AF900650

Green Energy and Technology

Climate change, environmental impact and the limited natural resources urge scientific research and novel technical solutions. The monograph series Green Energy and Technology serves as a publishing platform for scientific and technological approaches to “green”—i.e. environmentally friendly and sustainable—technologies. While a focus lies on energy and power supply, it also covers "green" solutions in industrial engineering and engineering design. Green Energy and Technology addresses researchers, advanced students, technical consultants as well as decision makers in industries and politics. Hence, the level of presentation spans from instructional to highly technical.
Indexed in Scopus.
Indexed in Ei Compendex.

Augustine O. Ayeni
Samuel Eshorame Sanni • Solomon U. Oranusi
Editors

Bioenergy and Biochemical Processing Technologies

Recent Advances and Future Demands

Springer

Editors
Augustine O. Ayeni
Chemical Engineering
Covenant University
Ota, Nigeria

Samuel Eshorame Sanni
Chemical Engineering
Covenant University
Ota, Nigeria

Solomon U. Oranusi
Biological Sciences
Covenant University
Ota, Nigeria

ISSN 1865-3529 ISSN 1865-3537 (electronic)
Green Energy and Technology
ISBN 978-3-030-96720-8 ISBN 978-3-030-96721-5 (eBook)
https://doi.org/10.1007/978-3-030-96721-5

© The Editor(s) (if applicable) and The Author(s), under exclusive license to Springer Nature Switzerland AG 2022
This work is subject to copyright. All rights are solely and exclusively licensed by the Publisher, whether the whole or part of the material is concerned, specifically the rights of translation, reprinting, reuse of illustrations, recitation, broadcasting, reproduction on microfilms or in any other physical way, and transmission or information storage and retrieval, electronic adaptation, computer software, or by similar or dissimilar methodology now known or hereafter developed.
The use of general descriptive names, registered names, trademarks, service marks, etc. in this publication does not imply, even in the absence of a specific statement, that such names are exempt from the relevant protective laws and regulations and therefore free for general use.
The publisher, the authors and the editors are safe to assume that the advice and information in this book are believed to be true and accurate at the date of publication. Neither the publisher nor the authors or the editors give a warranty, expressed or implied, with respect to the material contained herein or for any errors or omissions that may have been made. The publisher remains neutral with regard to jurisdictional claims in published maps and institutional affiliations.

This Springer imprint is published by the registered company Springer Nature Switzerland AG
The registered company address is: Gewerbestrasse 11, 6330 Cham, Switzerland

Contents

Part I
Energy Utilization, Conversion, and Management

Effective Moisture Diffusivity and Mathematical Modeling of the Drying Process for Cassava Stalk Biomass

Augustine O. Ayeni, O. Agboola, O. Oladokun, A. A. Ayoola, F. B. Elehinafe, E. E. Alagbe, M. Sika, and O. Azeta

1 Introduction

Moisture is the loss in weight when a substance undergoes a reversible process of drying, leading to constant weight. In such a process, the material does not undergo any noticeable changes except the loss in moisture. To avoid microbial growth in biological materials, the moisture content must be kept below 10% (Mercer 2008). Moisture measuring techniques are either direct or indirect. Direct methods involve gravimetric and chemical (Karl Fischer titration and coulometric). The indirect method can further be grouped under optical (infrared spectroscopy, infrared thermography, hyperspectral imaging), dielectric (electrical conductivity, microwaves, radiofrequency), nuclear (hydro-nuclear magnetic resonance, NMR,), and hygrometric (equilibrium relative humidity) (Zambrano et al. 2019). Lignocellulosic biomasses such as cassava stalk are excellent feedstocks for generating alternative sources of energy away from the conventional fossil fuels (Azeta et al. 2021). However, as a form of agricultural waste, cassava stalk must be properly disposed or utilized to reduce environmental pollution (Dick et al. 2020). Prior to biomass beneficiation processes, the drying process is essential. Most biological materials, such as lignocelluloses, must be dried to avoid hazardous environmental problems because these materials are easily exposed to microbial contamination or growth. Drying improves the shelf-life of biomasses. It also reduces weight, the volume, packing, storage, and transportation costs and can be stored conveniently at ambient temperatures (Uribe et al. 2011). Removal of moisture especially from

A. O. Ayeni (✉) · O. Agboola · O. Oladokun · A. A. Ayoola · F. B. Elehinafe
E. E. Alagbe · M. Sika · O. Azeta
Covenant University, College of Engineering, Department of Chemical Engineering, Ota, Nigeria
e-mail: augustine.ayeni@covenantuniversity.edu.ng

© The Author(s), under exclusive license to Springer Nature Switzerland AG 2022

A. O. Ayeni et al. (eds.), *Bioenergy and Biochemical Processing Technologies*, Green Energy and Technology, https://doi.org/10.1007/978-3-030-96721-5_1

lignocellulosic biomasses improves their caloric values which increases their fuel and energy properties. Analysis of drying experimental data through kinetic study allows us to understand the drying behavior, improving, and optimizing the process. During biomass transformation technologies, the drying process is crucial because energy is required which also decreases overall process yields (Anabel et al. 2018). During drying, moisture is evaporated along with mass and heat transfer occurring at the same time (Anabel et al. 2018). Therefore, to perfectly design a dryer, physical and thermal properties such as moisture diffusion, mass and heat transfer, and activation energy are very crucial (Mirzaee et al. 2009). Effective diffusivity is the rate of moisture movement. There are many parameters used in describing the drying process in which moisture diffusion is one (Absalan et al. 2016). Diffusivity is indicative of the diffusion mobility (of one substance with respect to other). It is derived from drying curves (Chen et al. 2012). Activation energy stands for the energy level of the molecules of water describing the diffusion and evaporation of moisture (Chen et al. 2012). How biological materials behave during the drying process can be explained by fitting experimental values into some drying models. Some of these models are either empirical or semi-empirical (Veras et al. 2012; Absalan et al. 2016).

In this study, the effect of drying temperatures (80, 100, and 120 °C) and drying intervals of 30 min to 8 h on the drying characteristics of cassava stalk was considered. The effective moisture diffusivity (diffusion coefficient) and the activation energy of the drying process were determined. In addition, the drying experimental data were fitted to three mathematical models of Page, Newton, and Midilli-Kucuk (Veras et al. 2012; Absalan et al. 2016).

2 Materials and Methods

2.1 Material Sourcing and Preparation

The cassava stalk biomass material was sourced from a local cassava farm in Ota Town, Ogun State, Nigeria. The harvested materials were sun-dried and further reduced in size through milling.

2.2 Drying Experiments

With the aid of the convection oven, the experiments were performed using temperatures of 80, 100, and 120 °C with sampling periods from 30 min to 8 h with time interval of 30 min (Table 1). The moisture contents for each sampling period were determined.

Table 1 Progression of the moisture content (%w/w) reduction with time for the cassava stalk drying process

	Temperature (°C)		
Time (h)	80	100	120
0.0	0.86	0.84	0.86
0.5	0.81	0.78	0.85
1.0	0.76	0.72	0.84
1.5	0.72	0.67	0.77
2.0	0.68	0.63	0.70
2.5	0.63	0.54	0.61
3.0	0.58	0.45	0.51
3.5	0.53	0.38	0.39
4.0	0.48	0.30	0.27
4.5	0.42	0.25	0.21
5.0	0.35	0.19	0.16
5.5	0.30	0.15	0.12
6.0	0.24	0.10	0.07
6.5	0.17	0.07	0.04
7.0	0.11	0.03	0.01
7.5	0.05	0.02	0.01
8.0	0.00	0.00	0.00

2.3 *Equations for Describing Effective Diffusivity and Activation Energy*

Using the Fick's second law of diffusion (Eq. 1), the obtained moisture content allowed the calculation of the moisture ratio, MR (Eq. 2). With the estimation of the moisture ratio, the effective moisture diffusivity and the energy of activation of the biomass drying were determined at each temperature.

$$\frac{\delta MR}{\delta t} = \Delta\left[Deff\left(\Delta \mathrm{MR}\right)\right] \tag{1}$$

where *Deff* = effective moisture diffusivity

$$\mathrm{MR}\left(\text{moisture ratio of biomass}\right) = \frac{M_t - M_e}{M_0 - M_e} \tag{2}$$

t is the time, M_t is the moisture content at any time, M_o is the initial moisture content, and M_e is the equilibrium moisture content.

Solving Eq. (2) leads to:

$$\text{MR} = \frac{8}{\pi^2}\sum_{n=0}^{\infty}\frac{1}{(2n+1)^2}\exp\left[-\frac{(2n+1)^2\pi^2 D_{\text{eff}}t}{4L^2}\right] \tag{3}$$

Equation (3) describes the drying process, where t is the drying time (h), L is the thickness of the sample, and n is the positive integer. Eq. 3 is appropriate to determine effective moisture for uniform moisture distribution and very negligible shrinkage (Doymaz and İsmail 2011). The effect of long period drying makes the first term $n = 0$ in Eq. 3 (Di Scala and Crapiste 2008; Chen et al. 2012).

Therefore,

$$\text{MR} = \frac{8}{\pi^2}\exp\left[-\frac{\pi D_{\text{eff}}}{4L^2}\right]t \tag{4}$$

Solving Eq. (4),

$$\ln(\text{MR}) = \ln\left(\frac{8}{\pi^2}\right) - \left[\frac{\pi^2 D_{\text{eff}}}{4L^2}\right]t \tag{5}$$

The slope = $\frac{\pi^2 D_{\text{eff}}}{4L^2}$ from Eq. 5 is estimated from the linear relationship of *Ln* (MR) and time that allows determining the effective diffusivity, D_{eff}. L was calculated from the average diameter of the sample.

The temperature dependence of the effective diffusivity can be represented by a form of Arrhenius relationship (Shi et al. 2008):

$$D_{\text{eff}} = D_o \exp\left(\frac{-E_a}{\text{RT}}\right) \tag{6}$$

In the Arrhenius equation, D_o ($m^2\,s^{-1}$) is known as the pre-exponential factor. The activation energy is Ea in kJ/mol^{-1}, the universal ideal gas constant is R ($8.314\ J\ mol^{-1}\ K^{-1}$), and T is the temperature of the drying air (K). Arrhenius equation explicitly explains the relation between effective diffusivity of moisture and temperature.

Simplifying Eq. (6):

$$\ln D_{\text{eff}} = \ln(D_o) - \frac{E_a}{\text{RT}} \tag{7}$$

The drying activation energy, Ea, can be calculated by plotting ln (D_{eff}) versus $1/T$ with slope ($-E_a/\text{R}$); then E_a can be determined.

2.4 *Mathematical Modeling of the Drying Kinetics*

Drying process for most biological materials is a complicated physical phenomena controlled by heat and mass transfer mechanisms (Gómez-de Lacruz et al. 2015). Different kinds of empirical, semi-empirical, and theoretical mathematical models are available to study, estimate, and predict optimum drying conditions for industrial applications (Gómez-de Lacruz et al. 2015). Drying parameters were determined from these mathematical models (of Page (Eq. 8), Newton (Eq. 9), and Midilli-Kucuk, Eq. 10) by fitting the experimental data of moisture ratio against time with the using non-linear regression method (minimization of sum of square error in Microsoft Excel). The best model describing the drying kinetics of the cassava stalk biomass was chosen by using the statistical tests of the highest correlation coefficient (R^2) and lowest average absolute error, AAE (Eq. 11).

$$\text{MR} = \exp\left(-kt^n\right) \tag{8}$$

$$\text{MR} = \exp\left(-kt\right) \tag{9}$$

$$\text{MR} = \text{a}\exp\left(-kt^n\right) + bt \tag{10}$$

$$\text{AAE} = \frac{1}{n}\sum_{i=1}^{n}\frac{\text{MR}_{\text{predicted}} - \text{MR}_{\text{measured}}}{\text{MR}_{\text{measured}}} \times 100\% \tag{11}$$

n = total number of data points

3 Results and Discussion

3.1 *Moisture Diffusion Activity*

Experimental data for cassava stalk were obtained for moisture contents as a function of time for the different temperatures (80, 100, and 120 °C). The drying data of moisture content were converted to moisture ratio (MR) using Eq. 2 and fitted to Eq. 5. The relationship between ln (MR) against time is provided in Fig. 1.

Drying is characterized by a decreasing drying rate (i.e., moisture decreases as time goes on). The falling drying rate period (time of moisture decrease) is characterized by the principle of internal diffusion. During constant drying rate period (principally controlled by capillary and gravity forces), drying occurs on the surface and the physical form of the cassava stalk biomass is affected. At 120 °C (393.15 Kelvin), moisture removal is expected to be high because water removal is higher at higher temperatures.

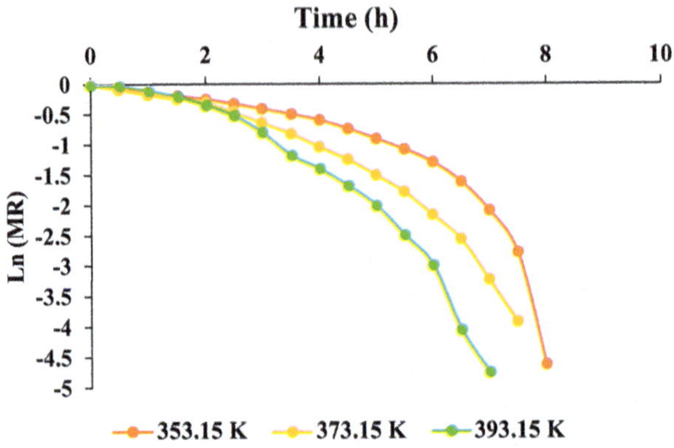

Fig. 1 Variation of the Ln (MR) with time for cassava stalk biomass at the different drying temperatures

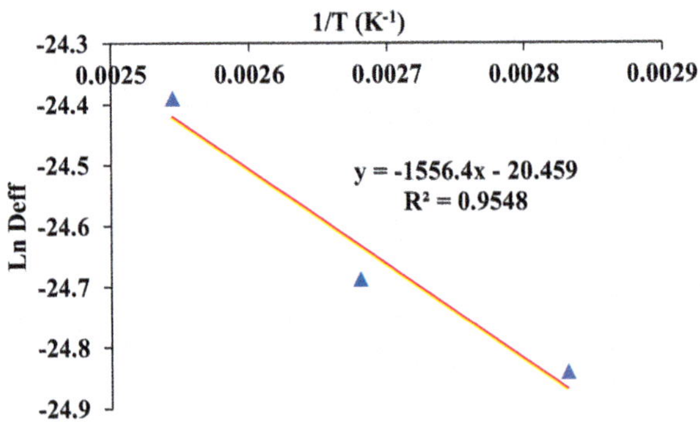

Fig. 2 Relationship between effective moisture diffusivity, Deff, and drying temperatures

Deff was determined for each heating rate (temperature of the drying process) and time by applying Eq. 5, which shows a linear relationship between ln D_{eff} and 1/T (Fig. 2) having R^2 of 0.9548. The effective moisture diffusivity (D_{eff}) for most biomass wastes can range between 10^{-16} and 10^{-6} m²/s (Anabel et al. 2018; Chen et al. 2012), which include the range of values obtained in this study (Table 2).

By applying Eq. 7 and using the value of the slope obtained from Fig. 2, the value of Ea, the activation energy was 12.93 kJ/mole, and the pre-exponential factor, Do, was estimated to be 1.30×10^{-9} m²/s. D_{eff} values increased as the temperature increased with a range of 2.56×10^{-11} to 1.63×10^{-11} (Table 2) for the cassava stalk biomass waste. It is expected that increased energy supply (at higher temperature) increases water molecules activity at increasing temperatures thereby leading tohigher moisture diffusivity (Absalan et al. 2016). The amount of energy required to initiate moisture removal during the drying process is described by the activation energy. Chen et al. (2012) obtained the effective moisture diffusivity values for poplar sawdust ranging from 9.38×10^{-10} to 1.38×10^{-9} m²/s by considering drying

Table 2 Regression models obtained from the relationship of Ln MR and time for estimating effective moisture diffusivity, Deff, using Eq. 7

Temperature, °C	Model	R^2	D_{eff} (m^2/s)
80	$y = -0.4022x + 0.5887$	0.7196	1.63×10^{-11}
100	$y = -0.4698x + 0.5221$	0.8957	1.90×10^{-11}
120	$y = -0.6332x + 0.7205$	0.8936	2.56×10^{-11}

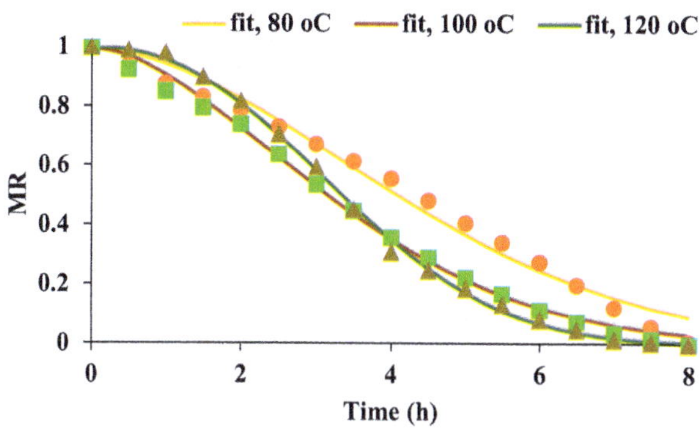

Fig. 3 Moisture ratio (MR) as a function of time for the drying process of the cassava stalk by the Page mathematical model at different temperatures. Markers represent the experimental data, ● (80 °C), ■ (100 °C), and ▲ (120 °C)

temperature of 60, 70, 80, and 90 °C, using a thermogravimetric analyzer. The activation energy in their work (Chen et al. 2012) was also calculated to be 12.30 kJ/mol. In addition, Anabela et al. (Anabel et al. 2018) obtained closed values for different regional wastes. The values obtained in this study for the Deff and Ea during the drying of cassava stalk are closer to results obtained for other biomass wastes reported in the literature.

3.2 Application of Drying Models to Experimental Data

The experiential data were fitted to the mathematical models of Page (Fig. 3), Newton (Fig. 4), and Midilli-Kucuk (Fig. 5) in order to evaluate the one which is best for representing drying kinetics of the cassava stalk biomass. The best model was selected statistically by considering the model with the highest coefficient of determination, R^2, and the least value of average absolute error (AAE). Table 3 gives the parameter values of the different models and the statistical analysis for each condition.

In Table 3, Page model provided the best fit at 120 °C drying temperature, Newton best fit occurred at the drying temperature of 100 °C, and Midilli-Kucuk provided the best fit at drying temperature of 100 °C. Averagely, considering the drying temperatures, 100 °C is the best temperature to be selected for effective

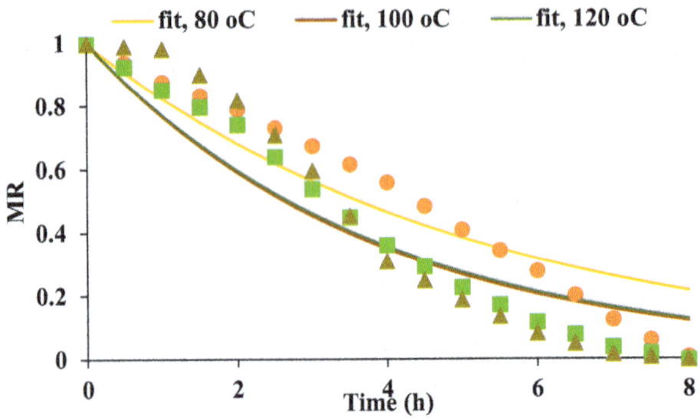

Fig. 4 Moisture ratio (MR) as a function of time for the drying process of the cassava stalk by the Newton mathematical model at different temperatures. Markers represent the experimental data, ● (80 °C), ■ (100 °C), and ▲ (120 °C)

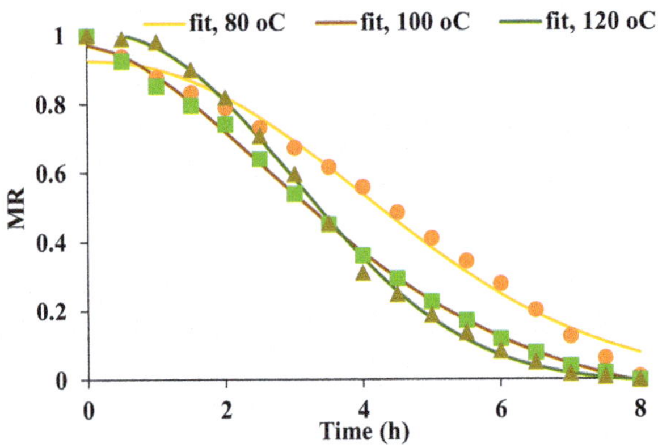

Fig. 5 Moisture ratio (MR) as a function of time for the drying process of the cassava stalk by the Midilli-Kucuk mathematical model at different temperatures. Markers represent the experimental data, ● (80 °C), ■ (100 °C), and ▲ (120 °C)

moisture removal during the drying process of the cassava stalk biomass. Obtained drying experimental data in this work showed that Midilli-Kucuk model provided the best fit out of the three models considered. The values of the mentioned tests for Midilli-Kucuk model were in the range of 0.9875–0.9985 for R^2 and 50.30–3.11 for AAE. In addition, Page model provided a better fit than the Newton model.

Table 3 Parameters for the drying mathematical models at different temperatures and their statistical results

Model	Temperature (°C)	k	n	a	b	R^2	AAE
Page							
	80	0.0524	1.82	–	–	0.9817	64.38
	100	0.096	1.72	–	–	0.9958	15.16
	120	0.0421	2.31	–	–	0.9985	8.55
Newton							
	80	0.1903	–	–	–	0.9312	155.55
	100	0.2625	–	–	–	0.9621	72.99
	120	0.2574	–	–	–	0.9390	184.50
Midilli-Kucuk							
	80	0.0265	2.18	0.93	−0.0001	0.9875	50.30
	100	0.0875	1.66	0.97	−0.0083	0.9985	3.11
	120	0.0466	2.24	1.01	−0.0012	0.9985	7.59

AAE Absolute average error measures the degree of closeness between the observed and predicted values. A lower value of AAE means the smaller error of model application

4 Conclusions

In this study, the diffusion controlling mechanisms and drying process of cassava stalk biomass experimental data allowed the evaluation of the effective moisture diffusivity, D_{eff}, and activation energy, Ea, under the prevailing experimental conditions. D_{eff} increased with increasing drying temperature indicating that more energy supply would increase the activity of water molecules when the biomass is dried at higher temperatures. In explaining the drying behavior of the cassava stalk biomass, the experimental data were fitted to three different drying models, and the models were compared according to their coefficient of determination and average absolute error. Considering the drying process, moisture diffusion is essential because it increases the understanding for the design, investigation, and the optimization of a dryer.

Acknowledgments The Management of Covenant University, Canaan land, Ota, Nigeria, is appreciated for sponsoring the authors to the first International Conference on Energy and Biochemical Engineering, Ota, Nigeria, and for the publication of this study.

References

Absalan G, Kianmehr MH, Arabhosseini A (2016) Effective moisture diffusivity and mathematical modeling of drying compost pellet. CIGR Journal 18:156-169.

Anabel F, Celia R, Germán M, Rosa R (2018) Determination of effective moisture diffusivity and thermodynamic, properties variation of regional wastes under different atmospheres. Case Studies in Thermal Eng. 12:248-257.

Azeta O, Ayeni AO, Agboola O, Elehinafe FB (2021) A review on the sustainable energy generation from the pyrolysis of coconut biomass. Sci. Afr. 13; e00909.

Chen D, Zheng Y, Zhu X (2012) Determination of effective moisture diffusivity and drying kinetics for poplar sawdust by thermogravimetric analysis under isothermal condition. Bioresour. Technol. 107:451-455.

Dick DT, Agboola O, Ayeni AO (2020) Pyrolysis of waste tyre for high-quality fuel products: A review. AIMS Energy 8:869 – 895.

Di Scala K, Crapiste G (2008) Drying kinetics and quality changes of drying of red pepper. LWT 41:789-765.

Doymaz I, İsmail O (2011) Drying characteristics of sweet cherry. Food Bioprod. Proc. 89:31–38.

Gómez-de Lacruz FJ, Casanova-Peláez PJ, López-García R, Cruz-Peragon F (2015) Review of the drying kinetics of olive oil mill wastes: Biomass recovery. Bioresour. 10:6055-6080.

Mercer DG (2008) Solar drying in developing countries: Possibilities and pitfalls. In G. L. Robertson, & J. R. Lupien (Eds.). Using food science and technology to improve nutrition and promote national development. International Union of Food Science & Technology http://www.iufost.org/publications/books/documents/Mercer.pdf)

Mirzaee E, Rafiee S, Keyhani A, Emam-Djomeh Z (2009) Determining of moisture diffusivity and activation energy in drying of apricots. Res. Agric. Eng. 55:114-120.

Shi JL, Pan ZL, McHugh TH, Wood D, Hirschberg E, Olson D (2008) Drying and quality characteristics of fresh and sugar-infused blueberries dried with infrared radiation heating. LWT Food Sci. Technol. 41:1962–1972.

Uribe E, Vega-Gálvez A, Di Scala K, Oyanadel R, Torrico JS, Miranda M (2011) Characteristics of Convective Drying of Pepino Fruit (*Solanum muricatum* Ait.): Application of Weibull Distribution. Food Bioproc. Technol. 4:1349-1356.

Veras AOM, Béttega R, Freire FB, Barrozo MAS, Freire JT (2012) Drying kinetics, structural characteristics and vitamin C retention of Dedo-de-Moca pepper (*Capsicum baccatum*) during convective and freeze drying. Braz. J. Chem. Eng. 4:741-750.

Zambrano MV, Dutta B, Mercer DG, Maclean HL, Touchie MF (2019) Assessment of moisture content measurement methods of dried food products in small-scale operations in developing countries: A review. Trends in Food Sci. Technol. 88:484-496.

Production and Characterization of Biochar and Hybrid Produced from the Co-carbonization of Corn Husk and Low-Density Polyethylene Wastes

Mubarak Adewale Amoloye, Sulyman Age Abdulkareem, and Adewale George Adeniyi

1 Introduction

Aside the global energy demand issues due to population rise, depletion, and unwanted consequences of the use of fossil fuel resources (Panwar et al. 2019), a major global concern especially in developing countries like Nigeria is found in municipal waste management (Adeniyi et al. 2020a). Ineffective municipal waste management has a lot of negative social, economic, and environmental effects. In Nigeria, agricultural and plastic wastes are major sources of municipal wastes raising a lot of concerns as a result of ineffective disposal methods and attitudinal behaviors of the populace. These wastes are found littering the streets and are eventually deposited in drainages causing blockages by dispersal agents (wind, water, and animals) (Adeniyi et al. 2020a). Usually, a quick way of disposal is to subject these wastes to open/direct burning. However, environmental concerns such as emission of greenhouse gases (GHG) are associated with this disposal method.

In addition, agricultural wastes could also emit methane (adding to global temperature rise) into the environment if they are a left to decompose (Huang et al. 2019) calling for a global concern. Co-carbonization, a form of slow pyrolysis, has been established in several studies (Adeniyi et al. 2020a, 2020b; Fan et al. 2020) as a waste utilization method for both agricultural and plastic wastes. It is a process which involves utilizing the synergistic effects of two or more materials for the production of a more valuable product under the influence of a driving force (temperature ranging between 350 and 450 °C) and in an environment which is less in supply of oxygen.

M. A. Amoloye (✉) · S. A. Abdulkareem · A. G. Adeniyi (✉)
Department of Chemical Engineering, University of Ilorin, Ilorin, Nigeria
e-mail: amoloye.ma@unilorin.edu.ng; adeniyi.ag@unilorin.edu.ng

© The Author(s), under exclusive license to Springer Nature Switzerland AG 2022

A. O. Ayeni et al. (eds.), *Bioenergy and Biochemical Processing Technologies*, Green Energy and Technology, https://doi.org/10.1007/978-3-030-96721-5_2

Corn (referred to as maize, sometimes) is an agricultural crop grown in large quantities globally and in Nigeria in particular. The versatility of the seeds in food processing industries and poultry feed manufacture is the reason for its high demand. Post-harvesting operation generates a lot of residues (biomass sources) such as cobs, leaves, stalks, husks, and a combination of all these residues which is referred to as stover or straw. Over the years, researches have paid more attention on upcycling of corn stover (Brewer et al. 2009; Chen et al. 2016, 2011; Peterson and Jackson 2014; Purakayastha et al. 2015; Sun et al. 2016; Yang et al. 2017), corn cobs (Du et al. 2015; Eduah et al. 2019; Godinho et al. 2019; Ioannidou et al. 2009;Liu et al. 2014), and corn stalks (Cen et al. 2016; Conti et al. 2016; Li et al. 2019; Lin et al. 2015; Liu et al. 2019; Zhang et al. 2019) to biochar with less attention paid to corn husk (CH). The husk houses the combination of seeds and corn cob. Although studies on yield and investigation of properties such as volatile matter content, ash content, pH, and electrical conductivity of biochar produce from CH have been reported elsewhere (Intani et al. 2016, 2018), few reports are available on the surface area, functional groups, and morphological properties of biochar produced from CH.

Low-density polyethylene (LDPE), a non-biodegradable material, belongs to the class of plastic wastes available in Nigeria. The sporadic rise in its utilization can be attributed to its high usage in the portable drinking water producing and packaging industries (Adeniyi et al. 2020a). Several advantages are accrued to the use of LDPE as its use provides a quick access to high-quality drinking water, provision of jobs for the populace in those industries to mention but a few. However, post-usage disposal methods of LDPE have been a daunting task as they are not degradable and burning emits several toxic gases into the environment (Adeniyi et al. 2020a). Several attempts have been reported in literature to harness the valuable products that could be recovered by degrading LDPE. One of such was reported in thermal and catalytic pyrolysis of LDPE to investigate what useful products could be obtained (Abdulkareem et al. 2019). The authors reported the constituents of pyrolysis oil produced from both pathways to be identical to hydrocarbon fuels obtainable in the refineries while also having a high yield of solid residue (34.89% for the thermal pathway, while 37.03% for the catalytic pathway). Moreover, research efforts have been geared towards harnessing the synergistic effects of agricultural wastes and plastics for biochar production (Adeniyi et al. 2020b, 2020c; Ighalo et al. 2021; Jin et al. 2019). Regardless of these efforts, the synergistic effect of co-carbonization of CH and LDPE has not been reported.

Therefore, the aim of this study was to investigate and compare the qualities of hybrid and biochar produced from the co-carbonization of CH/LDPE and carbonization of CH using similar methods described elsewhere (Adeniyi et al. 2020a, 2020b; Ighalo et al. 2021). The morphological properties, functional groups, surface areas, and pore volumes of both products were determined. The study, most importantly, should address issues of municipal waste management and upcycling of waste to wealth.

2 Methods

2.1 Sourcing and Preparation of CH and LDPE Waste

CH was obtained from a roadside roasted corn seller along the University of Ilorin road. The CH samples were dried in open air in order to reduce the moisture content. LDPE wastes (Sachet water waste) were handpicked within the university premises, cleaned with distilled water, and dried in open air. The combustion fuels used were mimosa and neem waste stalks and stems. They were also locally sourced within the university premises.

2.2 Experimental Set-up and Procedure

The experimental set up used in this study has been described in details in a recent study (Ighalo et al. 2021), while the procedures for operation are the same as in similar studies (Adeniyi et al. 2019a, 2019b, 2020d; Ighalo et al. 2021). The reactor operates in an autothermal mode (retort heating system) because the necessary heat of reaction is provided by the continuous partial oxidation of the combustion fuel by air within the reactor. The heat produced from the combustion of the fuel is sufficient to drive the endothermic reactions within the reactor to produce the biochar and hybrid (Nsamba et al. 2015).

2.3 Characterization of Products

In order to determine the qualities of the biochar and hybrid (corn husk biochar was referred to as CHB, while the hybrid from corn husk and LDPE waste was referred to as CH-LDPEH), characterization techniques such as scanning electron microscopy (SEM), Fourier transform infrared spectroscopy (FTIR), and Branueur–Emmett–Teller (BET) analysis were employed. The surface structure of the particles of the products from both experimental runs was studied using scanning electron microscopy (SEM PhenomProX). The SEM analysis was done with an acceleration voltage set at 20 kV. FTIR (Shimadzu, FTIR-8400S, Japan) was used to determine the functional groups and complexes present in both samples. The spectra were recorded using transmittance method in the 4000–650 cm^{-1} regions with 30 samples. Surface areas and pore volumes of both biochars were measured using a multipoint BET surface area and the DR (Dubinin–Radushkevich) method. The chars were characterized by N_2 adsorption test at 77 K.

3 Results and Discussion

3.1 Reactor Performance and Product Yield

The performance of the reactor was measured in terms of product yield, process time, and peak temperature. Product yields for both processes were computed by a material balance on the reactor using Eqs. 1, 2, and 3 as adopted from a similar study (Ighalo et al. 2021).

$$\text{Biochar}_{\text{yield}} = \left(\frac{M_{Bio-char}}{M_{\text{Raw}}} \right) \times 100\% \tag{1}$$

$$M_{Bio-char} = M_3 - M_2 \tag{2}$$

$$M_{\text{Raw}} = M_1 - M_2 \tag{3}$$

M_1 = Mass of combustion chamber + feed (in grams)

M_2 = Mass of combustion chamber (in grams)

M_3 = Mass of combustion chamber + product (in gram)

For the carbonization process $M_1 = 661$ g, $M_2 = 525$ g, and $M_3 = 593$ g, the CHB yield of 50% was obtained. Similarly, 59.91% CH-LDPEH yield was obtained for the co-carbonization process with $M_1 = 952$ g, $M_2 = 725$ g, and $M_3 = 861$ g, respectively (as summarized in Table 1). Higher CH-LDPEH yield is due to the presence of LDPE waste which contributed more carbon atoms in the reactor chamber during the co-carbonization process (Ighalo et al. 2021). The process time and peak temperatures for both processes were obtained from the temperature profiles (Figs. 1 and 2). Td represents the temperature profile within the carbonization chamber. As observed (Fig. 1), the process peaked at 351.9 ° C after the first 20 min of the carbonization process which dropped continually until it was observed that Ta, Tb, Tc, and Td were at ambient conditions (after 70 min) marking the end of the process. Ta, Tb, and Td are other points on the reactor where temperature measurements were taken to monitor the progression of the process. Though the reactor design was targeted for optimized biochar yield at low temperatures, it is also necessary to sustain the temperature between 350 ° *C and* 400 ° *C* for a longer period to achieve complete thermal decomposition of the feed (Ighalo et al. 2021). The >300 ° C region only lasted for about 20 min as observed (Fig. 1). Regardless of the appreciable yield of CHB, the peak temperature and the sustained period of the peak may

Table 1 Reactor performance summary

Index	Carbonization process	Co-carbonization process
Process time	70 min	120 min
Peak temperature	351.9 °C	332.8 °C
Biochar yield	50%	59.91%

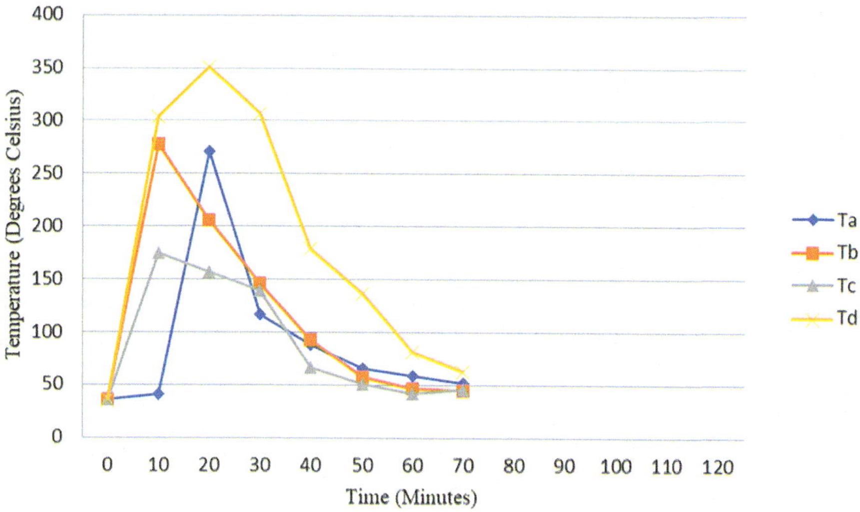

Fig. 1 Temperature profile for the carbonization process

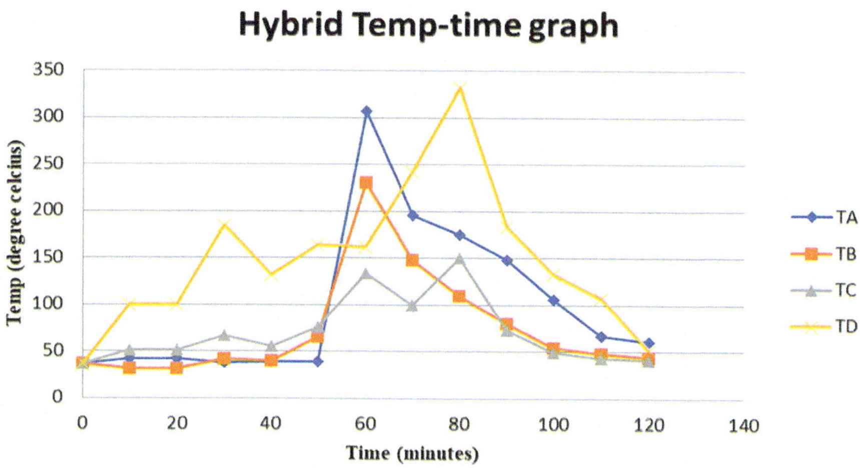

Fig. 2 Temperature profile for the co-carbonization process

have an influence on the quality of the CHB which was elucidated by the morphological and surface area measurements discussed in later sections. Similar studies (Adeniyi et al. 2020a; Ighalo et al. 2021) reported temperature regions sustained >300 ° C for longer periods; converse was the case in this study. This is attributed to thermal capacities (amount of heat that could be supplied) of the combustion fuels as the combustion fuels used in this study (Neem tree-Mimosa tree stalks and stems) were different from previous studies mentioned earlier.

Similar temperature profiles were obtained for the co-carbonization process (Fig. 2) in relation to Td maintaining the highest temperature all through the process. However, a lower peak temperature (332.8 ° C observed after 80 min) and longer process time of 120 min were observed for the co-carbonization process. The >300 ° C region only lasted for about 10 min as observed (Fig. 2). This is attributed to the self-regulating nature of the entire process. These peak temperatures attained by both processes were indicative that the combustion fuels possess the necessary thermal capacity to drive the endothermic reaction in the carbonization chamber for biochar production since it has been shown elsewhere that biochar has been produced at a temperature above 200 °C (Adeniyi et al. 2019b). However, they may not possess the necessary composition to sustain the process at the required temperature for a considerable length of time for complete thermal decomposition to occur. Further sections will discuss the influence of the temperature profile on the qualities of CHB and CH-LDPEH.

3.2 Product Morphology

Scanning electron microscope (SEM Phenom ProX) was used to study the morphology of both products. The accelerated voltage of the microscope was set to 20 kV for both the CHB and CH-LDPEH (Figs. 3 and 4).

The estimated sizes for both products are between 10 *and* 20 μm, respectively, as observed (Figs. 3 and 4). CHB micrograph showed a rough-like, heterogeneous surface bounded with some form of whitish gelatinous precipitate with very few visible pores. The same description could be repeated for the CH-LPDEH except for the disappearance of the very visible whitish precipitation into few dots of whitish precipitation with no visible pores on the surface. Usually, the presence of pores on char surfaces could be an indication of large surface areas. However, the absence of pores on the surfaces of both products could be as a result of incomplete thermal decomposition evident from the temperature profiles discussed earlier. The irregularities in the morphologies of both CHB and CH-LDPEH could be as result of presence of impurities, irregular surface, short microfibrils, and lumen present on the surface of the corn husk biomass which are typical of raw natural fibers (Carvalho-Mendes et al. 2015).

3.3 Product Surface Areas and Pore Volumes

The specific surface areas and pore volumes of CHB and CH-LPDEH were determined using the Branueur–Emmett–Teller (BET) method and the results summarized in Table 2. Both biochars were found to be relatively non-porous and had surface areas of 31.5587 m^2/g and 27.1052 m^2/g for the CHB and CH-LPDEH, respectively, which are relatively low when compared to similar studies reported in

Fig. 3 CHB micrograph

literature using the same method of biochar production (Adeniyi et al. 2019a, 2019b, 2020b, 2020a, 2020c; Ighalo et al. 2021). A further reduction in the specific surface area of CH-LPDEH was attributed to the presence of LDPE. The relatively low surface areas of these products could be as a result of the incomplete thermal degradation of biomass with compounds like tar entrapped within the pores (Ferreira et al. 2017; Yao et al. 2016) which was evident in the result discussed for the temperature profile of both processes. Porosity can be influenced by achieving high degradation of biomass lignin content, quick release of H_2 and CH_4, and the reaction of aromatic condensation as the temperature increases (Chen et al. 2012; Zhao et al. 2017). Kajina and Rousset (2018) reported biochar-specific surface area from sugarcane leaves to be higher than (253.2 m^2/g) that from coconut shell (25.8 m^2/g). Another study (El-Gamal et al. 2017) observed biochar from sugarcane bagasse had a higher specific surface area (185.2 m^2/g) than that produced from rice husks (154.7 m^2/g). These observations are suggestive of different responses of biomass composition to thermal degradation.

Fig. 4 CH-LDPEH micrograph

Table 2 Summary of the surface properties of CHB and CH-LPDEH

Index	BET surface area (m^2/g)	Micropore volume (cm^3/g)	Pore width (Å)
CHB	31.5587	0.126084	159.8085
CH-LPDEH	27.1052	0.161812	238.7915

3.4 *Product Functional Groups*

The changes in the chemical functionalities within a material are routinely investigated using the FTIR. The FTIR spectra (Figs. 5 and 6) show the functional groups present in both products. The observed peaks from both spectra are as summarized on Table 3. Similar spectra were observed for both products with the peak observed at 722 cm^{-1} on the CHB spectra not visible again on the CH-LDPEH spectra. The broadband at 3365 cm^{-1} and 3363 cm^{-1} corresponds to the hydroxyl group (O–H) stretching vibrations of cellulose. The peaks at 2850 cm^{-1} and 2920 cm^{-1} are attributed to simple alkanes in cellulose and hemicellulose components (Ferreira et al.

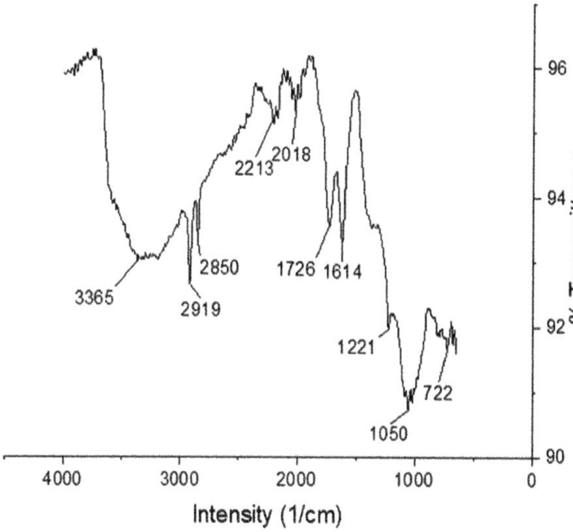

Fig. 5 FTIR spectra of CHB

Fig. 6 FTIR spectra of CH-LPDEH

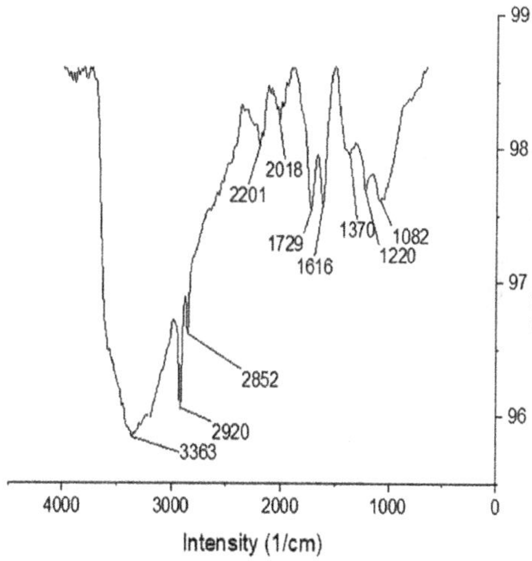

2017) and are characterized by absorptions due to C–H stretching and bending. 2213 cm^{-1} and 2201 cm^{-1} are peaks attributed to the presence of alkynes, which are compounds with carbon triple bond (–C ≡ C). The –C ≡ C– stretch appears as a weak band at these regions. Very few organic compounds show an absorption in this region. In the 1726 cm^{-1} and 1729 cm^{-1} regions, all carbonyl compounds absorb because of the stretching vibration of the C=O bond. This carbonyl band is useful for diagnostic purposes because of its high intensity and because few other functional groups are absorbed in this region. Also, the peaks at 1614 cm^{-1} and 1616 cm^{-1} are also assigned to the aromatic vibrations of C = O stretch in lignin (Adeniyi et al.

Table 3 FTIR peaks and possible assignments

Observed peak (cm^{-1})			
CHB	CH-LDPEHB	Functional group	Possible assignment
3365	3363	O–H stretching	Cellulose, hemicelluloses, and lignin
2850	2920	C–H stretching	Cellulose, hemicelluloses, and carboxylic acid
2213	2201	C ≡ C stretching	Lignin, hemicelluloses
1726	1729	C=O stretching	Lignin, hemicelluloses
1614	1616	C = O, C = C aromatic ring stretching vibrations	Lignin, hemicelluloses
722	–	C=H bending	

2019a). FTIR results of both products were found to contain more C=O and C–H functional groups. This inference is corroborated with what is found in other studies that biochars produced at lower temperatures are richer in C=O and C–H functional groups and could serve as nutrient exchange sites after oxidation (Mullen et al. 2010).

4 Conclusion

Co-carbonization of corn husk and low-density polyethylene waste for biochar production was carried out in a biomass gasifier with retort heating (BGRH) to investigate the quality of char produced to fill existing knowledge gap in this area. Product yields of 50 $\%$ *and* 59.91$\%$ were obtained for both corn husk biochar (CHB) and corn husk low-density polyethylene waste biochar (CH-LPDEH), respectively. Peak temperatures of 351.9 ° C and 332.2 ° C with short residence periods at these temperature regions were recorded for both carbonization and co-carbonization process. An observation attributed to the nature of the combustion fuel used for the process. The surface morphologies of both biochars were found to be rough, heterogeneous with the absence of pores which was more apparent from the (CH-LPDEH) micrograph. Low specific surface areas of 31.5587 m^2/g and of 27.1052 m^2/g were found for CHB and CH-LDPEH, respectively. FTIR results indicated both chars to contain more C=O and C–H functional groups. Biochars rich in these functional groups can serve as nutrient exchange sites.

Declarations Availability of Data and Materials: Not applicable.

Competing Interests: No potential conflict of interests was reported by the authors.

Funding: Not applicable.

References

Abdulkareem, S. A., Abu, T. O., & Eleburuike, N. A. (2019). Thermal and Catalytic Pyrolysis of Low- Density Polyethylene (LDPE) Wastes into Useful Fuel Oils : Comparative Studies. *ABUAD Journal of Engineering Research and Development (AJERD)*, *2*(1):36–41.

Adeniyi, A G, Ighalo, J. O., Onifade, D. V, & Popoola, A. O. (2020a). Production of Hybrid Biochar by Retort-Heating of Elephant Grass (Pennisetum purpureum) and Low Density Polyethylene (LDPE) for Waste Management and Product Development. *Journal of Materials and Environmental Science*, *2508*(12):1940–1952.

Adeniyi, A. G., Abdulkareem, S. A., Ighalo, J. O., Onifade, D. V., & Sanusi, K. S. (2020b). Thermochemical Co-conversion of Sugarcane Bagasse-LDPE Hybrid waste into Biochar. *Arabian Journal for Science and Engineering*.

Adeniyi, A. G., Ighalo, J. O., & Onifade, D. V. (2019a). Production of Bio-Char from Plantain (Musa paradisiaca) Fibers Using an Updraft Biomass Gasifier with Retort Heating. *Combustion Science and Technology*,*0*(0):1–15. doi:https://doi.org/10.1080/00102202.2019.1650269

Adeniyi, A. G., Ighalo, J. O., & Onifade, D. V. (2019b). Production of biochar from elephant grass (Pennisetum purpureum) using an updraft biomass gasifier with retort heating. *Biofuels*, *0*(0):1–8. doi:https://doi.org/10.1080/17597269.2019.1613751

Adeniyi, G. A., Ighalo, J. O., & Onifade, D. V. (2020c). Biochar from the Thermochemical Conversion of Orange (Citrus sinensis) Peel and Albedo : Product Quality and Potential Applications. *Chemistry Africa*, *3*:439–448. doi:https://doi.org/10.1007/s42250-020-00119-6

Brewer, C. E., Schmidt-Rohr, K., Satrio, J. A., & Brown, R. C. (2009). Characterization of Biochar from Fast Pyrolysis and Gasification Systems. *Environmental Progress & Sustainable Energy*, *28*(3):482–489. doi:10.1002/ep

Carvalho-Mendes, C. A., Adnet-Oliveira, F. A., Leite Moreira Amorim, C. M., Furtado, R. C. G., & De Sousa, A. M. F. (2015). Chemical, physical, mechanical, thermal and morphological characterization of corn husk residue. *Cellulose Chemistry and Technology*, *49*(9–10):727–735.

Cen, K., Chen, D., Wang, J., Cai, Y., & Wang, L. (2016). Effects of Water Washing and Torrefaction Pretreatments on Corn Stalk Pyrolysis : Combined Study Using TG-FTIR and a Fixed Bed Reactor. *Energy & Fuels*. doi:https://doi.org/10.1021/acs.energyfuels.6b02813

Chen, T., Liu, R., & Scott, N. R. (2016). Characterization of energy carriers obtained from the pyrolysis of white ash, switchgrass and corn stover — Biochar, syngas and bio-oil. *Fuel Processing Technology*, *142*:124–134. doi:https://doi.org/10.1016/j.fuproc.2015.09.034

Chen, X., Chen, G., Chen, L., Chen, Y., Lehmann, J., McBride, M. B., & Hay, A. G. (2011). Adsorption of copper and zinc by biochars produced from pyrolysis of hardwood and corn straw in aqueous solution. *Bioresource Technology*, *102*(19):8877–8884. doi:https://doi.org/10.1016/j.biortech.2011.06.078

Chen, Y., Yang, H., Wang, X., Zhang, S., & Chen, H. (2012). Biomass-based pyrolytic polygeneration system on cotton stalk pyrolysis : Influence of temperature. *Bioresource Technology*, *107*:411–418. doi:https://doi.org/10.1016/j.biortech.2011.10.074

Conti, R., Fabbri, D., Vassura, I., & Ferroni, L. (2016). Comparison of chemical and physical indices of thermal stability of biochars from different biomass by analytical pyrolysis and thermogravimetry. *Journal of Analytical and Applied Pyrolysis*. doi:https://doi.org/10.1016/j.jaap.2016.10.003

Du, C., Wu, J., Ma, D., Liu, Y., Qiu, P., Qiu, R., Liao, S., & Gao, D. (2015). Gasification of corn cob using non-thermal arc plasma. *International Journal of Hydrogen Energy*, *40*:12634–12649. doi:https://doi.org/10.1016/j.ijhydene.2015.07.111

Eduah, J. O., Nartey, E. K., Abekoe, M. K., Breuning-madsen, H., & Andersen, M. N. (2019). Phosphorus retention and availability in three contrasting soils amended with rice husk and corn cob biochar at varying pyrolysis temperatures. *Geoderma*, *341*:10–17. doi:10.1016/j.geoderma.2019.01.016

El-Gamal, E. H., Saleh, M., Elsokkary, I., Rashad, M., & El-latif, M. M. A. (2017). Comparison between Properties of Biochar Produced by Traditional and Controlled Pyrolysis. *ALEXANDRIA SCIENCE EXCHANGE JOURNAL, 38*(3).

Fan, S., Sun, Y., Yang, T., Chen, Y., Yan, B., Li, R., & Chen, G. (2020). Biochar derived from corn stalk and polyethylene co-pyrolysis : characterization and Pb (II) removal potential. *RSC Advances, 10*: 6362–6376. doi:https://doi.org/10.1039/c9ra09487c

Ferreira, D. S., Lazzarotto, P. I., Junges, J., Manera, C., Godinho, M., Osório, E., & Ferreira SD, Lazzarotto IP, J. J. (2017). Steam gasification of biochar derived from elephant grass pyrolysis in a screw reactor. *Energy Conversion and Management, 153*(153):163–174. doi:https://doi.org/10.1016/j.enconman.2017.10.006

Godinho, D., Nogueira, M., Bernardo, M., Dias, D., Lapa, N., Fonseca, I., & Pinto, F. (2019). Recovery of Cr (III) by using chars from the co-gasification of agriculture and forestry wastes. *Environmental Science and Pollution Research.*

Huang, Y., Zhao, Y. jie, Hao, Y. hong, Wei, G. qiang, Feng, J., Li, W. ying, Yi, Q., Mohamed, U., Pourkashanian, M., & Nimmo, W. (2019). A feasibility analysis of distributed power plants from agricultural residues resources gasification in rural China. *Biomass and Bioenergy, 121:* 1–12. doi:10.1016/j.biombioe.2018.12.007

Ighalo, J. O., Onifade, D. V., & Adeniyi, A. G. (2021). Retort-heating carbonisation of almond (Terminalia catappa) leaves and LDPE waste for biochar production: evaluation of product quality. *International Journal of Sustainable Engineering*, 1–9 doi:https://doi.org/10.1080/19397038.2021.1886371

Intani, K., Latif, S., Cao, Z., & Müller, J. (2018). Characterisation of biochar from maize residues produced in a self-purging pyrolysis reactor. *Bioresource Technology*. doi:https://doi.org/10.1016/j.biortech.2018.05.103

Intani, K., Latif, S., Kabir, A. K. M. R., & Müller, J. (2016). Effect of self-purging pyrolysis on yield of biochar from maize cobs, husks and leaves. *Bioresource Technology, 218*:541–551. doi:https://doi.org/10.1016/j.biortech.2016.06.114

Ioannidou, O., Zabaniotou, A., Antonakou, E. V, Papazisi, K. M., Lappas, A. A., & Athanassiou, C. (2009). Investigating the potential for energy, fuel, materials and chemicals production from corn residues (cobs and stalks) by non-catalytic and catalytic pyrolysis in two reactor configurations. *Renewable and Sustainable Energy Reviews, 13*:750–762. doi:https://doi.org/10.1016/j.rser.2008.01.004

Jin, Q., Wang, X., Li, S., Mikulčić, H., Bešenić, T., Deng, S., Vujanović, M., Tan, H., & Kumfer, B. M. (2019). Synergistic effects during co-pyrolysis of biomass and plastic: Gas, tar, soot, char products and thermogravimetric study. *Journal of the Energy Institute, 92*(1):108–117. doi:https://doi.org/10.1016/j.joei.2017.11.001

Kajina, W., & Rousset, P. (2018). *Coupled effect of feedstock and pyrolysis temperature on biochar as soil amendment.*

Li, J., Qiao, Y., Chen, X., Zong, P., Qin, S., Wu, Y., Wang, S., Zhang, H., & Tian, Y. (2019). Steam gasification of land, coastal zone and marine biomass by thermal gravimetric analyzer and a free-fall tubular gasifier : Biochars reactivity and hydrogen-rich syngas production. *Bioresource Technology, 289*:121–495. doi:https://doi.org/10.1016/j.biortech.2019.121495

Lin, X., Xie, J., Zheng, J., Liu, Q., Bei, Q., & Zhu, J. (2015). Effects of biochar application on greenhouse gas emissions, carbon sequestration and crop growth in coastal saline soil. *European Journal of Soil Science, 66*:329–338. doi:https://doi.org/10.1111/ejss.12225

Liu, L., Huang, Y., Zhang, S., Gong, Y., Su, Y., Cao, J., & Hu, H. (2019). Adsorption characteristics and mechanism of Pb (II) by agricultural waste-derived biochars produced from a pilot-scale pyrolysis system. *Waste Management, 100*:287–295. doi:https://doi.org/10.1016/j.wasman.2019.08.021

Liu, X., Zhang, Y., Li, Z., Feng, R., & Zhang, Y. (2014). Characterization of corncob-derived biochar and pyrolysis kinetics in comparison with corn stalk and sawdust. *Bioresource Technology, 170*:76–82. doi:https://doi.org/10.1016/j.biortech.2014.07.077

Mullen, C. A., Boateng, A. A., Goldberg, N. M., Lima, I. M., Laird, D. A., & Hicks, K. B. (2010). Bio-oil and bio-char production from corn cobs and stover by fast pyrolysis. *Biomass and Bioenergy*, *34*(1):67–74. doi:https://doi.org/10.1016/j.biombioe.2009.09.012

Nsamba, H. K., Hale, S. E., Cornelissen, G., & Bachmann, R. T. (2015). Sustainable Technologies for Small-Scale Biochar Production — A Review. *Journal of Sustainable Bioenergy Systems*, *5*:10–31.

Panwar, N. L., Pawar, A., & Salvi, B. L. (2019). Comprehensive review on production and utilization of biochar. *SN Applied Sciences*, *1*(2). doi:https://doi.org/10.1007/s42452-019-0172-6

Peterson, S. C., & Jackson, M. A. (2014). Simplifying pyrolysis: Using gasification to produce corn stover and wheat straw biochar for sorptive and horticultural media. *Industrial Crops and Products*, *53*:228–235. doi:https://doi.org/10.1016/j.indcrop.2013.12.028

Purakayastha, T. J., Kumari, S., & Pathak, H. (2015). Characterisation, stability, and microbial effects of four biochars produced from crop residues. *Geoderma*, 239–240:293–303. doi:https://doi.org/10.1016/j.geoderma.2014.11.009

Sun, J., He, F., Pan, Y., & Zhang, Z. (2016). Effects of pyrolysis temperature and residence time on physicochemical properties of different biochar types. *Acta Agriculture Scandinavica, Section B-Soil &Plant Science*. doi:https://doi.org/10.1080/09064710.2016.1214745

Yang, X., Wang, H., Strong, J. P., Xu, S., Liu, S., Lu, K., Sheng, K., Guo, J., Che, L., He, L., Ok, S. Y., Yuan, G., Shen, Y., & Chen, X. (2017). Thermal Properties of Biochars Derived from Waste Biomass Generated by Agricultural and Forestry Sectors. *Energies*, *10*:1–12. doi:https://doi.org/10.3390/en10040469

Yao, D., Hu, Q., Wang, D., Yang, H., Wu, C., Wang, X., & Chen, H. (2016). Hydrogen production from biomass gasification using biochar as a catalyst/support. *Bioresource Technology*. doi:https://doi.org/10.1016/j.biortech.2016.05.011

Zhang, H., Li, J., Yang, X., Song, S., Wang, Z., Huang, J., Zhang, Y., & Fang, Y. (2019). Influence of coal ash on CO2 gasification reactivity of corn stalk char. *Renewable Energy*, *147*:2056–2063. doi:https://doi.org/10.1016/j.renene.2019.10.009

Zhao, S., Ta, N., & Wang, X. (2017). Effect of Temperature on the Structural and Physicochemical Properties of Biochar with Apple Tree Branches as Feedstock Material. *Energies*, *10*. doi:https://doi.org/10.3390/en10091293

Energy and Fuels: Structural and Optical Characterization of Cu_2ZnSnS_4 Solar Absorber Layer for Photovoltaic Application Using SILAR Method

Abdulmutolib O. Olaoye, Thomas O. Daniel, Ebenezer O. Olabomi, Kazeem O. Olawale, and Akeem Mafe

1 Introduction

The global demand for energy is increasing in modern society. Due to the scarce (indium, gallium, tellurium) and hazardous (cadmium) components, CdTe and CIGS thin-film solar cell technologies have recently encountered production challenges on a larger scale (Wangperawong et al. 2011). As a result of its excellent band gap of 1.4–1.5 eV and a high 10^4 cm^{-1} absorption coefficient for the solar spectrum Cu_2ZnSnS_4 (CZTS) is a kesterite crystal structure and an appropriate material for affordable solar cells using thin films (Mitzi et al. 2011; Suryawanshi et al. 2013; Abermann 2013; Wallace et al. 2017). CZTS is a material of great interest in tackling these difficulties and P-type semiconductor that have good properties for photovoltaic applications.

Many researchers have reported the deposition of CZTS solar absorber layer via different techniques that include pulsed laser deposition, sputtering, evaporation, sol-gel processing, spray pyrolysis, electro-deposition, chemical vapor deposition, SILAR, and chemical bath deposition (Scragg et al. 2011; Mkawi et al. 2018; Shin et al. 2013; Deokate et al. 2019; Ahmed et al. 2012; Jiang et al. 2014; Cao et al. 2013). The study deposits CZTS thin-film solar absorber layer by employing the SILAR technique with the expectation of good homogeneity and crystallinity.

A. O. Olaoye (✉) · E. O. Olabomi · K. O. Olawale · A. Mafe
Science Technology Department, Physics with Electronics Unit, Federal Polytechnic Offa, Offa, Kwara State, Nigeria

T. O. Daniel
Department of Physics, Alex Ekwueme-Federal University Ndufu-Alike Ikwo, Abakaliki, Ebonyi State, Nigeria

© The Author(s), under exclusive license to Springer Nature Switzerland AG 2022

A. O. Ayeni et al. (eds.), *Bioenergy and Biochemical Processing Technologies*, Green Energy and Technology, https://doi.org/10.1007/978-3-030-96721-5_3

2 Material and Method

At room temperature, CZTS thin-film precursors were deposited via the SILAR technique. Glass substrates were treated in acetic acid for 5 minutes before the film deposition to remove oils and protein molecules from the surface. The glass substrates were also cleaned with ethanol and the detergent to remove possible organic contaminants. A total of four beakers are used in the experiment. The first beaker is filled with the cationic precursor, while the second beaker contains distilled water, which is used to get rid of any cations that have been weakly adsorbed from the surface of the substrate. The anionic solution is in the third beaker, and the precipitate on the substrate is removed with distilled water in the fourth beaker. Cu^{2+}, Zn^{2+}, and Sn^{2+} cations were obtained by dissolving 0.1 M of $CuSO_4$. $5H_2O$, 0.05 M of Zn $(CH_3(CO_2)_2)$, and 0.05 m of $SnSO_4$ solution in distilled water, while S^{2-} anionic was obtained by dissolving 0.1 M of CH_3CSNH_2 in distilled water. The glass substrate was immersed in a solution containing Cu^{2+}, Zn^{2+}, and Sn^{2+} ions (cations) until they were adsorbed onto the substrate surface and anionic S^{2-} precursor and then rinsed in distilled water after each immersion. For both precursors, a 30-second immersion duration was used, with a 10-second rinsing period. The films grown by nucleation as shown in Fig. 1. CZTS film in a thin layer was formed when the substrate has been immersed for 40 and 50 cycles, and the respective CZTS samples were identified as CZTS-40 and CZTS-50, respectively. To decrease film flaws and increase crystallite, the deposited substrates were annealed in a sulfur environment at 400 °C for 1 hour.

To estimate the film thickness of samples CZTS-40 and CZTS-50, surface profilometry (VEECO DEKTAK 150) was used. The structural characterization of CZTS films was investigated using an X-ray diffractometer (gracing incident at 40 mA, 45 kV with CuK and λ = 1.154060). The films' surface morphology was investigated by JOEL JSM-7600F Field Emission Scanning Microscopy (FESEM). The surface morphology of the thin films was examined using a JOEL JSM-7600F Field Emission Scanning Microscopy (FESEM) while XE-100Park system (XE Series SPM Controller) Atomic Force Microscopy (AFM). The optical characterization of the films produced on the glass substrate was measured using an Avantes UV-Visible spectrophotometer in the 350–800 nm wavelength range. AQUADPRO-301-6 four-point probe was used for electrical measurements.

3 Results and Discussion

3.1 Structural Analysis

Figure 2 shows the XRD pattern of CZTS thin films illustrating the preferential orientation growths of (011), (112), (004), and (220) (JCPDS No. 98–018-4358) for CZTS-40 while (002), (110), (112), (020), (004), (211), (114), (220), and (003) for

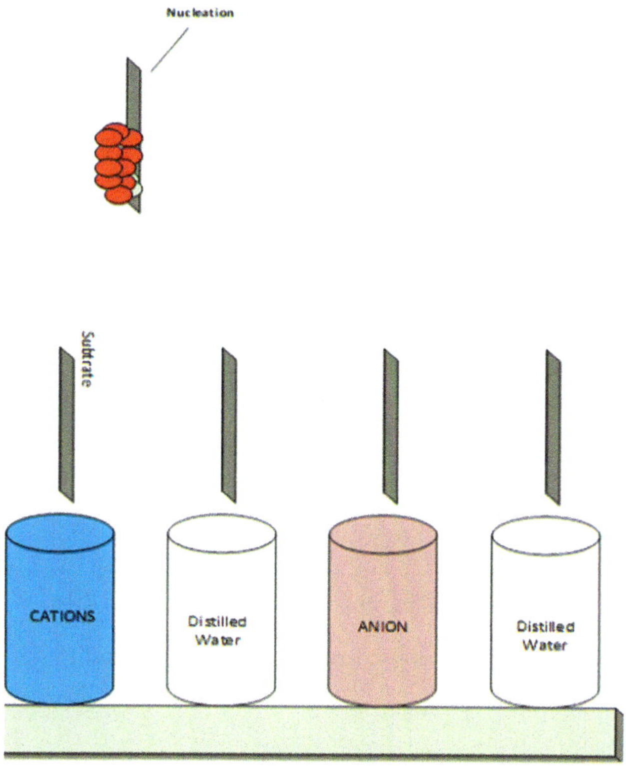

Fig. 1 Schematic presentation of CZTS thin films deposited using SILAR method

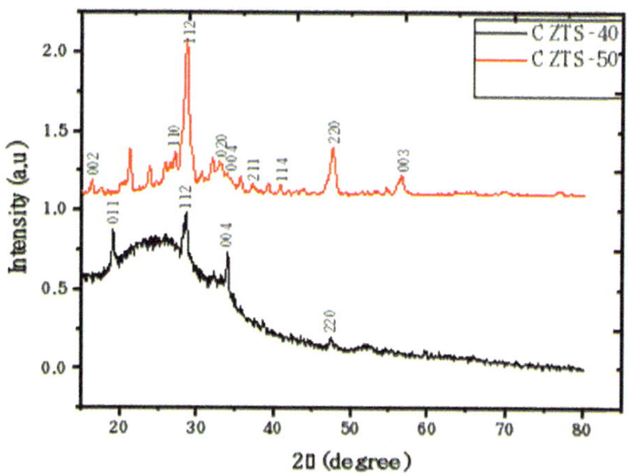

Fig. 2 XRD spectra of CZTS thin films deposited at different cycles

Table 1 FWHM, microstrain, dislocation density, and texture coefficient of the prepared films

CZTS	Peak	FWHM	2θ	Microstrain	Dislocation density (lines/m^2)	Texture coefficient
CZTS-40	112	0.3149	28.4849	51.723	1.579×10^{-5}	1.91
CZTS-50	112	0.3542	28.5548	51.596	1.998×10^{-5}	1.82

CZTS-50 (JCPDS No. 98–026-2389). From Table 1, the strongest peak occurs at 28.49 and 28.55 for CZTS-40 and CZTS-50 cycles, respectively, which is corresponding to (112) plane. The structure of this film is tetragonal. The lattice parameters for both samples were estimated to be $a = b = 5.433$ and $c = 10.604$ using Eq. 1. Likewise, the ratio of c/a is 1.952. The value indicates that the lattice is compressed in the c-axis direction (Nabeel et al. 2018).

$$\frac{1}{d^2} = \frac{h^2 + k^2}{a^2} + \frac{l^2}{c^2} \tag{1}$$

where (d) denotes the lattice spacing, (a) and (c) denote lattice parameters, and (hkl) denotes Miller indices.

The Scherrer's relation was used to determine the crystallite size (D) (Henry et al. 2018):

$$D = \frac{0.9\lambda}{\beta \tan\theta} \tag{2}$$

where (β) denotes full width at half maximum (FWHM), λ for X-ray wavelength, while θ for Bragg's angle, respectively. From the FWHM of the prominent peak (112), the crystallite size of CZTS-40 and CZTS-50 was determined to be 223.69 nm and 251.61 nm, respectively.

The microstrain of CZTS-40 is more than that of CZTS-50. The microstrain indicates the number of defects in the films is more in CZTS-40 than CZTS-50. The stress was caused by a mismatch between the films' crystalline lattices and the substrates or thin film deposition circumstances (Henry et al. 2018). The equation can be used to compute the microstrain:

$$\varepsilon = \frac{0.9}{4\tan\theta} \tag{3}$$

The dislocation density is a flaw in a crystal caused by a misalignment of the lattice in one part of the crystal, which can be calculated using Eq. 4:

$$\delta = \frac{1}{D^2} \tag{4}$$

The preferential crystal orientation plane with respect to reference samples can be calculated from the texture coefficient T_C, which is given by Eq. 5:

$$T_C = \frac{I_{(hkl)}}{I_{r(hkl)}} \left[\frac{1}{n} \sum \frac{I_{(hkl)}}{I_{r(hkl)}} \right]^{-1} \tag{5}$$

Tc is the orientation plane's texture coefficient, measure intensity of the orientation plane is $I_{(hkl)}$, $I_{r(hkl)}$ is the intensity of the identical peak in the reference, and n is the number of peaks taken into account.

Table 1 shows that the $T_{c(hkl)}$ values of the plane (112) for CZTS-40 and CZTS-50 are more than unitary, suggesting an excess of grain produced along the plane. The $T_{c(hkl)}$ value also indicates a greater degree of desired orientation along the plane (Olaoye et al. 2020).

3.2 *Morphological Analysis*

The morphologies of the films were examined using FESEM at a magnification of 10,000 x for the CZTS-40 and CZTS-50 are shown in Fig. 3a, c, respectively. ImageJ software was used to calculate the average grain size of the thin films (Rasband 2014), with average grain sizes of 7 nm and 11 nm for CZTS-40 and CZTS-50, respectively. The grain distribution histogram is shown in Fig. 3b, d. To investigate the grain distribution, the data were statistically examined using histograms and average grain size estimated based on the particle size distribution, using threshold approach for particle analysis (Julio). The films show the agglomeration of nanocrystallites. It is evident that a larger grain size of CTZS-50 compared to CZTS-40 enhances the decrease in grain boundaries and potential barriers, allowing trapped charge carriers to be released and increasing carrier conductivity. The decrease in grain boundary and reduction in deformation and defects in the crystals can also be linked to the growth in grain size with thickness, indicating a gain in the degree of grain perfection with the elimination of faults and the reduction of pores.

3.3 *Optical Analysis*

The optical spectra for the samples CZTS-40 and CZTS-50 were recorded and analyzed using UV-visible spectroscopy. It is evident from the UV-visible spectrum shown in Fig. 4. CZTS thin films take in a lot of light in the visible light spectrum. The optical bang gap for the thin films can be obtained from the optical transition using the Tauc plot (Eq. 5), while the absorption coefficient α was determined using Eq. 6:

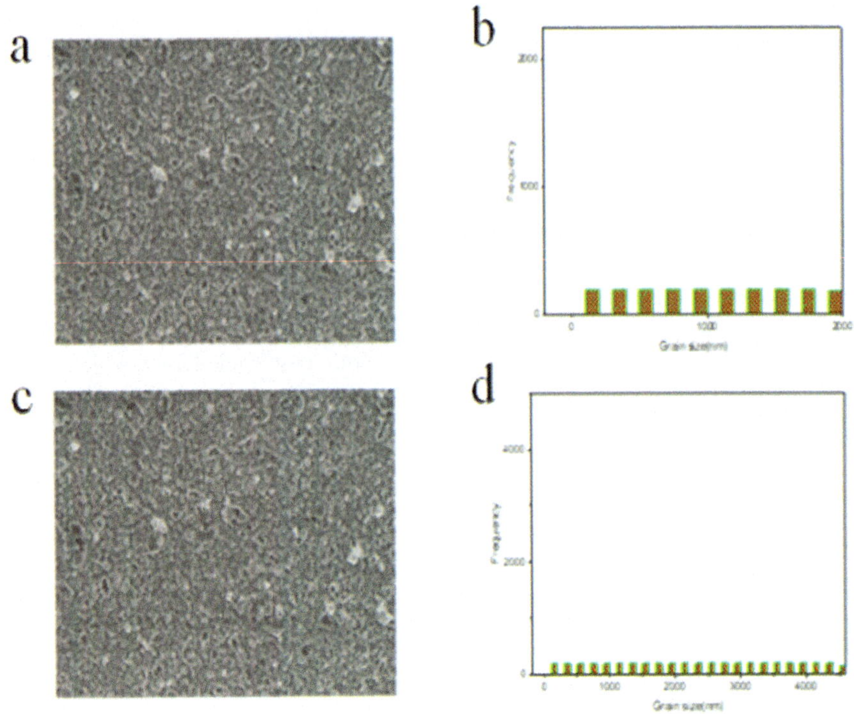

Fig. 3 FESEM images (**a**, **c**) and granular (particle) distribution (**b**, **d**) of the prepared films

$$\left(\alpha h\upsilon\right)^{\frac{1}{n}} = B\left(h\upsilon - E_g\right) \tag{6}$$

$$\alpha(\lambda) = \frac{1}{d}\ln\left(\frac{1}{T}\right) \tag{7}$$

where $h\upsilon$ is the photon energy and α is the absorption coefficient, B is a constant, E_g is the optical band gap, and n is a constant, which can be 2 for a direct transition and $\frac{1}{2}$ for an indirect transition. The plots of $(\alpha h\upsilon)^2$ versus $h\upsilon$ for CZTS-40 and CZTS-50 are shown in Fig. 4. Extrapolating the linear component of the $(\alpha h\upsilon)^2$ was used to determine the band gap of the samples and was revealed to be 1.60 eV and 1.54 eV for CZTS-40 and CZTS-50, respectively, which is in line with what was said before (Henry et al. 2016). The increase in film thickness causes the band gap to vary. The energy band gap was obviously less in thin films with a higher thickness. Both samples show band gaps that are near the ideal band gaps necessary for solar cells, indicating that they are potential photovoltaic materials.

Figure 4 shows the optical parameters of the deposit CZTS thin films. The band gap for CZTS-40 and CZTS-50 are 1.60 eV and 1.54 eV, respectively. We observe

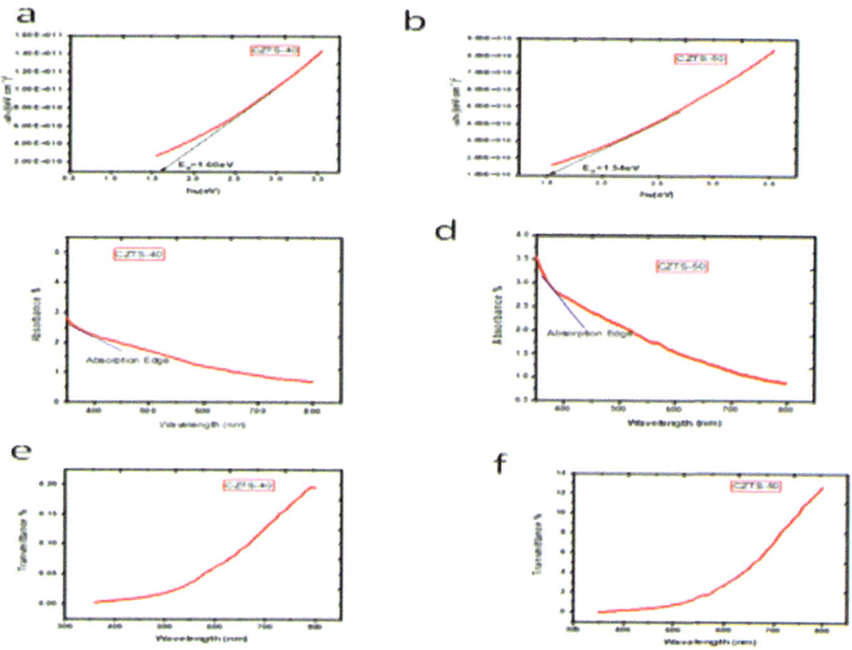

Fig. 4 Energy band (**a**, **b**), absorbance (**c**, **d**), and transmittance (**e**, **f**) plots of the prepared films

that as the number of deposition cycles grows, the energy band gap reduces. Both samples have a band gap that is near the suitable band gaps required for solar cells, which indicate potential materials for photovoltaic applications (Scragg et al. 2011).

The spectrum clearly shows CZTS thin films absorb more visible light, suggesting that there is an absorbent material. The coefficient of absorption in the visible area has been determined to be greater than 10^4 cm^{-1}, in line with prior findings (Yakuphanoglu et al. 2005).

The index of refraction and dielectric constant of semiconductor materials play a significant role in shaping the crystal's electrical and optical properties. The relative index of the films is calculated using the Herve-Vandamme equation:

$$n = \sqrt{1 + \left(\frac{A}{E_g + B} \right)^2} \tag{8}$$

It is observed from Table 2 that the films' index of refraction increases with a rise in thickness, which can be attributed to the high crystallinity and densification in CZTS-50. This implies that the incident light will be more reflected and generates reflectance losses in CZTS-50, which significantly lessens the efficacy of the photovoltaic device (Khan et al. 2019).

Table 2 Static dielectric, band gap energy, high-frequency dielectric, and refractive index for CZTS-40 and CZTS-50

Samples	Static dielectric	Band gap eV	Refractive index	High-frequency dielectric
CZTS-40	13.78	1.54	2.76	7.62
CZTS-50	13.90	1.50	2.78	7.73

Table 3 Electrical characterizations of CZTS thin films

Samples	Thickness (μm)	Sheet resistance (Ω/sq)	Resistivity (Ω.μm)	Conductivity (Ω^{-1} μm^{-1})
CZTS-40	25.0×10^{-2}	0.21912	5.478×10^{-2}	18.2548
CZTS-50	4.0×10^{-1}	0.01347	5.388×10^{-3}	185.5976

The reduction in the speed of light and dielectric absorbs energy from the electric field of the material is indicating the effect of dielectric by dipole motion (Uma Maheshwari and Senthil Kumar 2015). The electrical properties of the crystals are determined by high-frequency dielectric semiconductor materials, which were calculated using Eq. 9 (Akaltum et al. 2012):

$$\varepsilon_{\infty=n^2} \tag{9}$$

where n is the refractive index. It can be seen from Table 2 that the values of high-frequency dielectric constant increased from 7.62 and 7.73 with an increase in thickness.

Figure 4 shows the transmittance spectra of CZTS, low transmittance at photons of shorter wavelengths, which is higher at wavelength greater than 750 nm. It is observed that transmittance for CZTS-40 is lower than that of CZTS-50 with an increase in wavelength.

3.4 Electrical Analysis

At room temperature, the electrical properties of CZTS thin films were determined by a four-point probe. The films' electrical resistivity ranged from 10^{-2} to 10^{-3}. The decrease in resistivity is explained by the increase in grain size of CZTS thin films, shown in Table 3 (Shinde et al. 2012).

As observed the sheet resistance from Table 3 decreases as the thickness increases. Also, the sheet resistance increases because of the decline in crystallite size as being determined from XRD data analysis. The decrease is because of the Schottky barrier at the grain boundaries, which act as scattering center of the electronic conduction by trapping the electrons. The finding is in conformity with the relevant literature (Vilca-Huayhua et al. 2020).

4 Conclusion

CZTS thin films were fabricated using SILAR method at room temperature. The overwhelming preferred orientation of (112) planes is visible in the XRD patterns of the produced films, indicating the tetragonal structure of CZTS. The crystallite size of deposited CZTS thin films are 223.69 nm and 251. 61 nm for CZTS-40 and CZTS-50 respectively, whereas microstrain that indicates mismatch defect microstrain decreases from 51.723 to 51.596 numbers of deposition cycles increased. FESEM examined the surface morphology with particle sizes of 7 nm and 11 nm for CZTS-40 and CZTS-50, respectively. The energy band gap is decreasing from 1.60 eV to 1.45 eV for CZTS-40 and CZTS-50, respectively. The result reveals that the optical band gap decreases as the number of deposited cycles increases. The deposited thin films' solar absorber layers show great potential in the application of solar cell fabrication due to their energy band gaps. Due to an increase in the sheet resistance and a decrease in crystallite size, the Schottky barrier forms at grain boundaries, acting as a scattering point for electronic conduction by trapping electrons. In this research, CZTS thin films' solar absorber layer with optimum optical properties was successfully deposited at low temperature at the laboratory scale, implying that the fabrication process and process control can be achieved in large-scale manufacturing with low cost.

Acknowledgments The authors would appreciate Federal Polytechnic, Offa for supporting this work.

References

A. Wangperawong, J.S. King, S.M. Herron, B.P. Tran, K. Pangan-Okimoto, S.F. Bent, Thin Solid Films 519,2488 (2011)

A.B. Nabeel, A.S. Sabah, A.H.Sabreen, International Journal of Applied Engineering Research.13,(6),3379(2018)

A.O. Olaoye, R.S. Lawal, T.O. Daniel, M.S. Shehu, M.Q. Hamzah, M.A. Agam, Solid State Technology, 63, 1 (2020).

B. Shin, O. Gunawan, Y. Zhu, N.A. Bojarczuk, S.J. Chey, S. Guha, Photovolt. Res. Appl. 1, 72 (2013)

B. Uma Maheshwari, V. Senthil Kumar, Int. J. Energy Res. 39, 771-777 (2015)

C.A. Vilca-Huayhua, K.J. Paz-Corrales, F.F.H. Aragon, M.C. Mathpal, L. Villegas-Lelovsky, J.A.H. Coaquira, D.G. Pacheco-Salazar, Growth and vacuum post-annealing effect on the structural, electrical and optical properties of Sn-doped In_2O_3 thin films. *Thin Solid Films*. (2020) 709, 38207.

D.B. Mitzi, O. Gunawan, T.K. Todorov, K. Wang, S. Guha, Sol. Energy Mater. Sol. Cells 95(6), 1421(2011)

E.M. Mkawi, Y. Al-Hadeethi, E. Shalaan, E. Bekyarova, J. Mater. Sci.: Mater. Electron. 29(23), 20476 (2018)

F. Jiang, S. Ikeda, T. Harada, M. Matsumura, Adv. Energy Mater. 4, 1301381(2014)

F. Yakuphanoglu, M. Sekerci, A. Balaban, The effect of film thickness on the optical absorption edge and optical constants of the Cr (III) organic thin films. Optical Materials 27 (2005) 1369–1372.

J. Henry, K. Mohanraj, G. Sivakumar, Electrical and optical properties of CZTS thin films prepared by SILAR method. Journal of Asian Ceramic Societies (2016) pp 81-84.

J. Henry, K. Mohanraj, G. Sivakumar, Jordan Journal of Physics. 11, (2) 101 (2018)

J.J. Scragg, T. Ericson, T. Kubart, M. Edoff, C. Platzer-Bjorkman, Chem. Mater. 23, 4455 (2011)

M. Cao, L. Li, B.L. Zhang, J. Huang, L.J. Wang, Y. Shen, Y. Sun, J.C. Jiang, G.J. Hu, Sol. Energy Mater.Sol. Cells 117, 81(2013)

M.P. Suryawanshi, G.L. Agawane, S.M. Bhosale, S.W. Shin, P.S. Patil, J.H. Kim, A.V. Moholkar, Mater. Technol. 28(1-2), 98 (2013)

N.M. Shinde, D.P. Dubal, D.S. Dhawale, C.D. Lokhande, J.M. Kim,.J.H. Moon. Material Res. Bull., 47, (2012) 302-307.

R. J. Deokate, R. S. Kate, S. C. Bulakhe, https://doi.org/10.1007/s10854-018-00630-0 (2019).

S. Abermann, Sol. Energy. 94, 37(2013)

S. Ahmed, K.B. Reuter, O. Gunawan, L. Guo, L.T. Romankiw, H. Deligianni, Adv. Enegy Mater. 2, 253 (2012)

S.A. Khan, S. Irfan, Z. Zhuanghao, S.L. Lee, Influence of Refractive Index on Antireflectance Efficiency of Thin Films. Materials Communication. doi:https://doi.org/10.3390/ma12091483 (2019).

S.K. Wallace, D.B. Mitzi, A. Walsh, 2(4), 776 (2017)

W.S. Rasband, (2014), in: http://imagej.nih.gov/ij/1997-2014.

Y. Akaltum, M.A. Yildirim, A. Ates, M. Yildirim, Zinc concentration effect on structural, optical and electrical properties of $Cd_{1-x}Zn_xSe$ thin films. Materials Research Bulletin 47 (2012) 3390-3396.

Daily and Cumulative Biogas Yields from Selected Animal Dungs

Ochuko Mary Ojo

1 Introduction

Demand for energy is increasing with upsurge in population. The ever-increasing demands for energy, combined with the shortage of fossil fuels almost all over the world have generated a converted awareness on the utilization of renewable energy sources (Kavitha and Joseph, 2007). The exploration of substitute renewable energy sources is needed not to replace fossil fuels but also to safeguard the environment. Biogas production and its use as an alternative renewable source of energy is progressively gaining attention, particularly in the developing countries where access to modern sources of energy is limited (Abdulkareem, 2005).

Biogas is a gas produced from the anaerobic decomposition of organic waste. It contains mainly methane and carbon dioxide, but may also have traces of hydrogen sulphide and water vapour. Biogas is an affordable energy source suitable for application on Nigerian farms and villages where over 70% of the population lives (Itodo et al., 2007). Several researchers have studied biogas production from anaerobic digestion of animal and agricultural wastes (Sadaka and Engler, 2000; Bujoczek et al., 2000; Castrillon et al., 2002; Kivaisi, 2002; Gelegenis et al.,2007; Ojolo et al., 2007; Momoh and Nwaogazie, 2008; Li et al., 2009; Budiyono et al., 2010; Ofoefule et al., 2010; Yusuf et al., 2011; Ojo, 2017; Ojo et al., 2018 and Ojo et al., 2019, Ojo and Babatola, 2020). Biogas production is a viable solution to organic waste management. The primary aim of this study is to evaluate the production of biogas from three common animal waste namely cow dung, poultry droppings and swine dung. This was done by assessing the daily and cumulative biogas yields from the three selected feedstocks.

O. M. Ojo (✉)
Department of Civil Engineering, the Federal University of Technology, Akure, Nigeria
e-mail: omojo@futa.edu.ng

© The Author(s), under exclusive license to Springer Nature Switzerland AG 2022

A. O. Ayeni et al. (eds.), *Bioenergy and Biochemical Processing Technologies*, Green Energy and Technology, https://doi.org/10.1007/978-3-030-96721-5_4

2 Materials and Methods

Three 12-l capacity plastic digesters were used in this study. The materials used in constructing the digesters include 12-l capacity plastic containers, PVC pipes of different diameters, corks, 90° elbows, control valves and one-fourth inch rubber hose. Other materials used in the experimental set up include tyre tube, syringe tube, head and knob, gas hose and digital weighing scale.

2.1 *Digester Design Calculations*

A batch-fed anaerobic digestion system was adopted in this study. 12-l plastic containers were used in the construction of the digesters. The digesters consist of the digestion chamber and the gas chamber.

The volume of the digestion chamber was determined using Eq. 1.

$$\text{Volume of digestion chamber}\left(1\right) = \text{Daily feed in}\left(1/\text{day}\right)\times\text{Retention time}\left(\text{days}\right) \quad (1)$$

0.16 kg of feedstock was added to the digester on a daily basis for a retention period of 30 days and a waste to water ratio of 1:1 was used.

The total volume of digester's feed per day is given in Eq. 2 with the assumption that 1 kg of waste is equivalent to 1 l of water of substrate.

$$\begin{aligned}\text{Volume of Digester feed per day} &= 0.16\ \text{kg of waste} + 0.61\ \text{of water} \\ &= 0.321\ \text{of substrate}\end{aligned} \quad (2)$$

The digester volume is calculated using Eq. 3.

$$\begin{aligned}\text{Digester Volume} &= \text{Volume of digester feed per day}\times\text{retention time} \\ &= 0.32\times 30 = 9.6\,\text{litres}\end{aligned} \quad (3)$$

The volume of the gas chamber was taken as one-fifth the total volume of the digester and was given by Eq. 4.

$$\text{Volume of gas chamber} = 1/5\times\text{Total volume of digester} \quad (4)$$

$$\text{Volume of gas chamber} = 1/5\times 12 = 2.41$$

2.2 Anaerobic Digestion of Feedstocks

The feedstocks used in this study (cow dung, poultry droppings and swine dung) were collected from the teaching and research farm of the Federal University of Technology, Akure.

4.8 kg of waste was digested over the retention period of 30 days. The waste was mixed with water to form slurry in a ratio of one part of waste to one part of water to arrive at the substrate input of 0.0096 m^3/digester. The substrate input, S_d, for batch-type digester utilized in this study is given by Eq. 5.

$$S_d = \text{Biomass}(\text{B}) + \text{Water}(\text{W})\text{m}^3 / \text{batch} \tag{5}$$

$$S_d = 0.0048 + 0.0048 = 0.0096\text{m}^3 / \text{batch}$$

A retention time of 30 days was utilized in this study.

For a batch-type digester, the daily substrate input (batch input) equals the operating volume of the digester (V_0) (Kossmann et al. 2000).

Therefore,

$$V_o = S_d \; RT = 0.0096 = 30 \times 0.0032\,\text{m}^3$$

where V_0 is the operating volume of the digester.

Figure 1 shows the schematic diagram of the mini-digester and the gas collector. Each reactor was connected via its gas outlet to a tyre tube used as the gas collecting apparatus. The digestion process was done simultaneously for the three different substrates. The digesters were subjected to periodic agitation to ensure thorough mixing of the digester content while maintaining intimate contact between the microorganisms and substrate and to enhance complete digestion of the substrates. The volume of biogas yield was measured and recorded on a daily basis using a digital weighing scale.

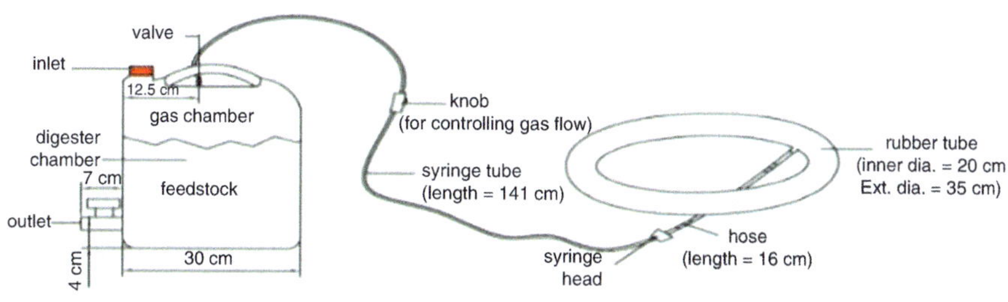

Fig. 1 Schematic representation of the mini-digester and the gas collector

3 Results and Discussions

3.1 Volume of Biogas Produced Daily for the Different Substrates

Figures 2, 3 and 4 show the average daily volume of biogas produced with the retention times for cow dung, poultry droppings and swine dung. The volume of biogas produced from cow dung began on the sixth day (as shown in Fig. 2) with a biogas volume of 0.0020 m^3; it increased steadily on the seventh and eighth day and is seen dropping on day 9. However, gas production fluctuated from day 10 up to day 23; then it began to increase until it attained its maximum volume at day 27 and then a decline in production rate occurred to day 30. 0.0032 m^3 of biogas was produced on day 30.

The volume of biogas production from poultry droppings (Fig. 3) began on day 1, it increased on day 5 to 0.0072 m^3 and it decreased to 0.0046 m^3 on day 8. Daily biogas produced fluctuated from 0.0072 m^3 on day 5 to 0.0070 m^3 on day 28. The maximum volume or peak volume of 0.0076 m^3 was attained on day 14. 0.0063 m^3 of biogas was produced on day 29 and 30, respectively.

The volume of biogas produced from swine dung (Fig. 4) began on day 10 with 0.0010 m^3 and increased gradually till day 14. Fluctuations in biogas production occurred till day 27, with maximum gas production of 0.0043 m^3 attained on days 22 and 25. Reduction in biogas produced occurred from day 27 to 0.0026 m^3 on day 30.

Comparing the daily volume of biogas produced for the different feedstocks, poultry droppings had the highest daily biogas production of 0.0076 m^3, and this occurred on day 14.

Fig. 2 Average daily biogas production against retention time (days) for cow dung

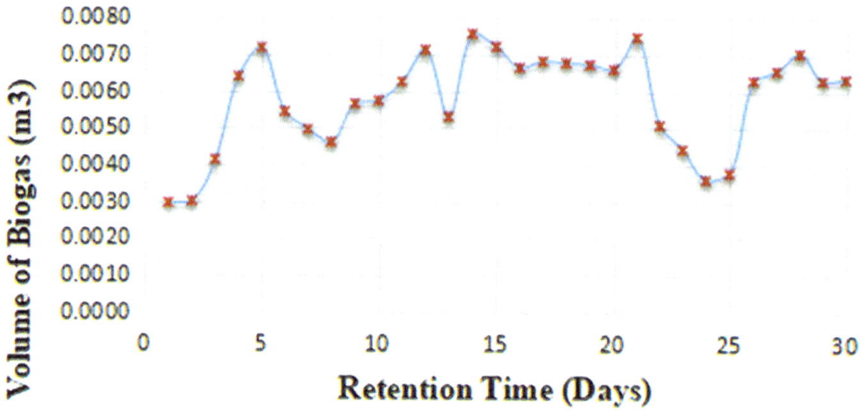

Fig. 3 Average daily biogas production against retention time (days) for poultry droppings

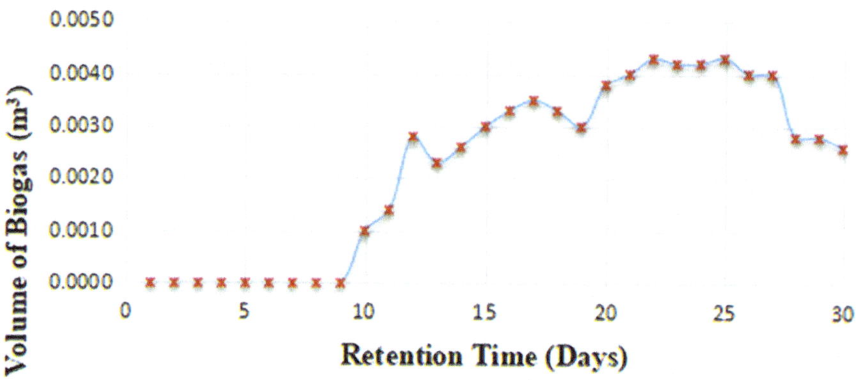

Fig. 4 Average daily biogas production against retention time (days) for swine dung

3.2 Cumulative Volume of Gas Produced for the Different Feedstocks

The cumulative volume of biogas produced for the different feedstocks is shown in Fig. 5.

Poultry droppings had the highest cumulative volume of biogas production of 0.1741 m^3 (0.0363 m^3/kg), while cow dung and swine dung, respectively, produced 0.1020 m^3 (0.0213 m^3/kg) and 0.0670 m^3 (0.0028 m^3/kg) of biogas. The single-substrate digestion of cow dung, poultry droppings and swine dung recorded average daily gas production of 0.0058 m^3, 0.0034 m^3 and 0.0022 m^3, respectively. Figure 5 shows that the poultry droppings produced more biogas than the cow dung and swine. This could be attributed to the nature of organics present in the substrates which also is a function of the feeds the animals were exposed to, as well as the slow rate of decomposition of the swine dung. The poultry feeds seem to be very rich in

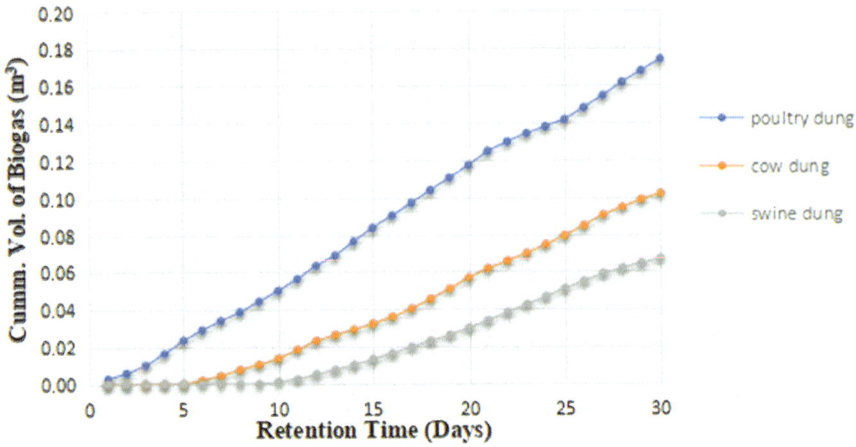

Fig. 5 Cumulative biogas production for the three substrates

organic nutrients. Cows also feed on organic-rich food; hence, their dungs contain more organic wastes which account for the higher biogas production relative to the swine dung.

The higher and faster biogas generation from the poultry dung could be attributed to the faster rate of decomposition of the substrate. The action of bacteria on this category of waste is fast relative to the swine dung and cow dung. The relative biogas yield from the substrates could be attributed also to the organic matter content of the wastes which confirms the submissions of earlier studies carried out on cow dung and poultry droppings (Ojolo et al., 2007, Ahmadu, 2009, Igboro, 2011, Ojo et al., 2018 and Ojo et al., 2019). The higher biogas production from poultry droppings could also be attributed to the available nutrient such as ammonia in the droppings. This is substantiated by Ojolo et al. (2007) and Ojo (2017) that ideal substrates for anaerobic digestion should contain adequate amount of carbon, oxygen, hydrogen and nitrogen. The drops in levels of gas production and subsequent peak could be associated to microbial succession in the digesters with respect to the retention time.

4 Conclusion

The cumulative biogas yield from 3 12-l capacity mini-digesters with 9.6 kg (1:1 waste to water ratio) slurry of cow dung, poultry droppings and swine dung was found to be 0.1020 m^3, 0.1741 m^3 and 0.0670 m^3, respectively. The results affirmed that poultry dung is one of the best feedstock for biogas production with its production profiles exceeding the production from the other two conventional feedstocks (cow dung and swine dung). Biogas is not just a renewable energy source for rural population but also an appropriate way of managing waste. The use of mini-digesters for biogas production should be considered by poor households who cannot afford the ever-increasing cost of fossil fuels.

References

Abdulkareem AS (2005) Refining biogas produced from biomass: An alternative to cooking gas. Leonardo J Sci 7: 1-8.

Ahmadu TO (2009) Comparative Performance of Cow Dung and Chicken Droppings for Biogas Production, M.Sc Thesis Submitted to the Department of Mechanical Engineering, Ahmadu Bello University, Zaria.

Budiyono IN, Widiasa S, Johari S. (2010) The kinetic of Biogas production Rate from Cattle Manure in Batch Mode. Int J Chem Biol 1: 39 – 45

Bujoczek G, Oleszkiewicz J, Sparling R, Cenkowski S (2000) High solid anaerobic digestion of chicken manure. J Agric Eng Res 76 : 51-60.

Castrillon L, Vazguez I, Maranon E, Satre H (2002) Anaerobic thermophilic treatment of cattle manure in UASB reactors. Waste Manage Res 20: 350-356.

Gelegenis J, Georgakakis D, Angelidaki I, Mavris V (2007) Optimization of biogas production by co-digesting whey with diluted poultry manure. Renew Energy 32(13): 2147-2160.

Igboro SB. (2011) Production of Biogas and Compost from Cow Dung in Zaria, Nigeria, unpublished PhD Dissertation Presented to the Department of Water Resources and Environmental Engineering, Ahmadu Bello University, Zaria Nigeria.

Itodo IN, Agyo GE, Yusuf P (2007) Performance evaluation of a biogas stove for cooking in Nigeria. J. Energy South. Africa, 18(3): 1-5.

Kavitha ES, Joseph K (2007) Biomethanation of vegetable wastes. J Inst Public Health Eng 3: 11-19.

Kivaisi AK (2002) Pretreatment of robusta coffee hulls and co-digestion with cow-dung for enhanced anaerobic digestion. Tanz J Sci 28:1-10.

Kossmann W, Pönitz U, Habermehl S (2000). Biogas Digest, Volume I: Biogas Basics, Information and Advisory Services on Appropriate Technology. Available online at: www.gtz.de/de/dokumente/en-Biogas-Volume I.pdf.

Li R, Chen S, Li X (2009) Anaerobic codigestion of kitchen waste and cattle manure for methane production. Energy Sources 31: 1848- 1856.

Momoh OLY, Nwaogazie LI (2008) Effect of Waste Paper on Biogas Production from Co-digestion of Cow Dung and Water Hyacinth in Batch Reactors J Appl Sci Environ Manage 12 (4): 95 – 98

Ofoefule AU, Uzodinma EO, Anyanwu CN (2010). Studies on the effect of Anaerobic Digestion on the microbial flora of Animal Wastes: Digestion and Modelling of Process Parameters. Trends Appl Sci Res 5 (1): 39-47.

Ojo OM (2017) Biomethanation of Water Hyacinth and Selected Animal Dungs for Biogas Production. Unpublished Ph.D Thesis in the Department of Civil Engineering, The Federal University of Technology, Akure.

Ojo OM, Babatola JO, Adesina AA, Akinola AO, Lafe O (2018) Synergistic Effect of co-digesting different mix ratios of Water Hyacinth and Cow-dung for Biogas production. FUTAJEET 12 (1): 54 – 59.

Ojo OM., Babatola JO, Akinola AO., Lafe O, Adelodun AA (2019) Co-Digestion of Water Hyacinth and Poultry Manure for Improved Biogas yield. ABUAD J Eng Res Dev 2 (1): 42 – 48

Ojo OM, Babatola JO(2020) Association between Biogas Quality and Digester Temperature for Selected Animal Dung-Aided Water Hyacinth Digestion Mixes. J Appli Sci Environ Manage 24 (6): 966-959

Ojolo SJ, Oke SA, Animasahun OK, Adesuyi, BK (2007) Utilization of poultry, cow and kitchen wastes for biogas production: A comparative analysis. J Environ Health Sci Eng 4: 223-228.

Sadaka S, Engler, C (2000) Effects of mixing on anaerobic composting of beef manure. In: Proceeding of ASAE Annual International Meeting, Technical papers: Engineering Solutions for a New Century, 9-12 July, pp. 4993-5001.

Yusuf MOL, Debora A, Ogheneruona DE (2011) Ambient temperature kinetic assessment of biogas production from co-digestion of horse and cow dung. Res Agric Eng 57: 97-104.

Energy and Fuels: Sol-Gel Synthesized Core-Shell $0.5Li_2MnO_3 \cdot 0.5LiNi_{0.5}Mn_{0.3}Co_{0.2}O_2$ Material: Effect of Mixed Fuel (Citric Acid and Ammonium Acetate) on the Structural Properties

Samuel O. Ajayi, Cyril O. Ehi-Eromosele, and Kolawole O. Ajanaku

1 Introduction

In material science, lithium-rich layered oxide (LRLO) cathode material is one of the recent research topic in recent years on the account of its outstanding thermal stability, high capacity, less toxicity, low cost of production, and high operating voltage (Ellis et al. 2010; Prakasha et al. 2017). They are generally represented by $xLiMO_2 \cdot (1 - x)Li_2MnO_3$ (M = $3d$-transition metals). Normally, LRLO materials are synthesized in form of composite where $LiMO_2$ R-3 m rhombohedral layered phase intergrowth with Li_2MnO_3 monoclinic layered phase forming an heterostructure (Ates et al. 2015; Sathiya et al. 2015; Li et al. 2016). Recently, studies have shown the structural performance of LRLO materials could be improved by synthesizing them in form of a core-shell configuration. The nickel-rich layered oxide which is accountable for high capacity of the material serves as the core, while the manganese-rich layer which is responsible for the thermal stability of the material serves as the shell in a hybrid system (Li et al. 2015, 2016; Ehi-Eromosole et al. 2020).

The solution combustion and the sol-gel methods were adopted due to their simplicity, time conserving, energy conserving, cost effectiveness, and purity and higher uniformity of reaction particles (Ma et al. 2016, 2017). The mixed fuel (citric acid monohydrate and ammonium acetate) was adopted to optimize the synthetic route which has not been reported for the synthesis of LRLO core-shell material in literatures to the best of our knowledge. In this study, citric acid monohydrate and ammonium acetate in 3:1 proportion were used since ammonium acetate is known to create pores and generate higher amount of gases during synthesis. The pores created affect the morphology and the structural properties of the synthesized material.

S. O. Ajayi (✉) · C. O. Ehi-Eromosele · K. O. Ajanaku
Department of Chemistry, Covenant University, Ota, Nigeria
e-mail: samuel.ajayi@covenantuniversity.edu.ng

© The Author(s), under exclusive license to Springer Nature Switzerland AG 2022

A. O. Ayeni et al. (eds.), *Bioenergy and Biochemical Processing Technologies*, Green Energy and Technology, https://doi.org/10.1007/978-3-030-96721-5_5

2 Materials and Methods

2.1 *Synthesis of 0.5Li_2MnO_3–0.5$LiNi_{0.5}Mn_{0.3}Co_{0.08}O_2$ (CS-SG-7525) Material*

The CS-SG-7525 material was synthesized using a combination of the solution combustion synthesis (SCS) to prepare the $LiNi_{0.5}Mn_{0.3}Co_{0.08}O_2$ core material and a sol-gel method to form the core-shell structure (Ehi-Eromosele et al. 2016; Zuo et al. 2017). The SCS route was optimized by varying the quantity of citric acid monohydrate and ammonium acetate in the ratio 3:1. In a usual procedure, 0.73 g of $Ni(NO_3)_2{\cdot}6H_2O$, 0.38 g of $Mn(NO_3)_2{\cdot}4H_2O$, 0.29 g of $Co(NO_3)_2{\cdot}6H_2O$ (Sigma-Aldrich ≥98%), 0.35 g of $LiNO_3$, 0.32 g of $C_2H_7NO_2$, and 0.66 g of $C_6H_8O_7.H_2O$ were dissolved in distilled water, and the solutions were further heated to 80 °C under magnetic stirring to form a gel. In addition, the gel was placed on a hot plate preheated to 300 °C, and combustion occurred forming the core precursor powder. Stoichiometric amounts of the shell coating Li_2MnO_3 precursor (0.42 g of $LiOH{\cdot}H_2O$ and 0.52 g of $Mn(NO_3)_2{\cdot}4H_2O$) were dissolved in distilled water. Then gradually, a stoichiometric amount of chelating agent (1.77 g of $C_6H_8O_7{\cdot}H_2O$) was added to form a sol. Then, the synthesized core precursor ($LiNi_{0.5}Co_{0.2}Mn_{0.3}O_2$) was added into the sol under magnetic stirring and heated at 80 °C to evaporate the solution to form a gel rapidly. The obtained gel was dried in the vacuum for 4 hours. Subsequently, it was annealed at 1000 °C for 10 hours to obtain final coated products. The coated product was cooled, grounded, and labeled CS-SG-7525 and was used for further characterization.

2.2 *Characterization of the CS-SG-7525 Material*

The quantitative elemental compositions (Li, Mn, Ni, Co) of the CS-SG-7525 material were precisely detected by inductively coupled plasma optical emission spectroscopy (ICP-OES, iCAP 7600DUO from ThermoFisher Scientific). The crystal structure and phase composition of the synthesized materials were determined by powder X-ray diffraction (XRD) using a STOE STADI MP diffractometer. The powder samples were loaded in 0.5 mm diameter glass capillaries with diffraction patterns collected in a transmission geometry using a Mo K_α radiation ($\lambda = 0.709320$ Å) and a MYTHEN 1 K detector. The local structure of the powder samples was studied by Raman spectroscopy using Horiba LabRam Evolution HR microscope equipped with a 632 nm HeNe solid-state excitation laser (17 mW), a 100× objective, and a 600 grooves per millimeter grating (Ehi-Eromosele et al. 2020). The morphology of the synthesized materials was examined by SEM/EDX (Zeiss Merlin GEMINI 2).

3 Results and Discussion

3.1 Structural Characterization of CS-SG-7525 Material

It was observed during the experimental study that the precursor solution formed a pink slurry gel after drying at 80 °C. The slurry gel further formed a darkish powder after combustion at 300 °C with gas evolution during the combustion process. This could be attributed to the addition on ammonium acetate which is known to generate pores during combustion process thereby leading to gas evolution. The chemical composition of the CS-SG-7525 material investigated by ICP-OES is presented in Table 1. The investigation shows the elemental ratio of Li:Mn:Ni:Co in CS-SG-W-1000 material is 1.14:0.47:0.18:0.07. These ratios are in excellent agreement with the target theoretical/nominal value stoichiometry within experimental error ($Li_{1.2}Mn_{0.52}Ni_{0.20}Co_{0.08}O_2$) with nickel in divalent state, cobalt in trivalent state, and manganese in tetravalent state (Koga et al. 2012). The Li/M ratio for the samples is higher than the theoretical/nominal value due to lower evaporative lithium loss during synthesis and annealing.

The XRD pattern obtained for the CS-SG-7525 material is presented in Fig. 1a. The diffraction peaks are narrow, sharp, intense, and clear and well-fitted indicating a high purity and excellent crystallinity of the synthesized material (Li et al. 2015; Kunjuzwa 2017; Ehi-Eromosele et al. 2020). The diffraction peaks observed could be indexed to α-$NaFeO_2$ hexagonal structure with rhombohedral symmetry (R3m). The diffraction peaks 003, 101, 006, 102,004, 108, and 110 were observed at 9°, 10°, 15°, 20°, and 25°, respectively. A clearly distinctive splitting of (006)/(102) and (108)/(110) doublet peaks indicating a good layered structure material was observed in all the XRD pattern. The broad and weak diffraction peak between 10–15° observed could be ascribed to the $LiMn_6$ cation ordering in the monoclinic Li_2MnO_3 domains (C2/m space group) (Ehi-Eromosele et al. 2020).

The full weight at half maximum (FWHM) of the (003) and (104) peaks is 0.08° and 0.16°, respectively, the intensity ratio is 1.25, and the crystallite size is 46.26 for CS-SG-7525 material as shown in Table 2. The small values of the FWHM suggest better crystallinity and homogenous distributions of cations in the synthesized materials. Little cationic mixing is observed for CS-SG-7525 material since the intensity ratio is higher than 1.20 (Shi et al. 2013; Ma et al. 2016; Deng et al. 2017; Huang et al. 2019). Ma et al. (2016) explained that the higher the synthesis annealing temperature, the better the crystallinity and the more pronounced the layered structure of the synthesized material. The highly crystallized layered structure with little amount of cation mixing observed for CS-SG-7525 material could be

Table 1 Chemical compositions of CS-SG-7525

	Li	Mn	Ni	Co	O	Li/M
Theoretical value	1.20	0.52	0.20	0.08	2.00	1.50
CS-SG-W-7525	1.10	0.51	0.16	0.06	1.95	1.51

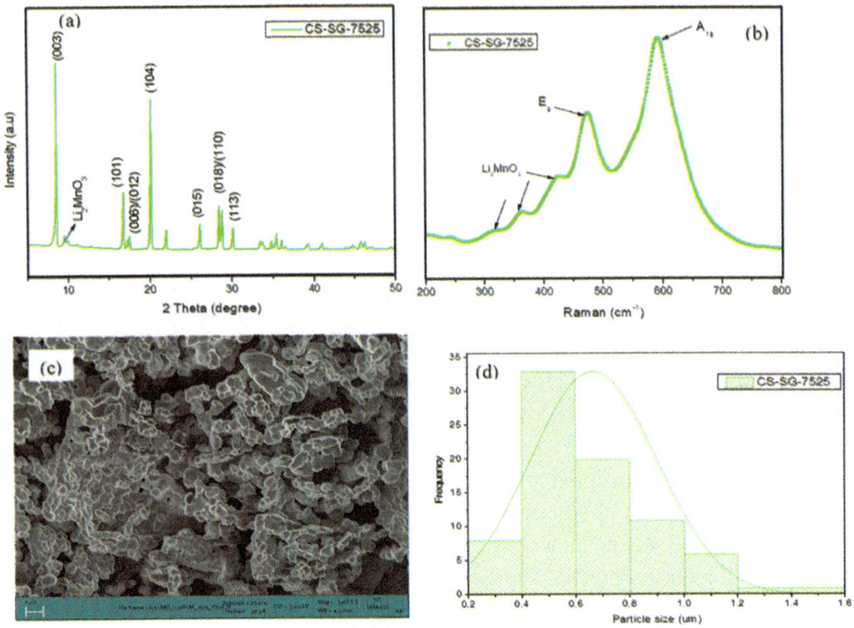

Fig. 1 (**a**) XRD pattern of CS-SG-7525 (**b**) Raman spectra of CS-SG-7525 (**c**) SEM micrograph of CS-SG-7525 (**d**) The particle size histogram of CS-SG-7525

Table 2 Structural parameters obtained from XRD data for CS-SG-7525

Cathode material	FWHM $_{(003)}$ (°)	FWHM $_{(104)}$ (°)	$I_{(003)}/I_{(104)}$	Crystallite size (nm)
CS-SG-7525	0.08	0.16	1.25	46.26

attributed to the synthesis high annealing temperature (Ma et al. 2016; Iftikhar 2016). Two strong peaks are observed at 500 cm^{-1} and 600 cm^{-1} for E_g and A_{1g} Raman modes, respectively, for the material. These peaks are attributed to rhombohedral phase (R3m space group) of a layered structure of core segment ($LiNi_{0.5}Co_{0.2}Mn_{0.3}O_2$). Some narrow and weak peaks are also observed between 300–440 cm^{-1}, and they can be attributed to the fingerprint vibration of monoclinic symmetry (C2/m space group) of the layered structure of the shell segment (Li_2MnO_3) (Deng et al. 2017). The Raman spectra validate the powder XRD data presented for CS-SG-W-7525 in Fig. 1a confirming its layered structure with monoclinic phase (C/2 m space group) and rhombohedral phase (R3m space group). The XRD and Raman spectra obtained in this study revealed a well-layered crystalline structure for the synthesized material.

The surface morphological characterization of the CS-SG-7525 material investigated by scanning electron microscopy (SEM) is presented in Fig. 1c. Figure 1c reveals that the material is made up of irregular particle shape with rough surfaces. The particle size of the materials ranges from 0.28 to 1.49 μm with average particle size of 0.67 μm as shown in Fig. 1d. The transmission electron microscopy (TEM) results of CS-SG-7525 are awaited to confirm the core-shell structure of this material.

4 Conclusion

The $0.5Li_2MnO_3{\cdot}0.5LiNi_{0.5}Mn_{0.3}Co_{0.2}O_2$ core-shell material was successfully synthesized using fuel mixture (citric acid monohydrate and ammonium acetate) in 3:1 proportion as chelating agent. The XRD and Raman spectra obtained in this study revealed a well-layered crystalline structure for the synthesized material. The material shows very low cationic mixing which is a desired structural characteristic in cathode materials. This material also revealed a chemical composition close to the nominal values indicating the suitability of the synthetic method. The material is made up of irregular particle shape with rough surfaces. The synthesized material could be used as cathode material in Li-ion batteries.

Acknowledgments The Alexander von Humboldt Foundation fellowship and the assistance of the technical staff of IAM-ESS, Karlsruhe Institute of Technology, Germany, are appreciated. The Covenant University's financial support is also sincerely appreciated.

Conflicts of Interest There are no conflicts of interest to declare.

References

Ates MN, Mukerjee S, Abraham KM (2015) A high rate Li-rich layered MNC cathode material for lithium-ion batteries. RSC Adv 5(35):27375-27386.

Deng YP, Yin ZW, Wu ZG et al (2017) Layered/spinel heterostructured and hierarchical micro/nanostructured Li-rich cathode materials with enhanced electrochemical properties for Li-ion batteries. ACS Appl Mater 9(25):21065-21070.

Ehi-Eromosele CO, Ita BI, Iweala EEJ et al (2016) Structural and magnetic characterization of $La_{0.7}Sr_{0.3}MnO_3$ nanoparticles obtained by the citrate-gel combustion method: Effect of fuel to oxidizer ratio. Ceram Int 42(1):636-643.

Ehi-Eromosele CO, Indris S, Bramnik NN et al (2020) In situ x-ray diffraction and x-ray absorption spectroscopic studies of a lithium-rich layered positive electrode material: Comparison of composite and core–shell structures. ACS Appl Mater 12(12):13852-13868.

Ellis BL, Lee KT, Nazar LF et al (2010) Positive electrode materials for Li-ion and Li-batteries. Chem Mater 22(3):691-714.

Huang L, Liu L, Wu H et al (2019) Optimization of synthesis parameters for uniform sphere-like $Li_{1.2}Mn_{0.54}Ni_{0.13}Co_{0.13}O_2$ as high performance cathode material for lithium ion batteries. J Alloys Compd 775:921-930.

Iftikhar M (2016) Synthesis and characterization of lithium-manganese rich cathode materials for lithium ion batteries (Doctoral dissertation, Quaid-i-Azam University Islamabad).

Kunjuzwa N (2017) Synthesis of cation substituted $LiMn_{2-x}M_xO_4$ (M= A1, Ni) cathode materials for a lithium ion battery: Improving energy storage, capacity retention, and lithium transport (Doctoral dissertation).

Li J, Shunmugasundaram R et al (2016) In situ x-ray diffraction study of layered Li–Ni–Mn–Co oxides: effect of particle size and structural stability of core–shell materials. Chem Mater 28(1):162-171.

Li L, Xu M, Chen Z et al (2015) High-performance lithium-rich layered oxide materials: Effects of chelating agents on microstructure and electrochemical properties. Electrochim Acta 174:446-455.

Ma Q, Li R, Zheng R et al (2016) Improving rate capability and decelerating voltage decay of Li-rich layered oxide cathodes via selenium doping to stabilize oxygen. J Power Sources 331:112-121.

Ma X, He H et al (2017) Synthesis of $Li_{1.2}Mn_{0.54}Co_{0.13}Ni_{0.13}O_2$ by sol–gel method and its electrochemical properties as cathode materials for lithium-ion batteries. J Mater Sci Mater Electron 28(22):16665-16671.

Prakasha KR, Sathish M et al (2017) Mitigating the surface degradation and voltage decay of $Li_{1.2}Ni_{0.13}Mn_{0.54}Co_{0.13}O_2$ cathode material through surface modification using Li_2ZrO_3. ACS Omega 2(5):2308-2316.

Sathiya M, Abakumov AM et al (2015) Origin of voltage decay in high-capacity layered oxide electrodes. Nat Mater 14(2):230-238.

Shi SJ, Tu JP, Tang YY et al (2013). Combustion synthesis and electrochemical performance of Li[$Li_{0.2}Mn_{0.54}Ni_{0.13}Co_{0.13}$]$O_2$ with improved rate capability. J Power Sources 228:14-23.

Zuo D, Tian G, Li X et al (2017). Recent progress in surface coating of cathode materials for lithium ion secondary batteries, J Alloys Compd 706:24-40.

Energy Use and Carbon Footprint in a University: Nigeria Case Study

A. O. Adelaja, O. A. Omotoriogun, A. A. Oluwo, and O. M. Oyekeye

1 Introduction

Enormous awareness has been created on the central role of energy and the associated impacts, arising from its use by humans, on the environment, and on the associated consequences for sustainable development since the Kyoto Protocol and United Nations Framework Convention on Climate Change. To develop sustainable policies and initiatives in this regard, data on energy supply and end-use are necessary. This data will facilitate the development and implementation of energy management policies for private or government-owned institutions. The typical university selected for this purpose is the University of Lagos, Akoka, Nigeria, where the authors are familiar, and access to some information is easier. In addition, no publicly known energy policy programme is implemented, so the information to be provided in this paper can serve as the starting point for putting one in place.

Energy consumption or demand and the related GHGs have been studied extensively in the domestic building, service, commercial, and industrial sectors. Ireland (Howley and O'Gallachoir 2004; Bosseboeuf et al. 2005; O'Leary et al. 2005; O'Gallachoir et al. 2007); Australia (Lenzen 1998); West Germany, France, and the Netherlands (Weber and Perrels 2000); the USA (Bin and Dowlatabadi 2005); the UK (Baiocchi and Minx 2010); India (Pachauri and Spreng 2002; Devi et al. 2009); Namibia (Vita et al. 2006), Syria (Hainoun et al. 2006); and the province of Granada in Spain (Carpio et al. 2018) are examples.

In Universities, however, analyses of energy consumption or carbon footprints have also been considered. An extensive literature review of energy consumption, mobility, and carbon footprints has been done by Helmers et al. (2021). Their study

A. O. Adelaja (✉) · O. A. Omotoriogun · A. A. Oluwo · O. M. Oyekeye
Department of Mechanical Engineering, University of Lagos, Akoka, Yaba, Lagos, Nigeria
e-mail: aadelaja@unilag.edu.ng

© The Author(s), under exclusive license to Springer Nature Switzerland AG 2022
A. O. Ayeni et al. (eds.), *Bioenergy and Biochemical Processing Technologies*, Green Energy and Technology, https://doi.org/10.1007/978-3-030-96721-5_6

compared the energy use, mobility, and carbon footprints of 20 universities across the globe and made some rankings based on the performance indices. Yanez et al. (2020) evaluated the carbon footprints of the University of Talca (UT), Chile, based on the sources and observed that though GHG is the most widely accepted means of estimation, there are other methodologies. They identified transportation of staff and students on- and off-campus and the operation of the boiler as a stressor. Syafrudin et al. (2020) studied the carbon footprint of Diponegoro University, Indonesia, based on academic activities in 2018. Electricity and transportation accounted for the most significant contributors to CF, and in terms of sectoral mapping, the faculty of engineering contributed the most. Other studies on the consumption and carbon footprints are as follows: University College Cork, Ireland (O'Gallachoir et al. 2007), Pennsylvania University, USA (Braham et al. 2007), Massey University, New Zealand (Kent and Venter 2006), the University of Lagos, Nigeria (Adelaja et al. 2008), and King Abdullah University of Science and Technology, Saudi Arabia (Adenle et al., 2017).

Others have considered remedies to cut carbon emissions by planting trees and biomass resources that could help as carbon sink either in national parks or in botanical gardens (Qasim, 2017; Agonafir and Worku, 2017). Efficient maintenance and replacing fossil fuel-based equipment with renewable powered ones was noted to cut down emissions considerable (Carpio et al., 2018; Nitkiewicz and Ociepa-Kubicka, 2018).

For the purpose of analysis, several predictive tools, either mathematical models or software, have been used for energy demand/consumption. Examples are the Bayesian approach (Xiao et al. 2007), genetic algorithm (GA) and artificial neural network (ANN) (Azadeh et al. 2007), multivariate linear regression model (Al-Ghandoor et al. 2008), Grey prediction with rolling mechanism (GPRM) (Akay and Atak 2007), and energy input and output approach (E-IO) (Chung et al. 2011).

In this paper, section "Introduction" presents the introduction, while section "Overview of University of Lagos" contains the overview of the University and energy consumption pattern on sectoral basis. The software employed for analysis is presented in section "The Software". Furthermore, sections "Projection and Growth in Energy Consumption in the University of Lagos (2009–2033)" and "Projection and Growth in Carbon Footprint in the University of Lagos (2009–2033)" deal with the projections in energy consumption and carbon footprint, respectively, while results and discussion are presented in section "Discussion". The paper ends in section "Conclusion" with conclusions.

2 Overview of University of Lagos

The University of Lagos, fondly called "Unilag", is one of the foremost Universities in Nigeria. The location at the centre of the cosmopolitan state of Lagos, the second most populous state in Nigeria according to the 2006 national census, the former national capital, and the habitat of most of the multinational companies make it a

University of the first choice for candidates seeking admission into a tertiary institution in Nigeria.

The institution is a federal government-established University founded in 1962. It comprises two campuses: the main campus is located at Akoka and the College of Medicine is located at Idi-Araba, both in Lagos State. The main campus is surrounded mainly by the Lagos lagoon and is situated on 3.25 km^2 of land. The enrolment was grown from 131 students in 1962 to the current over 40,000 students. The University comprises nine faculties and the College of Medicine. Table 1 shows the statistics of the student enrolment and staff population between 1996 and 2008. More than 117 programmes are offered, including master's and doctoral studies in Arts, Social Sciences, Engineering, Environmental Sciences, Pharmacy, Law, Sciences, Business Administration, and Education. Apart from the regular programmes, the school offers part-time programmes on the main campus under the Distance Learning Institute (DLI), Sandwich, and Human Resources Development Board (HRDB). All these programmes attract students from all walks of life to the University coupled with excellent scholars.

2.1 *Limitations*

Energy analysis in an economy with no energy policy poses a significant challenge to benchmarking analysis results on energy-related issues. This challenge is more complicated where energy data are not readily available and a "walk through audit" had to be conducted.

Table 1 Demography of student enrolments and staff in University of Lagos

Year	Academic staff population	Non-academic staff population	Total staff population	Total student population	Total school population
1996	573	2870	3443	24,242	27,685
1997	581	3106	3687	23,351	27,038
1998	727	2992	3719	26,629	30,348
1999	619	3036	3655	30,348	41,160
2000	814	2938	3752	35,174	38,926
2001	744	3430	4174	30,699	34,873
2002	723	3449	4172	39,228	43,400
2003	705	3320	4025	41,120	45,145
2004	734	3186	3920	40,869	44,789
2005	1024	3500	4524	39,914	44,438
2006	1069	3237	4306	39,759	44,065
2007	1316	3320	4636	39,084	43,720
2008	1287	na	na	38,829	40,116
2010	1290	na	na	49,402	50,692

na – these records are not available in published document of the academic planning unit

This involved interaction with every energy-consuming device available, checking the specifications in terms of wattage, obtaining electrical energy data from the facility management of the Works and Services Unit. Where accessibility was difficult, for example, in some staff, residence or estimation was necessary, questionnaires were used and analysed for gas, charcoal, and firewood usage in the eateries. This was necessary because there was no comprehensive database to consult.

Energy data in government agencies and tertiary institutions in Nigeria are generally inadequate in quality, depth, and scope of coverage. At the University of Lagos, various challenges were encountered in the course of gathering data. Some of these are inadequate qualitative and quantitative data of electrical energy consumption and faulty metering system on electricity consumption for the faculties, service areas, and the student hostels. In addition, the incessant power outages are not accounted for by the electricity utility division.

2.2 *Energy Consumption Pattern in University of Lagos*

There were six major forms of energy sources in the institution in 2006. These are firewood, charcoal, kerosene, liquefied petroleum gas (LPG), electricity, and diesel (Adelaja et al. 2008). Of the energy mix, electricity has the most significant contribution, about 97.4% (Table 2). Furthermore, electrical energy was most applicable for water heating, cooking, space cooling, refrigeration, computers, lighting, laboratory machines, electronics, miscellaneous machines/ equipment, etc. The University buildings and facilities are categorized into three: faculty and service, residential areas, and commercial centres.

The national energy policy document, though drafted in 2003, is not yet passed into law. Table 3 shows some energy data between 1996 and 2004. Between 1996 and 1998, there was an increase after which there was a decline until 2000. There was a peak in 2002 and nosedived until 2004. This was primarily due to breakdowns in the two 2250 kVA diesel-generating sets (which were synchronized) that complement the erratic supplies from the national grid.

It should be noted that the electrical energy consumed was not disaggregated into the proportions that come from the national grid and that which are contributed by the diesel generators. The electricity tariff for the University of Lagos was categorized under residential billing systems, and this was raised in 2002 to ₦/kWh 6.09 up from ₦/kWh 4.98. Other forms of energy utilized are diesel 1.05%, LPG 0.88%, kerosene 0.54%, charcoal 0.09%, and firewood 0.04%, respectively, on the same energy unit basis.

Table 2 University of Lagos energy mix in 2008

	Categorization				
Fuels	Faculty and service area	Residential area	Commercial centres	Total	Percentage %
Firewood (kg)	–	–	5, 022		
(MJ)	–	–	77, 839	77, 839	0.04
Charcoal (kg)	–	317	11, 692		
(MJ)	–	4, 914	181, 230	186, 144	0.09
Kerosene (litres)	–	8, 978	16, 625		
(MJ)	–	401, 785	743, 983	1, 145, 768	0.54
LPG (kg)	–	20, 554	18, 788		
(MJ)	–	972, 411	888, 837	1, 861, 248	0.88
Diesel (litres)	–	299, 25.2	19, 732		
(MJ)	–	1, 331, 689	884, 006	2, 215, 696	1.05
Electricity (kWh)	719, 254	725, 828	45, 530		
(MJ)	77, 679, 445	78, 389, 464	49, 172, 502	205, 241, 411	97.40

Table 3 Trends of energy consumption pattern in University of Lagos

Year	1996	1998	2000	2002	2004	2006	2008	2010
Energy consumption (GWh)	8.89	11.35	12.85	15.18	12.76	13.62	12.81	12.00
Energy per capita (MWh/person)	0.32	0.37	0.33	0.35	0.28	0.31	0.32	0.24
Electricity tariff (₦/kWh)	4.98	4.98	4.98	6.09	6.09	6.09	6.09	6.09

Currency conversion: $1.00 = ₦160.00 (2010)

The Demand Sector

By sector, energy consumption in the University can be divided into three significant areas earlier mentioned. The energy consumption on the sectoral basis is depicted in Table 4 for the base year 2008.

Table 4 indicates the proportion of the energy consumption for the base year 2008. This shows that the residential areas consume the most energy, about 70% of the total energy consumption, while the share of the commercial centre is merely 8.1%. 60% of the contribution from the residence is attributable to space cooling, refrigeration, and cooking (Fig. 5).

The faculty and service sector comprises the faculty buildings (offices, classrooms, and lecture theatres), administrative buildings (senate building, council chamber, cashier office, and student affairs), medical centre, indoor and outdoor lighting, etc. The energy consumption in the sector is presented in Fig. 1. From

Table 4 Energy consumption on sector basis for the base year 2008

Location	Average annual demand (GWh)	Percentage %
Faculty and service area	12.8	21.9
Residential area	40.9	70.0
Commercial Centre	4.7	8.1
Total	57.4	100

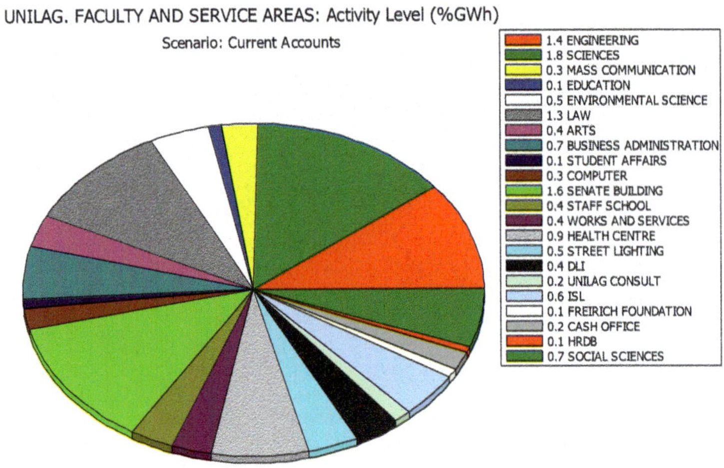

Fig. 1 Energy demand in faculty and service area for the base year 2008

Figs. 1, 2, 3, and 4, the Faculty of Sciences consumed 1.8 GWh, followed by Senate building 1.6 GWh, Engineering 1.4 GWh, Law 1.3 GWh, etc.

The large portion attributed to Sciences is mainly due to the space cooling and refrigeration and the laboratory equipment such as engineering laboratory, kitchen and office equipment such as electric kettle, microwave, photocopy machines, scanners, printers and computers. Space cooling, refrigeration, high wattage electric heating, and elevators are major contributors to the 12.8% ascribed to the Senate building.

The seven student hostels and the staff quarters comprise **the residential areas**, and the activities are presented in Fig. 2. In the residential area, Figs. 2 and 5 display energy data in the households, which are summed up on a street basis (where street refers to the street, avenue, or a close) to compress the input. The energy consumption in the street is summed up and referred to by the name of that street. Staff residences (Ozolua road 5.0 GWh and high rise building 5.0 GWh) and female student's hostels (Fagunwa 4.8 GWh, Makama 4.6 GWh, and Madam Tinubu 3.6 GWh) had high consumption. Space cooling and refrigeration are major contributors to consumption in staff residence, while high wattage electric heaters, pressing iron,

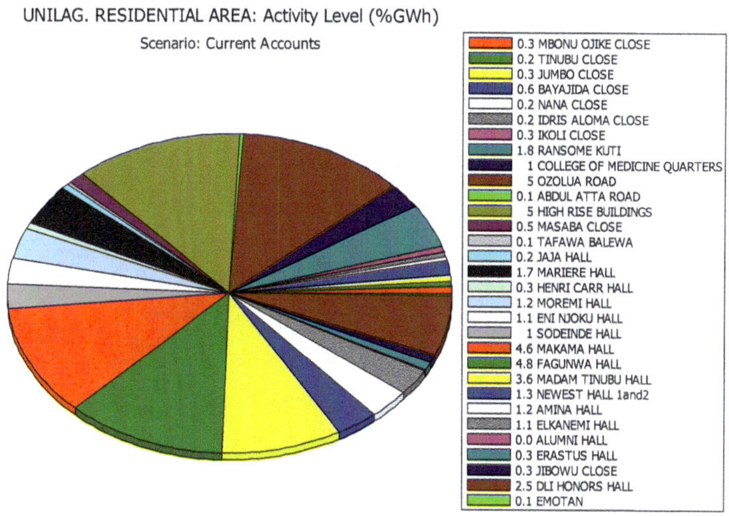

Fig. 2 Energy demand in the residential area for the base year 2008

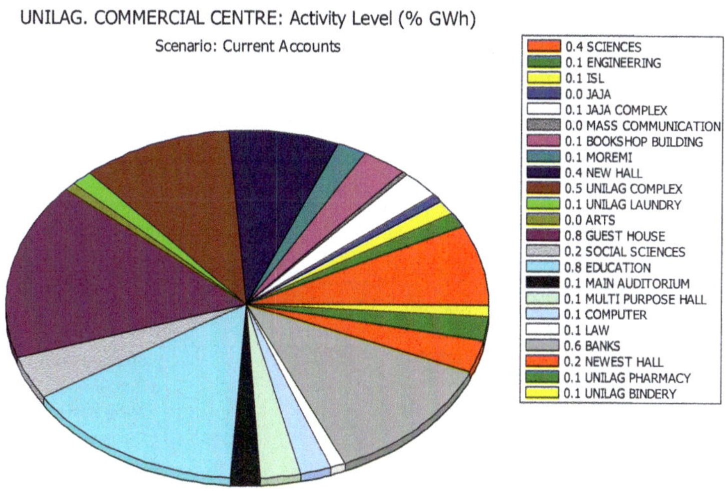

Fig. 3 Energy demand in the commercial centre for the base year 2008

electric stove, and hair driers are responsible for consumption in the three female halls (Fig. 5).

The commercial centres comprise all buildings associated with religion (two churches and a mosque) and business ventures such as the guest houses, business centres, pharmacy, laundry, bindery, banks, bookshop, post office, etc. A typical energy consumption distribution is shown in Fig. 3. Figures 3 and 6 reveal that activities such as conferences, lounge sing, etc. endeared the guest houses' 0.8 GWh. In comparison, Education Faculty's 0.8 GWh comes from office space

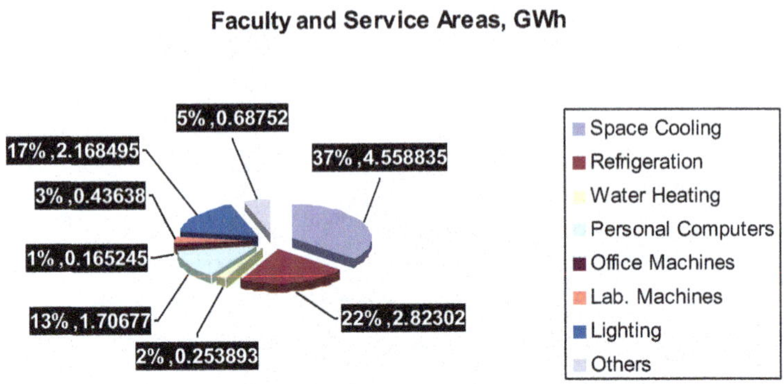

Fig. 4 End-use categorization for faculty and service areas for the base year 2008

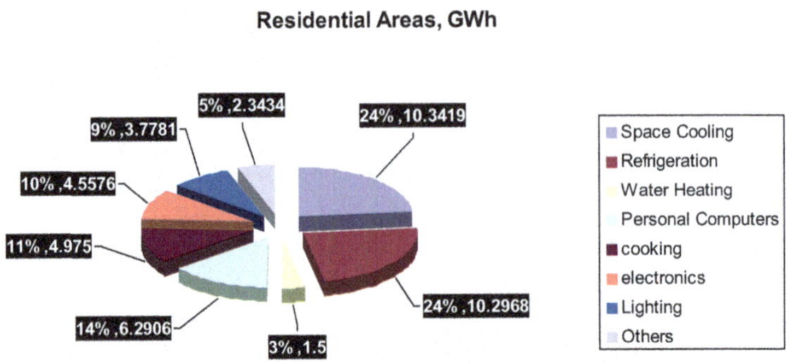

Fig. 5 End-use categorization for the residential centre for the base year 2008

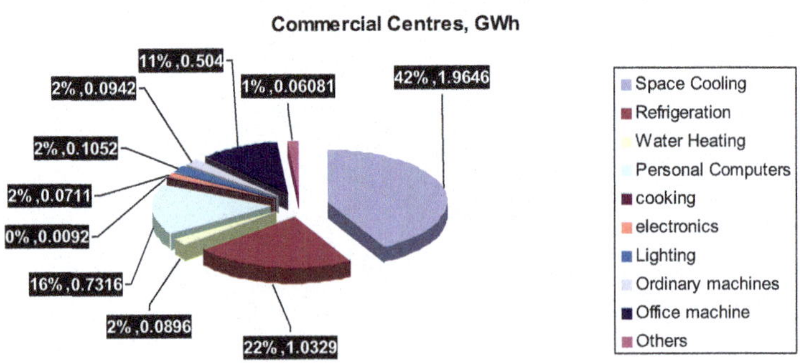

Fig. 6 End-use categorization for the commercial centre for the base year 2008

cooling, refrigeration of the numerous departments, and part-time programme offices and classrooms. The 0.6 GWh attributed to banks is the summation of consumption of the five banks situated in the institution.

End Use

The electrical energy end use is categorized as follows (Adelaja *et al.* 2008):

1. Water heating comprises energy use by various types of water heaters.
2. Cooking includes energy used for cooking (kerosene, gas, firewood, and electricity), bread toaster, microwave oven, etc.
3. Space cooling comprises energy utilized by air conditioners, ventilation equipment, etc.
4. Computers (laptops and desktops), printers, scanners, and servers.
5. Refrigerator comprises energy utilized by refrigerators, deep freezers, and water dispensers.
6. Lightening (incandescent, fluorescent, halogen, and stage lamps).
7. Laboratory equipment/machine: this can be in the form of electricity, gas, and kerosene.
8. Electronics appliances such as television, radio, VCRs, DVD players, decoder.
9. Miscellaneous machines include pumping, washing, drier machines, etc.

3 The Software

The LEAP software is used for the energy-environment simulation and is a scenarios-based tool. Its scenarios are based on comprehensive accounting of how energy is utilised, transformed, and generated in an economy, sector, region, and a country subject to a range of alternative assumptions regarding price, economy, population, technology, development, etc. The data structure of the software is flexible and enables the end user to specify some specific technical details. It can recognize the least cost scenarios, but the user experience and information will be needed to obtain market equilibrium or optimum plans. LEAP can be employed as a forecasting tool, to give energy information, store data, etc. It can function as a policy analysis instrument to model and evaluate the impacts of physical, environmental, and economic factors on alternative energy programme, investment options, and activities. It assists in assessing a wide range of projects, programme activities, technological innovations, and many other energy initiatives and proposes strategies that best address energy and environmental challenges. Figure 7 shows a typical LEAP window.

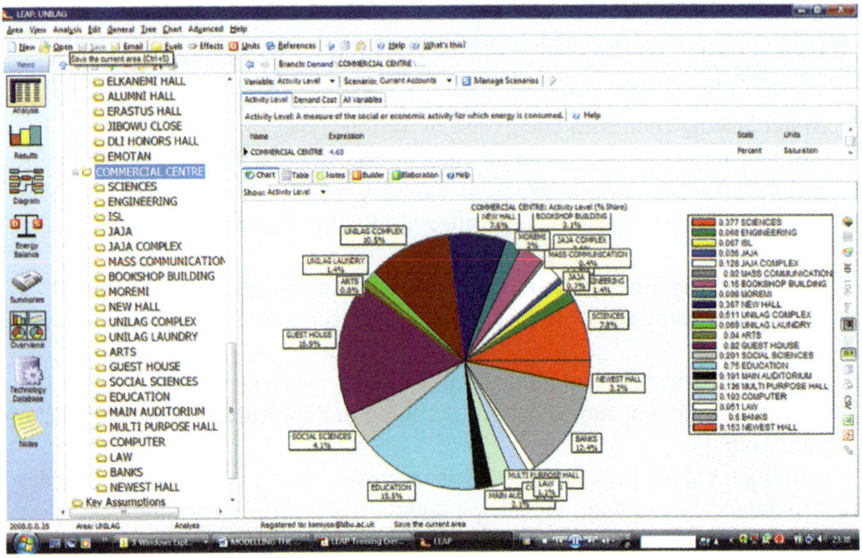

Fig. 7 A typical LEAP window

4 Projection and Growth in Energy Consumption in the University of Lagos (2009–2033)

LEAP is employed to forecast the 25-year period (2009–2033); the base year is taken as 2008 and the target year 2033. Calculating the average annual (compound) growth rate (AAGR) of the population between 1996 and 2007 gives 3.88%. This is necessitated by the fact that the population data for 2008–2010 are not available. However, not all the staff and students presented in Table 1 live in the University residence due to inadequate housing units. Considering that new hostels may be constructed in the near future to meet the inadequacy, there is the need to adjust the AAGR of the population accommodated in the University residence. For this reason, the population is multiplied by a factor of 0.722, and a new AAGR is obtained as 2.8%. The value obtained is utilized for the business as usual scenario growth rate. For the projections, three scenarios are considered based on different growth rates. For the 2.8% growth rate, the energy consumption projection is presented in Fig. 8, for the 3.8% growth rate in Fig. 9, while for the 1.8% growth rate, the result is illustrated in Fig. 10.

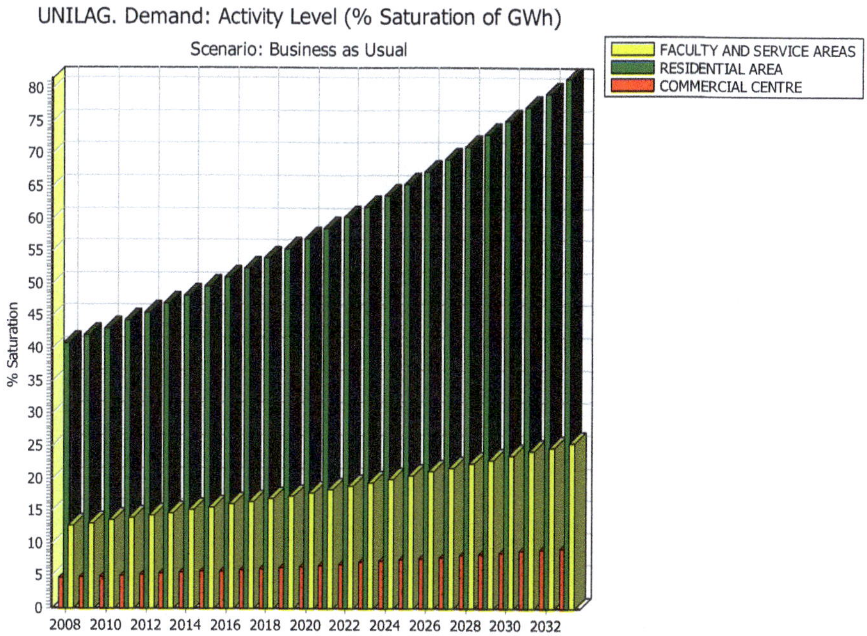

Fig. 8 Twenty-five years energy consumption projection for business as usual scenario

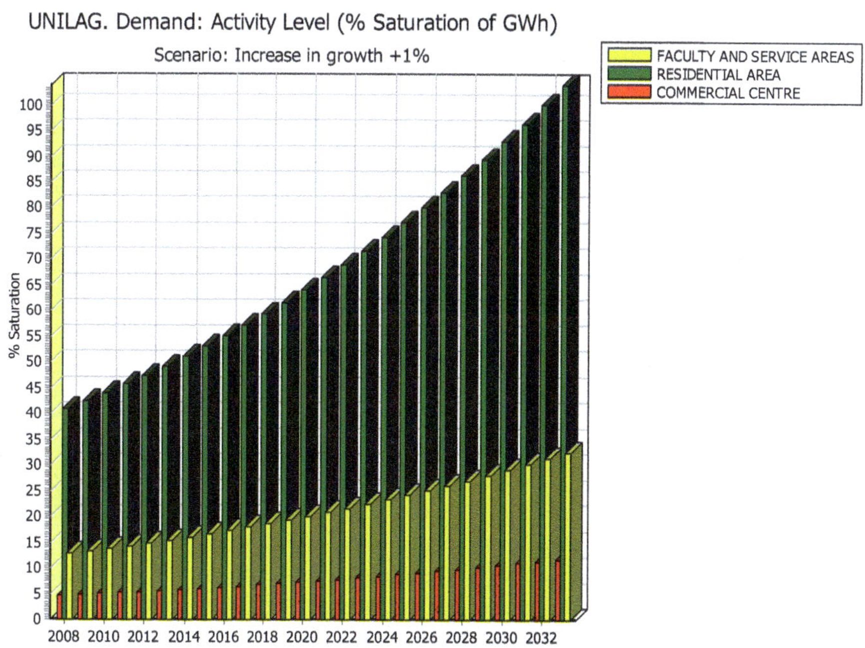

Fig. 9 Twenty-five years energy consumption projection for 3.8% growth rate scenario

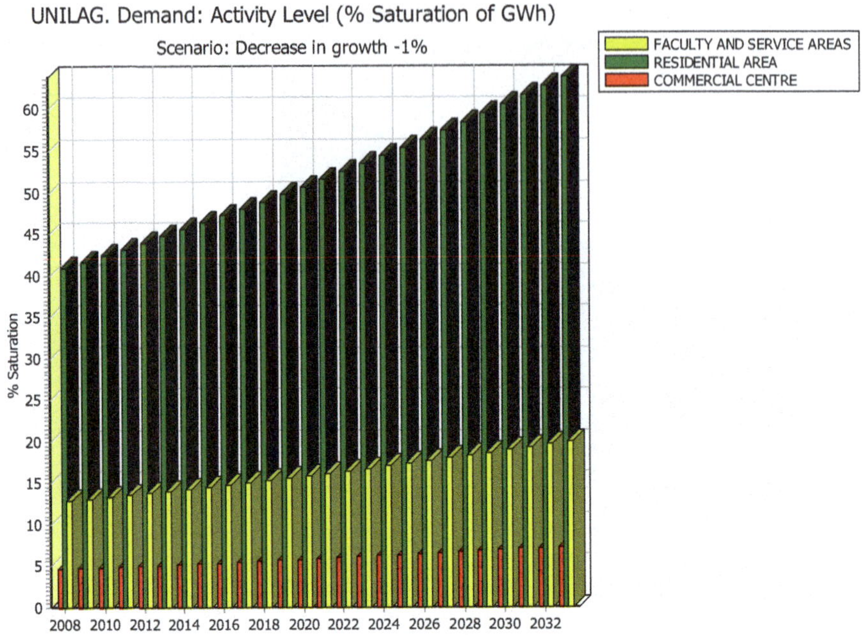

Fig. 10 Twenty-five years energy consumption projection for 1.8% growth rate scenario

5 Projection and Growth in Carbon Footprint in the University of Lagos (2009–2033)

Carbon footprint is an indication of the consequence of human activities on the environment with respect to the amount of GHGs produced. It evaluates the impact of institutions and their operations on global warming. Since the Kyoto Protocol, GHG's emission efforts have focused on reducing six atmospheric gases with recognised "direct" greenhouse effect on the global climate. These are carbon IV oxide (CO_2), nitrous oxide (N_2O), methane (CH_4), hydrofluorocarbon (HFC), perfluorocarbon (PFC), and sulphur hexafluoride (SF_b). Air pollutants or "indirect" GHGs such as non-methane volatile organic compounds (NMVOCs), nitrogen oxides (NO_x), sulphur dioxide (SO_2), and carbon monoxide (CO), however, contribute to global warming by producing GHGs through reactions with other chemical compounds.

These reactions lead to chemical alteration, affecting the atmospheric lifetime of other GHGs and affecting the absorptive characteristics of the atmosphere, such as cloud formation. While these gases have different global warming potentials (GWPs), they are commonly indexed to an equivalent amount of carbon dioxide (CO_2e) to make a simple comparison and evaluation. The trend in CO_2 emission in relation to energy patterns has varied in time according to the energy fuel mix, and the deciding factor relates to the share of electricity consumption. The loading of the

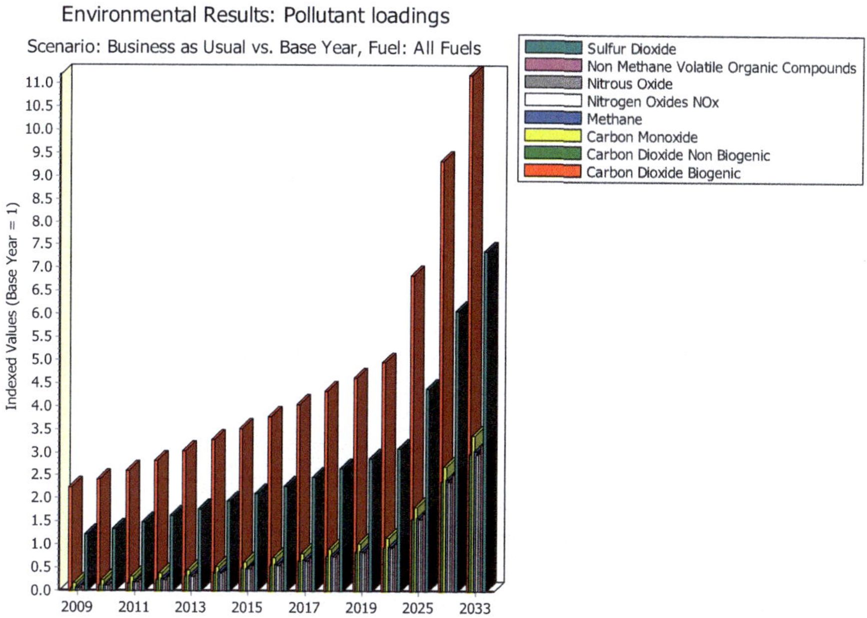

Fig. 11 GWP (CO_2e) of the GHGs in the energy mix for 2009–2033

pollutant in the sources projected over 25 years (2009–2033) using the current growth rate, i.e., business as usual, is shown in Fig. 11. The GWP (CO_2e) for the same period is given in Fig. 12 for the residence and commercial centre. Results of the faculty and service areas are not captured because they are very insignificant compared to the other two sectors.

6 Discussion

The growth rate utilized in the software is determined from demographic data; the population growth rate is estimated at 2.8%. Three scenarios are considered; high growth rate (3.8%) denoted by high, 2.8% growth rate as business as usual, and low growth rate (1.8%) as low.

Using these three scenarios, energy consumption and GWP of the GHGs at CO_2e are projected for 25 years. Figures 8, 9, and 10 show the growth in consumption for the faculty and service areas, residential areas, and commercial centres between the base and target years for the three scenarios, respectively. Expectedly, the high scenario shows high energy consumption with residential benchmarking 103% share, faculty and service area 30%, and commercial centres 12% percentage saturation, respectively, at the target year. For the business as usual, 82%, 25%, and 10% saturation, respectively, are projected, while for the low, it was 66%, 20%, and 7% saturation, respectively.

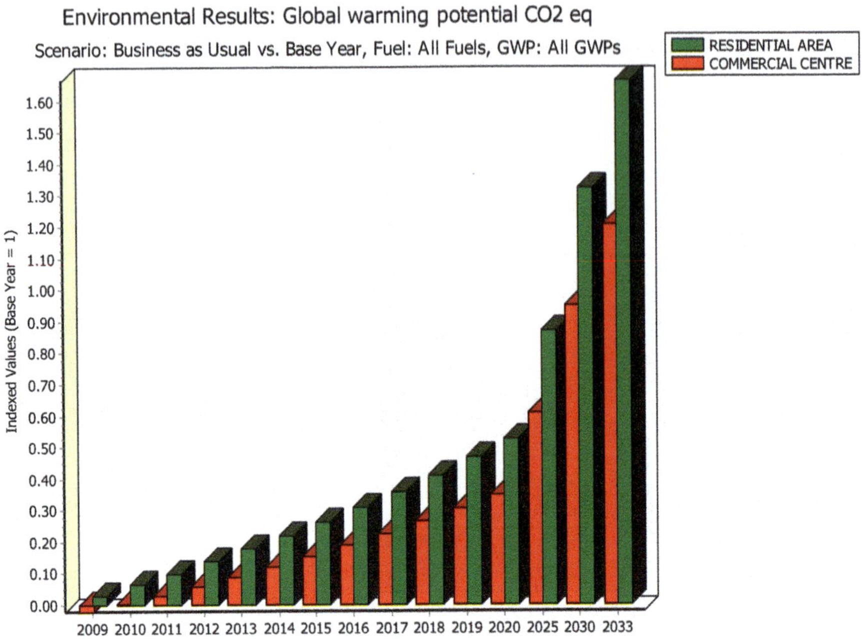

Fig. 12 GWP (CO_2e) of the GHGs in the energy mix for 2009–2033 on sectoral basis

For the carbon footprints, LEAP estimates the environmental loading of GHGs emitted from LPG, kerosene, firewood, charcoal, and diesel fuels utilized in the residential and commercial centres. This excludes the faculty and service areas since electricity is the form of energy applicable and no energy production. Figure 11 shows the trends of different GHGs emitted in the institution. A rise of 1100% in biogenic CO_2, 740% in SO_2, and 330% in CO, etc. is expected by the target year. Figure 12 shows that the residential area accounts for a more significant contribution of GWP compared with the commercial centres. For the CO_2**e**, by the target year, the residential area and commercial centre must have contributed 25% and 60%, respectively. In order to reverse this trend, cleaner energy options such as solar, wind (which may be available as a result of the nearness of the institution to the lagoon), and biomass are proposed as alternatives to the current fossil fuel-based options.

7 Conclusion

This paper reveals that despite the absence of a national energy policy to guide energy management in various institutions and agencies, energy analysis is imperative for regulating the balance between supply and demand either for the short, medium, or long term. A comprehensive data collection of the demographic and energy parameters were carried out during a walk-through energy audit in an educational institution, the University of Lagos, in 2006 and 2008. This information which excludes transportation helps estimate energy demand and carbon footprint for the institution between the base and target years. From the energy mix, GWP has the most contribution from biogenic CO_2, SO_2, and CO, respectively, while other GHGs are insignificant. Two sectors significantly contribute to the GWP; the residential area and commercial centres.

Acknowledgement The authors are grateful to the University of Lagos Works and Services and Academic Planning Unit for the energy consumption and demography data, respectively.

References

Adelaja, AO, Damisa, O, Oke, SA, Ayoola, AB, Ayeyemitan, AO (2008) A survey on the energy consumption and demand in a tertiary institution. Maejo Int. J. Sci. Technol. 2(2): 331-344.

Adenle, Y, Alshuwaikhat, H (2017) Spatial estimation and visualization of CO_2 emissions for campus sustainability: The case of King Abdullah University of Science and Technology (KAUST), Saudi Arabia. Sustainability 9: 15.

Agonafir, H, Worku, M (2017) Carbon stock in Gullele Botanical Garden: Implications for carbon emissions reduction, North Western Addis Ababa, Ethiopia. J. Environ. Earth Sci. 7: 40–52.

Akay, D, Atak, M (2007) Grey prediction with rolling mechanism for electricity demand forecasting for Turkey. Energy 32: 1670-1675.

Al-Ghandoor, A, Al-Hinti, I, Jaber, JO, Sawalha, SA (2008) Electricity consumption and associated GHG emissions of the Jordanian industrial sector: empirical analysis and future projection. Energy Policy 36: 258-267.

Azadeh, A, Ghaderi, SF, Tarverdian, S, Saberi, M (2007) Integration of artificial neural network and genetic algorithm to predict electrical energy consumption. Appl. Math. Comput. 186: 1731-1741.

Baiocchi, G, Minx, JC (2010) Understanding changes in the UK's CO_2 emissions: a global perspective. Environ. Sci. Technol. 44: 1177-1184.

Bin, S, Dowlatabadi, H (2005) Consumer lifestyle approach to US energy use and the related CO_2 emissions. Energy Policy 33: 197-208.

Bosseboeuf, D, Lapillone, B, Eichammer, W, Faberi, S. (2005). Energy efficiency monitoring in the EU-15. *ADEME*, ISBN-13: 978-2-86817-819-7.

Braham, W, Malkawi, A, Martin, M, Lee, J, William, S, Bernstein, E, Gabrielian, A. (2007). University of Pennsylvania carbon footprint. Report, 1-49. http://www.aashe.org/documents/resources/ PennGreenhouse GasReport.pdf.

Carpio, M, Martín-Morales, M, Zamorano, M (2018) Environmental and economic effects of using renewable energy in residential thermal installations according to 2030 targets: Case study in the Province of Granada (Spain). Constr. Technol. 93: 474–477.

Chung, W-S, Tohno, S, Choi, K-H (2011) Socio-technological impact analysis using an energy input- output approach to GHG emissions issues in South Korea. Appl. Energy 88: 3747-3758.

Devi, R, Singh, V, Dahiya, RP, Kumar, A (2009) Energy consumption pattern of a decentralized community in northern Haryana. Renew. Sustain. Energy Rev. 13: 194-200.

Hainoun, A, Self-Eldin, MK, Almoustafa, S (2006) Analysis of the Syrian long-term energy and electricity demand projection using the end-use methodology. Energy Policy 34: 1958-1970.

Helmers, E, Chang, C, Dauwels, J (2021) Carbon footprinting of universities worldwide: Part I—objective comparison by standardized metrics, Environ. Sci. Eur. 33: 30 1-25.

Howley, M, O'Gallachoir, BP (2004) Energy in Ireland 1990-2002, trend issues and indicators. Sustainable Energy Ireland Report 04-EPSSU-006-R/01.

Kent, C, Venter, M (2006) Massey University-energy management report for 2005.

Lenzen, M (1998) Primary energy and greenhouse gases embodied in Australian final consumption: an input-output analysis. Energy Policy 26(6): 495-506.

Nitkiewicz, T, Ociepa-Kubicka, A (2018) Eco-investments-Life Cycle Assessment of different scenarios of biomass combustion. Ecol. Chem. Eng. S. 25: 307–322.

O'Gallachoir, BP, Keane, M, Morrissey, E, O'Donnell, J (2007) Using indicators to profile energy consumption and to inform energy policy in a University-A case study in Ireland. Energy Build. 39: 913-922.

O'Leary, F, Howley, M, O'Gallachoir, BP (2005) Profiling energy and CO_2 emission in the services sector. Sustainable Energy Ireland Report 05-EPSSU-006-R/01.

Pachauri, S, Spreng, D (2002) Direct and indirect energy requirements of household in India. Energy Policy 30: 511-523.

Qasim, M (2017) Forest carbón stock assessment of the Musk Deer National Park, Azad Jammu and Kashmir (AJK). J. Trop. For. Environ. 7: 122–127.

Syafrudin, S, Zaman, B, Budihardjo, MA, Yumaroh, S, Gita, DI, Lantip DS (2020) Carbon footprint of academic activities: A case study in diponegoro university, IOP Conf. Ser. Earth Environ. Sci. 448, 012008

Vita, GD, Endersen, K, Hunt, LC (2006) An empirical analysis of energy demand in Namibia. Energy Policy 34: 3447-3463.

Weber, C, Perrels, A (2000) Modeling lifestyle effects on energy demand and related emissions. Energy Policy 28: 549-566.

Xiao, N, Zarnikau, J, Damien, P (2007) Testing functional form in energy modeling: an application of the Bayesian approach to US electricity demand. Energy Econ. 29: 158-166.

Yanez, P, Sinha, A, Vásquez M (2020) Carbon footprint estimation in a university campus: Evaluation and insights Sustainability 12, 18.

Overview on Advancement of Sustainable Heterogeneous Catalysts for the Production of Biodiesel

Vincent Efeovbokhan, Tolu Makinwa, Oluranti Agboola, and Olagoke Oladokun

1 Overview of Biodiesel

Biodiesel is a bioenergy source got from animal fats, vegetable oils and micro-algal oil (Abbaszaadeh et al. 2012). It is suitable for use in normal diesel engines. It came into effect in the 1990s as a result of global warming and emission of greenhouse gases from vehicles, together with its benefits over regular diesel in terms of toxicity and biodegradability (Ali 2017). Due to its similarity in physical characteristics with petro-diesel, it is utilized as a fuel in its neat state or mixed with petro-diesel, devoid of any adjustment to the engine (Robles-Medina et al. 2009), storage facility or distribution equipment (Gonzaga and Sobral 2012). Recently, the fast exhaustion of fossil fuel reservoirs has also made biodiesel an interesting substitute for diesel fossil fuel (Omar and Amin 2011). It is estimated that biodiesel along with other biofuels will substitute for 10% of Europe's diesel consumption and 5% of the total fuel demand in South East Asia (Phan and Phan 2008). Besides, biodiesel is environmentally friendly because it produced exhaust gas that is sulphur and aromatic compound free. It is biodegradable and innocuous (Marchetti et al. 2007; Nie et al. 2006). In addition, biodiesel possesses many other benefits, like having emission profile that is low (this includes potential carcinogens). It is a renewable resource, and it has no contribution to the upsurge in the levels of carbon dioxide in the atmosphere, it has a greater cetane number, reduced toxicity and, as it can be used in sensitive areas and hence, reduced the intensity of the greenhouse effect (Abbaszaadeh et al. 2012; Yunus-Khan et al. 2014; Balat and Balat 2010; Demirbas 2009). Nonetheless, biodiesel production is very expensive, and this is the main

V. Efeovbokhan · T. Makinwa · O. Agboola (✉) · O. Oladokun
Department of Chemical Engineering, Covenant University, Ota, Nigeria
e-mail: oluranti.agboola@covenantuniversity.edu.ng

© The Author(s), under exclusive license to Springer Nature Switzerland AG 2022

A. O. Ayeni et al. (eds.), *Bioenergy and Biochemical Processing Technologies*, Green Energy and Technology, https://doi.org/10.1007/978-3-030-96721-5_7

hindrance to its commercialization. Literature has the report that roughly 70–95% of the whole biodiesel production cost is associated with raw materials cost (Azócar et al. 2010). The other limitations of biodiesel over diesel fuel include higher emission of nitrogen oxides and lower energy content – about 12% less than diesel – possess a greater cloud and pour point, higher viscosity and greater corrosion rate (Yunus-Khan et al. 2014). For the purpose of mitigating the cost of production, waste cooking oil (WCO) has been considered an attractive and hopeful feedstock; this could result in the reduction of the cost of biodiesel production to 60–70% as it is considered a low-cost raw material (Math et al. 2010). Apart from the fact that the production of biodiesel from WCO will circumvent the competition of the same oil resources for food and fuel, it will, in addition, solve the glitches related to the discarding of WCO (Farooq et al. 2013). However, to obtain a realistic conversion of the WCO to biodiesel, the transesterification reaction is usually implemented with the aid of an appropriate catalyst system.

In general, homogeneous acid or base catalysts are employed in the production of biodiesel from diverse feedstocks. Conventional homogeneous catalysts can be acid or base types. They usually have numerous benefits such as high catalytic activity and slight reaction conditions in the temperature range of 313.15 K to 338.15 K and atmospheric pressure. The difference between homogeneous catalyst and heterogeneous catalyst is depicted in Table 1. Nonetheless, the utilization of homogeneous catalysts results in numerous glitches, such as reactor corrosion, soap production, strain in recovering the catalyst and the production of large quantities of wastewater, thus increasing the total cost of biodiesel production (Farooq et al. 2013). Figure 1 shows some of the drawbacks of using homogeneous catalysts for biodiesel production, making heterogeneous catalysis a prospective technology for biodiesel production.

This review discusses the sustainable advancement of heterogeneous catalysts for the production of biodiesel.

Table 1 Difference between homogeneous catalyst and heterogeneous catalyst

Homogeneous catalyst	Heterogeneous catalyst
The catalyst and the reactant of homogeneous catalytic system possess same phase that is intensive to water content and fatty acid	The catalyst and the reactant of heterogeneous catalytic system possess different distinctive solid phases
Its separation is usually difficult	Its separation is usually easy
Its selectivity is high	Its selectivity is low
The process is not cost effective and its recycling process is difficult	The process is cost effective and its recycling process is easy
The system is unstable at lower temperature and pressure	The system is robust at high temperature and pressure
The system possess short life and extensive purification procedure	The system possess long life and less purification procedure

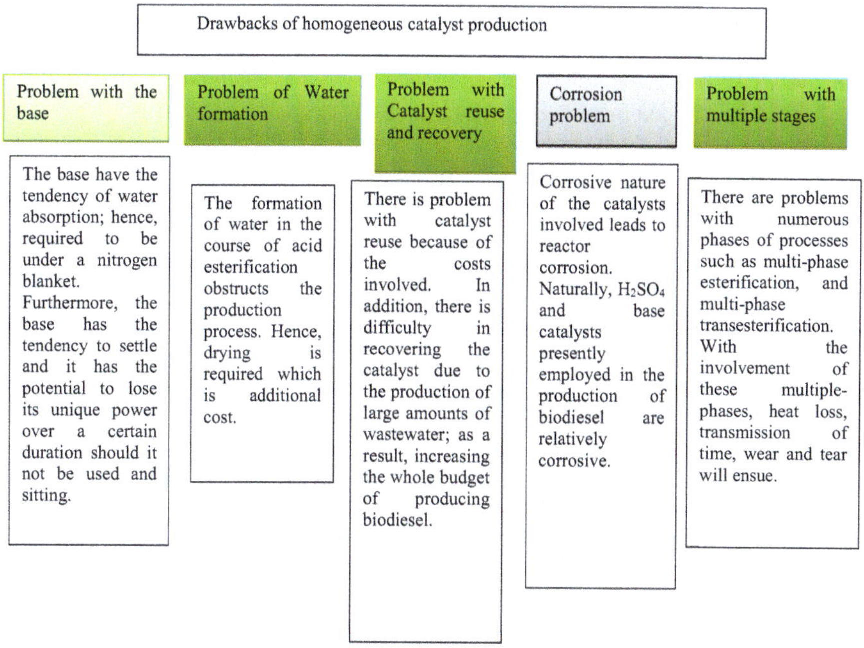

Fig. 1 The potential drawbacks of homogeneous catalyst

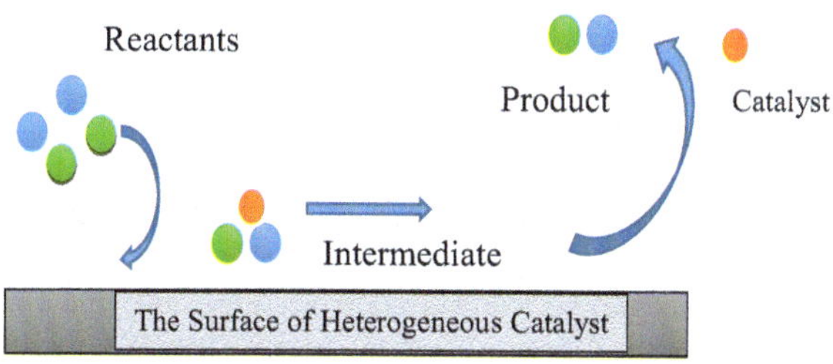

Fig. 2 Result of a simple heterogeneous catalysis that ensued during catalytic process

2 Overview of Heterogeneous Catalysts

Heterogeneous catalysts are those which phase is different to the reactants. The result of a simple heterogeneous catalysis that ensued during catalytic process is shown in Fig. 2. Numerous research works on these heterogeneous catalysts have been and are still being carried out to find solution glitches related to homogeneous catalysis in the production of biodiesel (Talha and Sulaiman 2016). Heterogeneous catalysts could be either acidic or basic. In recent times, basic catalysis is now the most extended process in the biodiesel production (Borges and Díaz 2012). Generally, solid base catalysts have high activeness than their acidic counterparts in that they require relatively less extreme reaction conditions – temperature and time

(Hara 2009). However, the uses of these basic catalysts come with their disadvantages. For example, the presence of water and free fatty acids in the oil has great impact on transesterification, resulting in side reactions which reduce the yield and lead to complications in recovering the biodiesel (Leung et al. 2010).

Metal oxides are the most usually basic heterogeneous group studied. Examples of such are magnesium oxide, calcium oxide and strontium oxide, just to mention a few (Borges and Díaz 2012). Within these catalysts, calcium oxide is the most preferred due to its higher activity, longer life time and ability to run under moderate reaction conditions (Kouzu et al. 2008a). In addition, it can be obtained naturally as calcium has numerous biological waste sources.

2.1 Natural Sources of Heterogeneous Base Catalysts

For the purpose of ensuing the sustainability of biodiesel synthesis, interest has peaked on the utilization of waste heterogeneous catalysts (Talha and Sulaiman 2016). There are several natural sources of calcium widely used catalysts in transesterification, to eliminate waste and reduce the cost of production (Chouhan and Sarma 2011). Generally, $Ca(NO_3)_2$, $CaCO_3$ and $Ca(OH)_2$ are raw materials from which calcium oxide (CaO) is synthesized (Cho et al. 2009). These raw materials are major compounds found in calcium-rich materials such as shells, bones and horns, which often times form a huge bulk of human waste.

The use of waste crab shells as a calcium oxide catalyst source has been used to synthesize esters from palm olein (Boey et al. 2009). Calcium oxide obtained from waste eggshells was also found to be competent catalyst, producing a yield of 97% (Viriya-empikul et al. 2010b). Calcined oyster and chicken shells have been evaluated to be suitable catalysts in the conversion of oils to fatty acid methyl esters (Boey et al. 2009). CaO obtained from cow bones, after calcination, has also been tested and found to follow similar trends in performance with the conventional CaO in the production of biodiesel (Ayoola et al. 2018a). The bones of animals were employed as economical catalyst for the transesterification of palm olein, giving an optimal biodiesel yield of 96.78% (Obadiah et al. 2012). Calcium oxide gotten from chicken eggshells and ostrich eggshells were utilized for biodiesel production, with yields of 96% and 94%, respectively (Buasri et al. 2014). Some authors have utilized calcined scallop shells as catalyst in the transesterification of waste oil with methanol for the production of biodiesel (Sirisomboonchai et al. 2015). White bivalve clam shell gotten from the sea shore was also used to synthesize calcium oxide for transesterification (Niju et al. 2016). The ash of the river snail shell was used to obtain calcium oxide (Kaewdaeng et al. 2017). The catalytic activities of calcium oxide samples obtained from the shells in biodiesel production have likewise been looked into (Buasri et al. 2013). Heterogeneous catalyst in the form of calcium oxide was obtained from the *Pomacea* specie shell and used as a catalyst for biodiesel production (Margaretha et al. 2012). Calcium oxide from eggshells can also been employed as a catalyst for biodiesel production synthesis (Ayoola et al. 2018b).

2.2 Preparation of Heterogeneous Base Catalysts

Typically, purchased heterogeneous base catalysts do not require special preparation before use. However, in some cases, they are made to undergo some form of treatment to improve their activity. For example, CaO may be calcined at temperatures up to 900 °C for some time to combat poisoning that may have occurred from the absorption of atmospheric water and carbon dioxide (Zhang et al. 2010). Calcinations may also be used as a form of preparation treatment to get rid of carbonates and hydroxyl groups from the surface of the catalysts, as well as to improve the transesterification reaction (Granados et al. 2009). However, calcium oxide obtained from natural sources requires extensive preparation procedures before it can be used for catalysis in transesterification. This is because that natural sources of calcium often occur in the form of carbonates, nitrates and hydroxides, mixed with other compounds and organic matter (Cho et al. 2009). Hence, cleaning procedures along with extended high-temperature conditions (calcinations) are required to convert these natural sources to calcium oxide and maximize its activity. Reports state that for calcium oxide prepared from alternative sources, the catalytic activity is maximized at temperatures between 900 °C and 1100 °C (Zhu et al. 2006).

In a study involving calcium oxide derived from cow bones, the waste cow bones were initially washed, then broken and milled to small particles, cooked at 100 °C for 4 hours using a pressure cooker, removed and dried for 1 hour at 105 °C, further grinded into powder and finally calcined at to 800 °C for another 4 hours to ensure maximum conversion to CaO (Ayoola et al. 2018a). Obadiah et al. (2012) prepared calcium oxide from sheep bones by a series of steps beginning with crushing the bones in a hydraulic press as 100 psi. The bone chips were pressure cooked at 1000 °C and 15 psi for duration of 4 hours, with the water being changed intermittently. After boiling, the bones were dried at 105 °C for 16 hours and grinded finely with a hammer mill into particle sizes (<2 mm). Lastly, the finely milled bone powder was heated from 200 °C to 1000 °C to monitor the effect of calcination temperature on the transformation to calcium oxide. It was found that the sample with highest calcium oxide component was obtained at 800 °C.

In the synthesis of calcium oxide from chicken and ostrich egg shells (Tan et al. 2015), the eggshells were initially soaked in distilled water and washed to eliminate contaminants and their white inner membranes removed. The shells were allowed to dry in an oven at 100 °C, pounded into chips and grinded into fine powder by a mechanical grinder. Calcination was performed on the fine powder at 1000 °C in a tubular furnace for 4 hours to covert calcium carbonate to calcium oxide. Calcium oxide obtained from waste scallop shells was prepared by crushing grinding the shells to powder and calcination at diverse temperatures from 600 °C to 400 °C for 2 hours, to determine the optimum calcination temperature for calcined scallop shell (Sirisomboonchai et al. 2015). The optimum calcination temperature was found to be in the span of 900 °C and 1100 °C. Another study performed the calcinations at a temperature of 1000 °C with good conversion (Buasri et al. 2013).

A different method of preparing calcium oxide with high activity is the calcination-hydration-dehydration method (Niju et al. 2016). Here, waste shells were washed and rinsed thoroughly with distilled water and dried in an oven for 1 day at 105 °C. The dried pieces were then crushed, grinded and calcined at 900 °C for 4 hours. At this point, the calcium component in the shells had oxidized to calcium oxide. This product was then refluxed in water for 6 hours at 60 °C to convert to calcium oxide to calcium hydroxide. Following this, the refluxed solid articles were calcined again at 600 °C for duration of 3 hours to convert the calcium hydroxide to calcium oxide. Calcium oxide from river snail shell was obtained by washing, drying and calcinations for 4 hours at 800 °C (Kaewdaeng et al. 2017). Waste shells were calcined at temperatures between 700 °C and 1000 °C for 4 hours, grinded and sieved to obtain white calcium oxide powder for catalysis (Buasri et al. 2013). Specie of shell was calcined for 2 hours at 900 °C to convert calcium carbonate to calcium oxide catalyst (Margaretha et al. 2012).

In some cases, calcium oxide may be impregnated (doped) with another element or compound to increase its basic strength (Liao and Chung 2013). Calcium oxide was impregnated with 25% of KBr solution by a method known as wet impregnation for 2 hours under continuous stirring (Mahesh et al. 2015). After which, the sample was heated at 900 °C for 4 hours and allowed to cool. Alumina-calcium oxide (60 wt% of CaO on Al_2O_3) was synthesized using a one-step synthesis procedure (Yalman 2012). Following the preparation, the sample was calcined at 700 °C for 6 hours, sieved and stored. In all cases, calcium oxide is stored in desiccators to prevent catalyst poisoning as a result of absorption of atmospheric water and carbon dioxide (Kaewdaeng et al. 2017).

2.3 *Biodiesel Production*

Different set-ups and several combinations of reaction factors exist for the production of biodiesel by transesterification. Reaction factors include reaction temperature, the ratio of moles of alcohol to oil, weight of catalyst used, agitation intensity, the time allowed for the reaction to take place and pretreatment, among others. Many researchers have studied different processes of biodiesel production.

Tan et al. (2015) researched on a method of biodiesel production by employing a process called two-step transesterification reaction. This was done in light of the type of oil used, which is WCO and the fact that it possessed a high acid value. The first step was an acid esterification which was performed using 50 ml of oil, 25 ml of methanol and 0.05 ml of concentrated H_2SO_4. The methanol and acid were blended together by using a stirrer in a conical flask, while the oil was preheated to 65 °C in another conical flask. The blend was added to the preheated oil, and the reaction was left to proceed for 1 hour at 65 °C in an orbital shaker. The product was then separated out and purified before being passed to the next step. The second step of this procedure involved the base transesterification of oil product with methanol and calcium oxide catalyst in a 500 ml screw-capped glass bottle enclosed in an incubator with orbital shaker. Here five factors were varied: catalyst weight % per volume of oil – 1, 1.5, 2, 2.5, 3; reaction temperature in °C – 25, 50, 65, 75, 85; proportion of methanol to oil – 6:1, 9:1, 10:1, 12:1, 14:1; time in hours – 1, 2, 3, 4, 5; and orbital speed in rpm – 50, 100, 150, 250, 350. For each run, the calcium oxide was first activated by pouring it into methanol and shaking for 1 hour at 65 °C to form methoxide which was later added to the esterified oil and vigorously stirred.

The result showed that the maximum condition for heterogeneous biodiesel production was a combination of the following factor selections: 1.5 weight% per volume of oil, 12:1 ratio, 65 °C temperature, 2 hours reaction time and 250 rpm agitation speed. In addition to these findings, this study also compared the catalytic activity of calcium oxide attained from chicken eggshells and ostrich eggshells. Although both catalysts gave a high biodiesel yield, it was discovered that production using catalyst obtained from ostrich eggshells at the optimum condition gave a higher biodiesel yield of 96% compared to that from chicken eggshells which gave a yield of 94%. This variance was accredited to the ostrich eggshells having a larger surface area which enhances transesterification and the higher basicity of ostrich eggshells. The yield obtained here is higher than that obtained in other studies which employed the use of eggshells from other birds with values ranging from 92 to 96% (Viriya-empikul et al. 2010a; Chen et al. 2014; Wei et al. 2009b). The yield in this study is also higher than that from a study that used ostrich eggs as well with 30.08 methanol weight % of oil, 3.79 wt% of catalyst and temperature of 60 °C for a duration of 4 hours (Satar 2014). The reasons for discrepancy stated here were the differences in the catalyst preparation method and the molar ratio. The properties of the biodiesel obtained fell within the range set by the ASTM. A major gap in this study was the absence of reflux equipment to condense methanol that may have evaporated during the reaction, as temperatures 65 °C and above were tested. This may have affected the final yield of biodiesel obtained.

Another study involved the transesterification of waste oil by employing calcium oxide catalyst obtained from the calcinations of scallop shells (Sirisomboonchai et al. 2015). Here, to ensure the methanol remained liquid for maximum contact with the triglycerides in the waste cooking oil, a closed reflux reactor was employed for the transesterification. Prior to the reaction, the calcined scallop shell (now calcium oxide) was mixed with the methanol in the batch reactor at 40 °C for 1 hour with an agitation speed of 500 rpm. The oil was then introduced to the reactor for transesterification. Three factors were varied here, with levels for each factor. These factors and their respective levels are methanol-to-oil ratio (3:1, 6:1, 12:1), catalyst in wt% (1, 2, 5, 10) and time in minutes (30, 60, 120, 180). The reaction temperature for each run was fixed at 65 °C. Monitoring the time of the reaction and at constant molar ratio and catalyst loading, it was observed that for the initial 30 minutes, the methyl ester yield was not so high due to the time taken for the triglycerides to disperse in the methanol. However, beyond 30 minutes, the generation of the esters increased rapidly till 120 minutes where it remained constant, indicating the reaction equilibrium had been reached. As reports shed more light on the contribution of excess reaction time to the backward reaction and saponification reaction (Leung and Guo 2006), 120 minutes was chosen as the maximum reaction time for this study. Looking into the impact of the catalyst weight on the reaction, a proportional upsurge in the yield of esters with an upsurge in the weight of catalyst was noticed.

However, on increasing the weight from 5% to 10%, the rate of increase decreased. This was attributed to the fact that excess amount of catalyst than required for the reaction could contribute to saponification side reaction (Eevera et al. 2009). Hence, 5 wt% was chosen as the optimal weight percentage of catalyst to oil. Observing the trend while varying molar ratio, the ester yield greatly increased on upsurge in the molar ratio from 3:1 to 6:1. From 6:1 to 12:1, there was increase in the yield but only slightly. Beyond 12:1, the yield decreased. This was due to the decrease in the overall concentration of catalyst by dilution in the excess methanol. In addition, reports suggested that the optimum molar ratio for a transesterification process should be selected considering the cost of the materials, separation and purification (Leung et al. 2010). Hence, the ratio 6:1 of methanol to oil was picked as the optimal molar ratio. This study also provided a comparison between calcium oxide obtained by the calcinations of waste scallop shells and lab grade calcium oxide. While the use of the lab grade calcium oxide at the optimum conditions discovered gave 85% yield of biodiesel, the calcined scallop shells gave a maximum yield of 80%. Overall, both yields are low relative to other works done on this topic. This can faulted to the use of a high acid value oil feed without any form of pretreatment to decrease the percentage of free fatty acids (FFA), a major gap in this study.

A similar study was conducted to produce biodiesel from palm oil (Buasri et al. 2014). Here, calcined scallop shell was also used as the catalyst for transesterification. The reaction was done in a flask, furnished with a condenser (water-cooled), on a hot plate magnetic stirrer. A design of experiment was conducted using the *Taguchi* method. In the experiment, the effects of four factors were considered. For each factor, three levels were considered. The factors and their levels are reaction time (hours) – 3, 4, 5; reaction temperature (°C) – 60, 65, 70; loading of catalyst in weight percent – 8, 10, 12; and molar ratio of methanol to oil – 6:1, 9:1, 12:1. At the end of the experimental runs, nine combinations were selected based on the fatty acid methyl ester (FAME) % conversion. These combinations gave the most economic span of conditions that would produce biodiesel of high quality. Of the nine, the optimum condition generated was one which gave a conversion of 95.44% using a molar ratio of 9:1 and catalyst loading of 10 wt% at a temperature of 65 °C for 3 hours. It was also recorded in the study that the calcined scallop shell performed as well as commercial calcium oxide, and this waste scallop shell can be utilized as a suitable source of efficient and economical calcium oxide catalyst. The ester produced was light brown and met the requirements of the Thai biodiesel standard. A major gap observed in this study was the lack of esterification or any other suitable pretreatment step for the oil to reduce the amount of free fatty acids in the palm oil.

The biodiesel production process that employs calcium oxide gotten from river snail ash has also been observed (Kaewdaeng et al. 2017). Here, transesterification reaction was done in a 200 ml lab-scale reactor with a molar ratio of methanol to waste cooking oil (6:1, 9:1 and 12:1). Catalyst weight percentages of 1, 2 and 3 were considered as well. Keeping the molar ratio constant at 6:1, and observing the effect of catalyst loading at the maximum duration of 3 hours, the highest yield of biodiesel (95.91%) was realized using 1 wt%. Additional increase in the catalyst loading to 2 wt% and 3 wt% caused a reduction in the yield, subsequently the role of excess catalyst in impeding the distribution of the triglycerides and methanol in the reaction. With an upsurge in the molar ratio to 9:1, the highest yield (95.7%) was observed after just 1 hour with a catalyst loading of 1 wt%. An extended duration for the reaction resulted in a decrease in the percentage yield of biodiesel. This was credited to the reversible nature of the transesterification reaction and the possibility of saponification side reaction (Hatthason et al. 2013). A further increase of the molar ratio to 12:1 at 1 wt% catalyst loading for duration of 2 hours yielded 91.4% biodiesel, while a catalyst loading of 2 wt% at the same conditions yielded 93.6% biodiesel. This showed that an upsurge in the catalyst loading could result in an upsurge in the biodiesel yield, provided that sufficient reagent is available to dilute the effects of the catalyst's impeding tendencies. Further upsurge in the loading of catalyst to 3 wt% would, however, result in the reduction of the biodiesel yield. Comparing the yield from varying the molar ratio at the same reaction conditions, it was found that the optimum ratio was 6:1, and continuous increase in the proportion led to a gradual decrease in the yield. Overall, the runs gave good results, with the maximum biodiesel yield at 95.7%. The chemical composition of the product was found to have a FAME content of 98.12%. Other characteristics of the esters were obtained and discovered to be within requirements. As with others, the absence of a pretreatment step for the waste cooking oil and reflux device during transesterification is a major gap in this study.

Another study on biodiesel production using waste sources of heterogeneous catalyst was carried out on the *Pomacea* species shell (Margaretha et al. 2012). Transesterification reaction was performed on palm oil in a flask that has reflux condenser, magnetically stirred heating mantle. The reaction temperature, time and agitation speed were left constant at 60 °C, 4 hours and 700 rpm, respectively. The mass palm oil used for each run was 200 g. The weight of catalyst was varied from 1 wt% to 5 wt% of weight of oil. The proportion of oil to methanol was varied between 1:5 and 1:11. For comparison purposes, sodium hydroxide (NaOH) was used as a catalyst under the same reacting conditions. Using the CaO obtained from the shell, it was observed that a rise in catalyst weight resulted in a rise in the yield of biodiesel. This resulted from the upsurge in the availability of active sites, causing an upsurge in the yield of biodiesel (Hattori et al. 2000). As a result of the reversible nature of the transesterification reaction, more methanol above the stoichiometric quantity will drive the forward reaction. This was observed to be true owing to the upsurge in methanol-to-oil ratio, which was from 5:1 to 7:1. An additional increase, however, decreased the yield of biodiesel due to side reactions (Kouzu et al. 2008b). The maximum yield (95.61%) was obtained with a 4 wt% loading of catalyst and methanol-to-oil proportion of 7:1. This result is similar to a study on heterogeneous catalysed transesterification (Chen et al. 2011). It is however higher than the results obtained in another work (Buasri et al. 2012). This maximum yield however was slightly lower than the yield (97.24%) obtained from catalysis using NaOH, confirming that homogeneous catalyst gives higher yields compared to heterogeneous catalysts.

An investigation on the heterogeneous catalysis of biodiesel production by employing calcined animal bones was also carried out (Obadiah et al. 2012). Catalyst weights from 5 wt% to 25 wt% were studied. Methanol-to-oil proportions of 1:1 and 18:1 were considered. The reaction occurred at 65 °C for 4 hours. The optimum condition observed was with a proportion of 18:1, catalyst weight of 20 wt%, reaction temperature of 65 °C and duration of 4 hours, at which the % conversion was 96.78%. A major gap identified in this study is the lack of extensive factor variations (temperature, time and agitation intensity). The optimum conditions prescribed here may not be accurate due to the fixed temperature and reaction time used in the study. In addition, there was no control experiment provided to compare the efficiency of calcium oxide obtained from animal bones to commercial calcium oxide or other catalysts purchased from the market or stores. A biodiesel yield of 96.5% was attained from soybean oil by transesterification with a methanol-to-oil molar ratio of 6:1, at 65 °C, for 5 hours, using heterogeneous catalyst from oyster shells (Nakatani et al. 2009). Viriya-empikul et al. (2012) got 94.1% yield of biodiesel from palm oil by transesterification at 60 °C, using 10 wt% catalyst loading and a molar proportion of 15:1, for 2 hours. A 3-hour transesterification reaction produced 95% biodiesel with a methanol-to-oil ratio of 9:1, 3 wt% loading of catalyst and reaction temperature of 65 °C (Wei et al. 2009a).

Switching our focus from waste heterogeneous catalysts to commercial catalysts, a study was conducted on biodiesel production from WCO by employing calcium oxide (Aworanti et al. 2013). Here, a blender with a tight lid was served as the reactor. Three factors were considered here: methanol-to-oil molar ratio, catalyst weight and reaction time. The maximum condition was observed to be at a molar ratio of 9:1, 3.5 wt% catalysts and a reaction time of 1 hour. At this optimum condition, the maximum conversion was found to exceed 90%, and the yield of biodiesel was approximately 96%. In addition, other characteristics like viscosity, density, acid value, pour point and flash point complied with the standards set by the ASTM. A study was carried out on the use of magnesium oxide (MgO) as a catalyst in biodiesel production (Lopez et al., 2005). Triacetin was used in place of regular alcohol for transesterification reaction at 600 °C for 8 hours. The biodiesel conversion was extremely low here at 18%. A high biodiesel yield of 91% was attained in an investigation into biodiesel production using MgO catalyst at supercritical conditions (300 °C) with a high molar ratio of methanol to oil (39.6:1) (Di-Serio et al. 2005). Another study produced biodiesel with a yield of 92% using 5.0 wt% of MgO catalyst, 12:1 molar proportion in 1 hour (Di-Serio et al., 2008). Although, next to calcium oxide, magnesium oxide is the most preferred base heterogeneous catalyst for biodiesel production, it is hardly used owing to its low surface area for contact leading to low conversions (Chouhan and Sarma 2011).

Strontium oxide (SrO) has also been highlighted as a suitable base heterogeneous catalyst for transesterification, though available literature on it is extremely limited (Zabeti et al. 2009). However, one study observed the transesterification reaction using 3 wt% SrO, methanol to oil proportion of 12:1 at 65 °C for 30 minutes for a biodiesel yield of 90% (Liu et al. 2008). KBr impregnated CaO has also been studied as a catalyst for transesterification of waste oil (Mahesh et al. 2015). The oil was pretreated prior to transesterification with 0.5 grams of activated charcoal to 100 ml of oil. The transesterification reaction was performed in flask furnished with a reflux condenser. The flask was positioned in a constant temperature water bath. The temperature of the reaction was kept constant at 65 °C, while the other factors were varied. The varied methanol-to-oil molar ratio was from 9:1 to 15:1; the amount of catalyst was varied from 2 to 6 weight% of oil and the time from 1 hour to 3 hours. The optimum FAME yield here was 83.6% with optimum conditions: molar ratio of 12:1, catalyst quantity of 3 wt% and duration of 2 hours.

2.4 Recovery and Regeneration of Base Heterogeneous Catalysts

Although heterogeneous catalysts produce lower yields of biodiesel, a major advantage is their reusability (Margaretha et al. 2012). That is, heterogeneous catalysts can be recovered, regenerated and put to use again, and different methods exist for each purpose. In one study, the catalyst was recovered using filtration method through a size 42 Whatman filter paper, washed with distilled water and washed again with acetone (Obadiah et al. 2012). The catalyst was dried, after which it was ready for reuse. Results showed that the catalyst could be recovered and reused for up to 5 cycles, with the fifth cycle giving a conversion of 83.7%. Another study separated out the catalyst by centrifugation at 5000 rpm for 10 minutes (Tan et al. 2015). The recovered catalyst was then washed with n-hexane to get rid of adsorbed material. The washed catalyst was dried and re-calcined at a temperature of 700 °C for reuse. Reports showed that the catalyst was also able to be reused for up to 5 cycles, with the fifth cycle attaining a biodiesel yield of 75%. Sirisomboonchai et al. (2015) also recovered their calcium oxide catalyst by centrifugation at 6000 rpm for 15 minutes. Upon reactivation, it was recorded that the catalyst could be effectively reused for up to 4 cycles. However, after the fourth cycle, the biodiesel yield reduced from 86% to 62%.

Margaretha et al. (2012) discussed and studied three methods of dealing with spent CaO catalyst after recovery. These three methods are reusing the spent catalyst without any form of treatment or reactivation, washing the spent catalyst with methanol to eliminate adsorbed biodiesel and oil and re-calcination of the spent catalyst. Three cycles were tested using these three methods to determine the influence of each method on the biodiesel yield per cycle. Using the first stated method, the biodiesel yield dropped from 95% in the first cycle to 38% in the second cycle, emphasizing the great degree of deactivation of the catalyst. The third cycle did not produce any biodiesel. Observing the trend using the second method, the biodiesel dropped from 96% to 90% and then to 75% in the first, second and third cycle, respectively. This method gave better yields in the second and third cycles than the first method. The third method gave yields above 90% in the first, second and third cycles. Hence, re-calcination of recovered catalyst is the best method of catalyst re-activation for reuse in biodiesel production.

From the foregoing, and generally therefore, heterogeneous catalysts are not corrosive in nature, and they are friendly, environmentally. One of their characteristics is easy separation from the products through filtration, and their disposal glitches are not much when compared with those stumble upon by the homogeneous catalysts (Yoosuk et al. 2010). Furthermore, heterogeneous catalysts have the capacity to be recycled and employed many times, as a result of the enhanced separation of the catalyst from the final products (Farooq et al. 2013). This offers a potential economical avenue for the production of biodiesel. Heterogeneous catalysts uphold easy recovery and a less expensive green process (Robles-Medina et al. 2009). They can withstand high free fatty acid and moisture content. Effective and cost-effective heterogeneous catalysts assist in minimizing the overall cost of biodiesel production (Robles-Medina et al. 2009; Marchetti et al. 2007). Homogeneous catalytic systems are devoid of active molecules usually found in heterogeneous catalysts; hence, catalytic activity could be weakened. Since heterogeneous catalysts possess several advantages, their characteristics make them sustainable.

2.5 *Sustainability of Heterogeneous Catalysts*

Making heterogeneous catalysts sustainable is a great step in the direction of green chemistry. Nonetheless, approaches to the actual green catalysts entail more than synthesizing them. Environmentally friendly synthesis and degradation of the catalysts after the duration of their lifespan, their excellent stability, insignificant leaching of vigorous components, the utilization of non-harmful solvents and the utilization of environmentally friendly reactants are all significant mechanisms that provide the catalyst and their reaction certainly green characteristics (Pálinkó, 2018). With regards to sustainability, it is essential to look for low-cost means of synthesizing heterogeneous catalysts that utilize earth abundant elements (Lee et al., 2014). Hence, sustainable heterogeneous catalysts that is highly effective is an important factor that will contribute to the final cost of biodiesel.

3 Conclusion

Biodiesel is regarded as an environmentally friendly biofuel, which is a promising substitute to diesel fuel. As a result of the challenges in the synthesis of present catalysts, the switch toward cost-effective paths for synthesizing sustainable catalysts for renewable biofuel fuels has increased. Heterogeneous catalytic processes play a central role in the production of biodiesel fuel. Hence, the role of heterogeneous catalysts is critical in every aspect of biodiesel production: from the conversion process of biofuels systems to making the whole process more effective, with low energy demand and realistic, economically.

Acknowledgement The authors would like to acknowledge the financial support provided by Covenant University for the publication of this review.

References

Abbaszaadeh A, Ghobadian B, Omidkhah MR, Najafi G (2012) Current biodiesel production technologies: A comparative review. Energy Conversion and Management 63:138-148

Ali UJ (2017) Production of Biodiesel from Used Cooking Oil. IJIRST:90-95

Aworanti OA, Agarry SE, Ajani AO (2013) Statistical Optimization of Process Variables for Biodiesel Production from Waste Cooking Oil Using Heterogeneous Base Catalyst. British Technology Journal:116-132

Ayoola AA, Igho BE, Fayomi SO (2018b) Data on calcium oxide and cow bone catalysts used for soybean biodiesel production. Data in Brief:512-517

Ayoola AA, Ojewunmi ME, Rasheed B, James AM (2018a) Data on CaO and eggshell catalysts used for biodiesel production. Data in Brief:1466-1473

Azócar L, Heipieper HJ, Navia R (2010) Biotechnological processes for biodiesel production using alternative oils. Applied microbiology and biotechnology 88 (3):621-636

Balat M, Balat H (2010) Progress in biodiesel processing. Applied energy 87 (6):1815-1835

Boey PL, Maniam GP, Hamid SA (2009) Biodiesel production via transesterification of palm olein using waste mud crab (Scylla serrata) shell as a heterogeneous catalyst. Bioresource Technology:6362-6368

Borges ME, Díaz L (2012) Recent developments on heterogeneous catalysts for biodiesel production by oil esterification and transesterification reactions: A review. Renewable and Sustainable Energy Reviews 16:2839-2849

Buasri, Chaiyut N, Loryuenyong V, Rodklum C, Chaikwan T, Kumphan N (2012) waste frying oil using potassium hydroxide supported on Jatropha curcas fruit shell as solid catalyst. Applied Sciences:641-653

Buasri A, Chaiyut N, Loryuenyong V, Worawanitchaphong P, Trongyong S (2013) Calcium Oxide Derived from Waste Shells of Mussel, Cockle, and Scallop as the Heterogeneous Catalyst for Biodiesel Production. The Scientific World Journal:1-8

Buasri A, Worawanitchaphong P, Trongyong S, Loryuenyong V (2014) Utilization of Scallop Waste Shell for Biodiesel Production from Palm Oil - Optimization Using Taguchi Method. APCBEE Procedia 8:216-221

Chen GY, Shan R, Shi JF, Yan BB (2014) Ultrasonic-assisted production of biodiesel from transesterification of palm oil over ostrich eggshell-derived CaO catalysts. Bioresource Technology:428-432

Chen YH, Huang YH, Lin RH, Shang NC, Chang CY, Chang CC, Chiang PC, Hu CY (2011) Biodiesel production in a rotating packed bed using K/γ-Al2O3 solid catalyst. Journal of Taiwan Institute of Chemical Engineers:937-944

Cho YB, Seo G, Chang DR (2009) Transesterification of tributyrin with methanol overcalcium oxide catalysts prepared from various precursors. Fuel Process Technology:1252-1258

Chouhan APS, Sarma AK (2011) Modern heterogeneous catalysts for biodiesel production: A comprehensive review. Renewable and Sustainable Energy Reviews:4378-4399

Demirbas A (2009) Progress and recent trends in biodiesel fuels. Energy conversion and management 50 (1):14-34
Di-Serio M, Tesser R, Dimiccoli M, Cammarota F, Nasatasi M, Santacesaria E (2005) Synthesis of biodiesel via homogeneous Lewis acid catalysts. Journal of Molecular Catalysis:111-115
Di-Serio M, Tesser R, Pengmei L, Santacesaria E (2008) Heterogeneous catalysts for biodiesel production. Energy Fuels:207-217
Eevera T, Rajendran K, Saradha S (2009) Biodiesel production process optimization and characterization to assess the suitability of the product for varied environmental conditions. Renewable Energy:762-765
Farooq M, Ramli A, Subbarao D (2013) Biodiesel production from waste cooking oil using bifunctional heterogeneous solid catalysts. Journal of Cleaner Production 59:131-140
Gonzaga FB, Sobral SP (2012) A new method for determining the acid number of biodiesel based on coulometric titration. Talanta 97:199-203
Granados L, Poves Z, Alonso M, Marisca LR, Galisteo C, Moreno-Tost R (2009) Biodiesel from sunflower oil by using activated calcium oxide. Applied Catalysis B:25-30
Hara M (2009) Environmentally benign production of biodiesel using heterogeneous catalysts. ChemSusChem:129-135
Hatthason S, Pairintra R, Krisnangkura K (2013) Biodiesel Production from Coconut Oil Using CaO of Waste Cockle Shell as a Catalyst. Agricultural Science Journal:361-364
Hattori H, Shima M, Kabashima H (2000) Alcoholysis of ester and epoxide catalyzed by solid base. Studies in Surface Science and Catalysis:3507-3510
Kaewdaeng S, Sintuya P, Nirusin R (2017) Biodiesel production using calcium oxide from river snail shell ash as catalyst. Energy Procedia 138:937-942
Kouzu M, Kasuno T, Tajika M, Sugimoto Y, Yamanaka S, Hidaka J (2008b) Calcium oxide as a solid base catalyst for transesterification of soybean oil and its application to biodiesel production. Fuel 87:2798-2806
Kouzu M, Kasuno T, Tajika M, Yamanaka S, Hidaka J (2008a) Active phase of calcium oxide used as solid base catalyst for transesterification of soybean oil with refluxing methanol. Applied Catalysis A: General:357-365
Lee AF, Bennett JA, Manayil JC, Wilson K (2014) Heterogeneous catalysis for sustainable biodiesel production via esterification and transesterification. Chem Soc Rev 43 (22):7887-7916. doi:https://doi.org/10.1039/c4cs00189c
Leung DYC, Guo Y (2006) Transesterification of neat and used frying oil: optimization for biodiesel production. Fuel Process Technology:883-890
Leung DYC, Wu X, Leung MKH (2010) Areview on biodiesel production using catalyzed transesterification. Applied Energy:1083-1095
Liao C, Chung T (2013) Optimization of process conditions using response surface methodology for the microwave - assisted transesterification of jatropha oil with KOH impregnated CaO as catalyst. Chemical Engineering Research and Design:2457-2464
Liu R, Wang X, Zhao X, Feng P (2008) Sulfonated ordered mesoporous carbon for catalytic preparation of biodiesel. Carbon:1664-1669
Lopez DE, Bruce DA, Lotero JGE (2005) Transesterification of triacetin with methanol on solid acid and base catalysts. Applied Catalysis A: General:97-105
Mahesh SE, Ramanathan A, Begum KMMS, Narayanan A (2015) Biodiesel production from waste cooking oil using KBr impregnated CaO as Catalyst. Energy Conversion and Management:442-450
Marchetti J, Miguel V, Errazu A (2007) Possible methods for biodiesel production. Renewable and sustainable energy reviews 11 (6):1300-1311
Margaretha YY, Prastyo HS, Ayucitra A, Ismadji S (2012) Calcium oxide from Pomacea sp. shell as a catalyst for biodiesel production. International Journal for Energy and Environmental Engineering:1-9
Math M, Kumar SP, Chetty SV (2010) Technologies for biodiesel production from used cooking oil—A review. Energy for sustainable Development 14 (4):339-345

Nakatani N, Takamor H, Takeda K, Sakugawa H (2009) Transesterification of soybean oil using combusted oyster shell waste as a catalyst. Bioresource Technology:1510-1513
Nie K, Xie F, Wang F, Tan T (2006) Lipase catalyzed methanolysis to produce biodiesel: optimization of the biodiesel production. Journal of Molecular Catalysis B: Enzymatic 43 (1-4):142-147
Niju S, Begum KMMS, Anantharaman N (2016) Enhancement of biodiesel synthesis over highly active CaO derived from natural white bivalve clam shell. Arabian Journal of Chemistry:633-639
Obadiah A, Swaroopa GA, Kumar SV, Jeganathan KR, Ramasubbu A (2012) Biodiesel production from Palm oil using calcined waste animal bone as catalyst. Bioresource Technology:512-516
Omar WNNW, Amin NAS (2011) Biodiesel production from waste cooking oil over alkaline modified zirconia catalyst. Fuel Processing Technology 92 (12):2397-2405
Pálinkó I (2018) Heterogeneous catalysis: A fundamental pillar of sustainable synthesis. In: Green Chemistry. Elsevier, pp. 415-447
Phan AN, Phan TM (2008) Biodiesel production from waste cooking oils. Fuel 87 (17-18):3490-3496
Robles-Medina A, González-Moreno P, Esteban-Cerdán L, Molina-Grima E (2009) Biocatalysis: towards ever greener biodiesel production. Biotechnology advances 27 (4):398-408
Satar KB (2014) Biodiesel production from waste cooking oil via heterogeneous catalyst using ostrich eggshel catalyst. Universiti Malysia, Kota Samarahan, Malaysia
Sirisomboonchai S, Abuduwayiti M, Guan G, Samart C, Abliz S, Hao X, Kusakabe K, Abudula A (2015) Biodiesel production from waste cooking oil using calcined scallop shell as catalyst. Energy Conversion and Management:242-247
Talha NS, Sulaiman S (2016) Overview of Catalysis in Biodiesel Production. ARPN Journal of Engineering and Applied Sciences:439-448
Tan YH, Abdullah MO, Nolasco-Hipolito C, Taufiq-Yap YH (2015) Waste ostrich- and chicken-eggshells as heterogeneous base catalyst for biodiesel production from used cooking oil: Catalyst characterization and biodiesel yield performance. Applied Energy:58-70
Viriya-empikul N, Krasae P, Puttasawat B, Yoosuk B, Chollacoop N (2010b) Waste shells of mollusk and egg as biodiesel production catalysts. Bioresource Technology:27-32
Viriya-empikul N, Krasae P, Puttasawat B, Yoosuk B, Chollacoop N, Faungnawakij K (2010a) Waste shells of mollusk and egg as biodiesel production catalysts. Bioresource Technology:3765-3767
Wei Z, Xu C, Li B (2009a) Application of waste eggshell as low-cost solid catalyst for biodiesel production. Bioresource Technology:2883-2885
Wei Z, Xu C, Li B (2009b) Application of waste eggshell as low-cost solid catalyst for biodiesel production. Bioresour Technol 2009;100:2883–5. Bioresource Technology:2883-2885
Yalman E (2012) Biodiesel Production from Safflower Using Heterogeneous CaO Based Catalysts. Iznir Institute of Technology, Izmir
Yoosuk B, Krasae P, Puttasawat B, Udomsap P, Viriya-empikul N, Faungnawakij K (2010) Magnesia modified with strontium as a solid base catalyst for transesterification of palm olein. Chemical Engineering Journal 162 (1):58-66
Yunus-Khan TM, Atabani AE, Badruddin IA, A B, Khayoon MS, Triwahyono S (2014) Recent scenario and technologies to utilize non-edible oils for biodiesel production. Renewable and Sustainable Energy Reviews:840-851
Zabeti M, Wan M, Wan D (2009) Activity of solid catalysts for biodiesel production: a review. Fuel Process Technology:770-777
Zhang J, Chen S, Yang R, Yan Y (2010) Biodiesel production from vegetable oil using heterogeneous acid and alkali catalyst. Fuel:2939-2944
Zhu H, Wu Z, Chen Y, Zhang P, Duan S, Liu X (2006) Preparation of biodiesel catalyzed by solid super base of calcium oxide and its refining process. Chinese Journal of Catalysis:391-396

Nigerian Marginal Oilfield Development Program: PIA and Current Issues

Rachael E. Josephs, Charles Y. Onuh, Oyinkepreye D. Orodu, Oluwasanmi A. Olabode, Yuven T. Nchila, and Christian N. Dinga

1 Introduction

The oil and gas industry is charged with the exploration, production, refining and distribution of petroleum, natural gas and its products around the world. Currently, the world is heavily dependent on hydrocarbon as a major source of energy. Daily consumption was approximately about 100.87 million barrels of oil equivalent globally in 2019 and an estimated 92.2 million barrels in 2020 (Ogolo, 2021; US Energy Information Agency, 2020). Therefore, in the future, it will be necessary to maintain a stable supply of this very important commodity. Companies are beginning to invest in hydrocarbon exploration and production to ensure that depleted hydrocarbon reserves are replenished (Ogolo, 2021).

Marginal fields are estimated to currently represent between 30% and 40% of world oil production and are gaining importance due to the natural decline in production from large mature fields (Onolemhemhen, 2017). Most of the offshore oilfields that have been discovered in recent years have been classified as marginal (Zhao et al., 2016). In general, most fields with low-abundant reserves, poor economic performance, high cost, and technical difficulty in development are called marginal fields or small fields (Akinpelu, 2009; Kue, 2006). On the other hand, in the Nigerian context, marginal fields are defined as fields which have been neglected by the leaseholders. Other features include long distance away from infrastructure, costly development during low oil price, limited volume of reserves, etc. (Otombosoba, 2017).

Marginal oil field reservoirs located in Nigeria are cautiously estimated to hold reserves of over 2.3 billion barrels of Stock Tank Oil Initially in Place (STOIIP)

R. E. Josephs (✉) · C. Y. Onuh · O. D. Orodu · O. A. Olabode · Y. T. Nchila · C. N. Dinga
Petroleum Engineering Department, Covenant University, Ota, Nigeria
e-mail: rachael.joseph@covenantuniversity.edu.ng

© The Author(s), under exclusive license to Springer Nature Switzerland AG 2022

A. O. Ayeni et al. (eds.), *Bioenergy and Biochemical Processing Technologies*, Green Energy and Technology, https://doi.org/10.1007/978-3-030-96721-5_8

spread across 183 fields (Akinwale, 2016). These fields are discoveries found in areas which are managed by International Oil Companies (IOCs), and no development effort has been carried out owing to their perceived limited economic value due to size, volume/reserves, and economic/technical unavailability of current technology for their development (Otombosoba, 2017). These fields may be dubbed uneconomic when produced by the IOCs; however, they may be profitable if explored by indigenous entrepreneurs owing to their low operating cost as explained by Dagogo (2018). National Oil Companies (NOCs) and IOCs are advancing in skills and capabilities to unlock the potential of marginal fields and small-scale developments (Onolemhemhen, 2017).

Marginal fields are categorized in accordance with a mixture of specific conditions: hydrocarbon resources, oil price, technical feasibility, operating company's profile, commercial structure, and regulatory aspects. They can largely be defined as oil and gas fields that are assumed uneconomic to develop due to size of reserves, complexity, financial and market conditions, or remoteness of nearby infrastructure (Humphrey, 2016). They are recovered by the Federal Government with the aim of reallocating them to indigenous companies with technical and financial capability and willingness to develop them (Ezeani, 2017). They can also be fields with low production at the end of their economic life cycle when revenues are below operating expenses (Humphrey, 2016). The 1969 Petroleum Act confers on the Petroleum Resources Minister (The President) the authority to award oil mining and prospecting licenses. As such, "Marginal Fields" are fields which the president may from time to time identify as such. According to Onolemhemhen et al. (2017) to limit the capricious classification of fields as marginal, DPR (1996) issued guidelines that list the traits that must be in place for a field to be classified as such. They are as follows:

1. Low STOIIP.
2. Considerable distance to available production facilities.
3. Reservoirs that possess unusual crude oil characteristics (e.g., crude oils characterized by high viscosity and low API gravity) that are not recoverable through conventional methods.
4. Acreage which has not been considered for development due to their marginal economics under prevalent financial and market conditions.
5. Fields with one or multiple wells that are yet to be developed by the operating companies including undiscovered fields and unappraised discoveries.
6. Producing fields that have become uneconomic owing to their attainment of near or exceeded abandonment limits.

Irrespective of geographical and technical characteristics, a key factor which influences the commerciality of marginal fields is the fiscal regime under which they are operated (Ezekiel, 2020). The fiscal regime is what guarantees reasonable return on investment (Adetoba, 2012). Ever since oil was discovered in 1956, the Nigerian government has continuously proposed a number of fiscal regulations to improve the petroleum industry. A number of them among others include the 1959 Petroleum Profit Tax Act (PPTA), Petroleum Act of 1969, and Nigerian National Petroleum Corporation (NNPC) Act 1977. More recently, the Local Content Act known as the

"Nigerian Oil and Gas Industry Content Development Act" (NOGICDA) 2010 and the Petroleum Industry Act (PIA) were introduced.

The complex regulatory framework that governs the planning and permitting of oil and gas projects is built on interdependent conditions that create bottlenecks in projects. For marginal field developers with limited financial resources, navigating this complex regulatory process frequently delays or even halts projects. Thus, this study seeks to carry out an extensive review of marginal oilfield development within the framework of the ensuing industry fiscal terms in Nigeria. A proper understanding of marginal field development in light of these fiscal systems will greatly impact decision making for the NOCs and contribute to the body of knowledge in oil field economics.

2 Theoretical Framework and Review of Existing Literature

Around the world, there have been various definitions developed for the classification of marginal fields. Hui et al. (2012) defined marginal fields as such fields which when developed by conventional technologies yield an internal rate of return lower than the standard in the industry but higher than the industry discount rate. Table 1 presents marginal field definition in some selected countries. In Nigeria however, categorizing a field to be marginal is the sole prerogative of the Federal Government through the discretion of the president (Petroleum Amendment Act, 1996). From Table 2, marginal fields can be defined as fields producing between 100 and 25, 000 barrels per day.

The Nigerian Marginal Field initiative was enacted when item 23 of the Petroleum Act (1996) mandated Oil Mining License (OML) holders to farm-out fields which have not been produced for 10 years or more to Nigerian owned Exploration and Production (E & P) companies with the consent of the president (Uwaga, 2008). Most farm-out agreements take the form of a sublease. Through the DPR, the OML holder acquires the license from the Minister; then the interest in the license is farmed out by the farmor, i.e., the holder of the license, to the farmee, i.e., the party

Table 1 Definition of marginal field in selected countries

USA	Fields with stripper oil wells and producing 10 barrels per day or below for a 12-month period and stripper gas wells producing 60,000 cubic feet per day or below at its peak flow rate
Netherlands	Gas fields which hold less than 4 billion cubic meters of reserves
United Kingdom	Oil fields estimated to contain 20 million Boe
Malaysia	Oilfields producing 30 million barrels of oil equivalent or less
China	An offshore oilfield of about 6.29 billion barrels of proven OOIP was classified as marginal
Egypt	Recoverable reserves equal to or less than five million barrels of oil

Sources: (Hui et al. 2012; Adeogun, 2015)

Table 2 Royalty rates as stipulated by the Marginal Fields Operations Regulations 2005

Production (bopd)	Royalties (%)
Below 5000	2.5%
Between 5000 and 10,000	7.5%
Between 10,000 and 15,000	12.5%
Between 15,000 and 25,000	18.5%

Table 3 Overriding royalty rates stipulated by the Marginal Fields Operations Regulations 2005

Production (bopd)	Royalties (%)
Below 2000	2.5%
Between 2000 and 5000	3.0%
Between 5000 and 10,000	5.5%
Between 10,000 and 15,000	7.5%
Above 15,000	7.5%

obtaining the license. Based on the agreement of both parties, the farmee is required to pay a percentage overriding interest to the farmor in addition to the royalties to be paid to the government who owns the mineral rights. Tables 2 and 3 present the percentage royalties and overriding royalties due for each corresponding production volume in barrels of oil per day (bopd) based on the Marginal Fields Operations Regulation of 2005 (Fig. 1).

Farm-out agreements may also be conducted through bidding rounds for marginal fields by the President (through the DPR), provided that such fields have been left unproduced, undeveloped or unattended to for more than 10 years. Bidding rounds are usually conducted in accordance with set guidelines which include information on payments, eligibility, timeline of the process, method of selection, and proposals. Once the winners are selected, they must pay a one-time signature bonus in order to complete the procurement process. The farmee is then required to negotiate necessary terms and conditions with the farmor. Since the inception of the Marginal Field Program (MFP), the DPR has announced three bidding rounds. The first bidding round took place in 2001, and indigenous companies were invited to bid for 24 out of the 116 fields identified. The 24 fields were awarded to 31 Nigerian companies in 2003 (Akinwale, 2016). The second was announced in 2013 by the Federal Government, and companies were invited to put in bids. Out of the 31 fields that put on offer, 16 were located onshore, while the remaining 15 were located in the continental shelf. Unfortunately, that bid never took place despite the fact that the guidelines to oversee the process had already been released (Ezekiel, 2020).

The third bidding round which commenced on the first of June 2020 was the first successful bid round in 20 years. The Guidelines for Award and Operations of Marginal Fields in Nigeria (the "Guidelines") 2020 document the criteria and procedure for the process (Okwurionu, 2020). At the end of the bid round in June 2021,

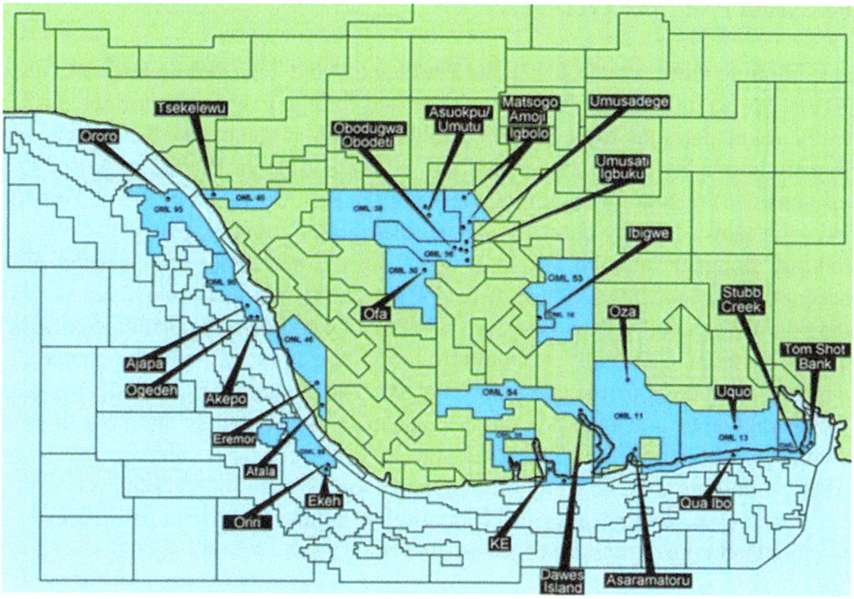

Fig. 1 Map of Niger Delta showing some previously awarded marginal fields (DPR 2014)

161 companies were shortlisted as potential awardees, out of 591 firms that submitted statements of interest forms for the 57 fields on offer (DPR, 2021). Wood Mackenzie estimated the total resources on offer to be around 800 MMbbl oil and 4.5 Tcf gas. Twenty-five of the largest oilfields are projected to generate over $38 billion in revenue over the life of the fields (Anderson, 2020). Ezekiel (2020) stated that about 251 marginal fields have been identified in Nigeria so far with about 2.3 billion barrels recoverable oil reserves.

2.1 Fiscal Regulations Applicable to Marginal Oil Fields in Nigeria

Fiscal regimes are applied by host governments of oil and gas producing nations to allocate, manage and regulate revenues accruing from oil and gas exploration. It is a key part of the decision-making process used by government and investors. According to the Nigerian Association of Petroleum Explorationists (NAPE), the economics of marginal fields are not considered to be robust enough under the prevailing “fiscal regime”; thus they remain unproduced. It is therefore necessary to examine the fiscal regulations applicable to the Nigerian petroleum industry as it is a major profitability determinant for marginal oil and gas field development.

Petroleum Profit Tax Act (PPTA)

According to Gbegi et al. (2017), the Petroleum Profit Tax (PPT) contributes about 70 and 95 percent of government revenue and foreign exchange earnings, respectively. It mandates tax to be paid on taxable income of companies that are engaged in petroleum activities (Abdul-Rahamoh, 2013). Petroleum Profit Tax Act of 1959 (as amended) regulates the administration of the PPT. Section 8 of the PPTA stipulates that any company carrying out petroleum operations is obligated to render returns, alongside audited accounts within an allowed time after the end of its accounting period. In Section 2 of the PPTA, petroleum operations refer to "winning or obtaining oil in Nigeria by or on behalf of a company, by drilling, extracting or other similar operations, excluding refining at a refinery, or in the course of a business carried out by the company engaged in such operations, and all operations incidental thereto and any sale of or disposal of chargeable oil by or on behalf of the company" (Gbegi, 2017). The applicable PPT rate for deep offshore and inland basin operations is 50%, while onshore and shallow water operations is 85%. The tax is usually paid in advance in 12 monthly installments (Lawal, 2013). However, marginal field operations are to enjoy a lower PPT rate of 55%.

Local Content Act

Local content comprises all government measures, projects, or interventions geared towards increasing employment, services, production, and overall monetary value to the country's local industry value chain (Acheampong, 2016). Table 4 presents the main objectives for introducing the Local Content Act in Nigeria's petroleum industry. The Nigerian Oil and Gas Industry Content Development Policy (also called the Nigerian Content Act, NCA), passed in April 2010, is the template used for the efficient execution of local content policy in Nigeria. One of its unique features is that it properly defines the term Nigerian content. NCA makes it necessary to give prior consideration to indigenous servicing companies which have demonstrated technical capacity and ownership of equipment whose personnel is predominantly Nigerian to execute jobs in the oil and gas industry. In addition, it presents a template for the classification of companies and lists the responsibilities of institutions charged with the effective delivery of the Nigerian content and a value matrix for measuring local input (Balouga, 2012). Section 2 of the Act makes it mandatory for regulatory authorities, contractors, operators, alliance partners, and other body desiring to engage in any oil and gas business related to consider local content. According to Adedeji (2016), the target for local content policy was to increase local content development by 45% in 2007, 70% in 2010, and above 80% by 2020 (Adedeji, 2016). Prior to enacting the NCA, foreign firms carried out most of the technical activities and operations in the upstream sector of the oil and gas industry. This was due to a perceived inadequacy of indigenous technical expertise, labor, and financial resources (Ezeani, 2017).

Table 4 Objective of the Nigerian government for some legislations in the petroleum industry

Policies	Objectives	Sources
Local Content Act	Growth in the upstream and downstream sectors of the industry Integration of the oil and gas industry into the main economy Diversification of the sources of investment within the industry Improved indigenous participation and transfer of technology and skills Increase in the employment proportion and provision of services in the oil industry Increase in the production capacity, etc.	Acheampong et al., 2016; Ayomike & Okeke, 2015
Petroleum Profit Tax Act	Exercise control over the public asset Regulate the number of participants in the industry and discourage rapid depletion of reserves Provide a veritable source of income to perform its obligations to the citizenry Create an instrument for the re-distribution of labor, technical expertise, and capital needed to build the industry and the country Promote a cleaner and healthier environment by the imposition of environmental tax	Abdul-Rahamoh et al., 2013
Petroleum Industry Bill	Reform the Nigerian National Petroleum Corporation (NNPC) Overhaul the current petroleum laws and fiscal system Revamp the petroleum revenue system and encourage sustainable prosperity within host communities Create a progressive fiscal framework that encourages investment	Sayne et al., 2012; Iledare, 2010
Marginal Field Program	Diversify investment sources and capital inflow into the country's oil industry Foster technology transfer by promoting indigenous participation in the sector Increase oil and gas reserves by engaging in aggressive exploration and development Provide opportunity to employ high-level technically competent Nigerians in the oil and gas business Increase production capacity; promote sharing of assets/facilities in order to ensure optimum utilization of available capacities	Ezekiel, 2020; Atsegbua, 2005

Petroleum Industry Act (PIA)

In 1999, President Olusegun Obasanjo established the OGIC (Oil and Gas Implementation Committee). The conclusions of the committee were approved by the government in 2007 forming the backbone of the then petroleum industry bill.

In 2012, a new draft was submitted to the National Assembly by the Minister of Petroleum Resources (De Montclos, 2014). The 2012 Draft of the PIB sought to remodel the Nigerian National Petroleum Corporation (NNPC) and to overhaul the existing fiscal systems (Iledare, 2010). Presently, NNPC acts as neither a meaningful operator nor a commercial entity. Thurber et al. (2010) noted that NNPC plays the role of de facto regulator in its transactions with IOCs as such conflict of interest is inevitable. The paper cited the implementation of local content as an example. NNPC is a regulator of local content, but it can also provide it. NNPC is also charged with cost regulation in joint venture arrangements and simultaneously serves as the major partner in these ventures. Other examples are in the sale of crude oil to the nation's four refineries; NNPC acts as a regulator, buyer, and commercial agent. As a government parastatal, NNPC manages government's share in Nigeria's oil operations, making it unanswerable to shareholders (Sayne, 2012).

The restructuring section in the PIA 2021 creates separate operations for the upstream, midstream, and downstream sectors. The Upstream Regulatory Commission ("the Commission") is to oversee the upstream sector, while the midstream and downstream sectors would be under the oversight of the Nigerian Midstream and Downstream Petroleum Regulatory Authority (the "Authority"). The Minister of Petroleum ("the Minister") is vested with general oversight powers over the petroleum industry; the Commission and the Authority are required to report to him. Section 53 of "the Act" directs the Minister to incorporate a Nigerian National Petroleum Company Limited with the Corporate Affairs Commission ("CAC") within 6 months of the Commencement of the PIB (PIA, 2021). The NNPC Ltd. like every incorporated entity will run as a commercial organization. Its share in joint venture arrangements with international firms will be taken over by the National Petroleum Assets Management Agency (NAPAMA) (De Montclos, 2014).

The Petroleum Profit Tax (PPT) which has been the most significant instrument for petroleum taxation will be replaced by Corporate Income Tax (CIT) and National Hydrocarbon Tax (NHT) in the PIB (Iledare, 2010). Hydrocarbon tax is a new form of tax which is only applicable to upstream petroleum operations. The proposed NHT rate is 0% for deep-water and frontier basins and 15% to 30% for onshore and shallow waters. In the case of the NHT, only 80% of foreign expenses are deductible, while in the case of CIT which is 30% of net profit, 100% of the relevant costs incurred are deductible except costs designated as unrecoverable. However, the NHT expense which was earlier deducted is not allowed to be deducted for CIT purposes. NHT is only imposed on oil revenue. The tax rate for gas is 5% and 2.5% for gas utilized in the country. The seventh schedule of the PIA stipulates different categories of royalties based on price, location, and production. The royalty is determined on a sliding scale and not a jumping scale; paragraph 10, subparagraph 2 stipulates royalties as seen in Table 5. A special provision for marginal fields producing less than 10,000 barrels per day within a month is made in subparagraph 4; the first 5000 bopd is charged at 5%, while the next 5000 bopd is charged at 7.5%; marginal fields producing above 10,000 bopd is charged under the provisions of subparagraph 2 stated earlier. Under the PIA, no overriding royalties shall be paid and no royalty by price for frontier acreages.

Table 5 Current regulations and the corresponding provisions of the PIB 2020 for the Nigerian petroleum industry

Current regulations	Terms	PIB proposed regulations	Terms
Petroleum profit tax	1. 85% onshore & 50% offshore 2. Marginal fields – 55%	1. Nigerian hydrocarbon tax 2. Corporate income tax 3. Host community levy 4. Withholding tax	1. 0% for deep-water/frontier basins; 15% for onshore/shallow water (PPL) and 30% for onshore/shallow water (PML) 2. 30% on net profit 3. 3% of its operating expenditure 4. 10% on dividends paid to shareholders
Nigerian National Petroleum Corporation (NNPC)	–	1. Nigerian upstream regulatory commission ("the commission") 2. Nigerian midstream and downstream petroleum regulatory authority ("the authority") 3. Nigerian National Petroleum Company Limited (NNPC ltd.)	–
Royalties and rent	1. Onshore and shallow water – 16.67–20% 2. Deepwater – 12% 3. Marginal fields – 2.5–18%	1. Location-based royalties 2. Price-based royalties 3. Production-based royalties 4. Marginal field royalties (producing less than 10,000 bopd within a month)	1. Onshore areas – 15%; shallow water – 12.5%; frontier basins – 7.5%;deep offshore – 10% 2. Below US$20 per barrel – 0%; above $20 and up to US $60 – 2.5%; above $60 up to US $100 – 4.0%; above $100 and up to US $150 – 8.0%; above US$150 per barrel – 10% 3. Monthly production in deep offshore <50,000 bpd – 5%; > 50,000 bpd – 7.5% 4. First 5000 bopd – 5%; next 5000 bpd – 7.5%
Oil prospecting license (OPL) Oil exploration license (OEL) Oil mining lease (OML)	–	1. Petroleum prospecting license (PPL) 2. Petroleum exploration license (PEL) 3. Petroleum mining lease (PML)	–
Others	–	1. Frontier exploration fund 2. Cost recovery limit	1. 30% of profit oil and gas 2. 80% of available hydrocarbons

Sources: PIA 2021; Okon, 2006

Furthermore, the PIA mandates a fund for frontier exploration, consisting of 30% of profit oil and gas from NNPC Limited received from its Production Sharing Contracts (PSCs), profit sharing contracts, and risk service contracts. It classifies a frontier basin as one in which no previous discovery or exploratory activity has been carried out or as determined by the commission. Some examples include Dahomey, Bida, Anambra, Benue. Sokoto, and Chad (KPMG, 2021). The Petroleum Host Community Fund (PHC Fund) was newly introduced in the PIA to address social and economic development in the oil-producing community. It is taken as 10% of the contractor's profit (Adeogun, 2015). However, this amount was changed to about 3%. The PIA (2021) also stipulates an "incorporated joint venture company" (IJV), in which NNPC Limited and other participants in the joint operating agreements with respect to upstream petroleum operations may voluntarily restructure their joint operating agreement to a joint venture carried out as a limited liability company, each referred to as an "incorporated joint venture company" (IJV) (PIA, 2021). Some other important provisions of the PIA are presented in Table 4.

3 Discussion

The marginal oil field development program is a laudable initiative of the Nigerian government which offers immense economic, social, and technical benefits such as increase in oil and gas production, engagement of indigenous manpower, transfer of technology and technical expertise, etc. Ezeani (2017) concluded that the major instruments employed to incentivize local participation in the oil and gas industry are the allocation of marginal field licenses alongside local content implementation. As stated earlier, some of the challenges facing the Nigerian marginal field program include economic viability, high cost, and technical difficulty. One of the ways to enhance the economic viability of marginal fields is to acknowledge the huge potential of the gas market and harness natural gas as a strategic asset rather than an undesirable byproduct of crude oil production (Humphrey, 2016). In order to reduce high cost of production and technical difficulties for marginal fields, the following should be encouraged: knowledge sharing, capacity building, joint venture partnerships, etc. However, these objectives cannot be achieved without the right fiscal legislations in place. The complex regulatory framework that governs the implementation of oil and gas projects can create bottlenecks which are not favorable for marginal field developers with limited financial resources. Ekeh (2015) opined that only 8 out of 24 marginal fields awarded in 2001 were producing a cumulative of about 53,000 barrels per day (bpd) as at 2013. One of the factors sighted as responsible for this was the tax regime and regulatory framework.

History was made on the first of July 2021, when the Nigerian Senate and the House of Representatives passed the Petroleum Industry Bill (PIB) after two decades. It was later signed into law by President Muhammadu Buhari on August 16, 2021. Without the Petroleum Industry Act (PIA), fiscal terms will be based on the Marginal Fields Operations (Fiscal Regime) Regulations 2005 and the Petroleum

Profits Tax Act 1959. Take, for instance, the current farm in agreement between IOCs and indigenous companies which requires royalties be paid to the IOCs for acreage farmed out to Marginal Field Operators (MFOs). The PIA demands that the IOCs completely surrender areas being operated to the indigenous companies, as such no overriding royalties will be paid. MFOs will be allowed to directly acquire the PML for those fields (Otombosoba, 2018). The PPT for marginal fields under the current fiscal regime is placed at 55%; this is lower than the 85% for mature fields (Otombosoba, 2018). Notwithstanding the PIA proposes even lower tax ranging from 15% to 30%. Also, the royalties and taxes are generally lower and should enhance the economic viability of marginal fields.

Local content policies are developed to increase the use of local manpower and resources to increase indigenization and foster sustainable development (Bello, 2017). The Petroleum Industry Bill (PIB) has been viewed as a growth enabler in the development of the Nigerian Content (Saidu, 2014). For marginal field development, using indigenous manpower could significantly reduce cost and create opportunities for improved expertise. This could make it more attractive to indigenous investors.

4 Conclusion

The Nigerian marginal field development program is a commendable initiative introduced to promote indigenous participation in the oil and gas industry, foster technological transfer, and increase oil and gas reserves and production capacity among others. Following the recent passage of the PIB and successful completion of the 2020 marginal field bid round by the DPR, there is need for further technical and economic evaluation to investigate the potential and profitability of the marginal field development initiative under the prevailing fiscal regulations. Although the PIA does not fully stipulate special provisions for marginal fields, it can be concluded that the PIA positively impacts marginal field development for the operators due lower taxes and royalties.

Acknowledgments The authors would like to appreciate the management of Covenant University for providing an enabling environment to carry out this research and assistance in publication.

References

Abdul-Rahamoh, O. A., Taiwo, F. H., & Adejare, A. T. (2013). The analysis of the effect of petroleum profit tax on Nigerian economy. *Asian Journal of Humanities and Social Sciences (AJHSS)*, *1*(1), 25-36

Acheampong, T., Ashong, M., & Svanikier, V. C. (2016). An assessment of local-content policies in oil and gas producing countries. *The Journal of World Energy Law & Business*, *9*(4), 282-302.

Adedeji, A. N., Sidique, S. F., Abd Rahman, A., & Law, S. H. (2016). The role of local content policy in local value creation in Nigeria's oil industry: A structural equation modeling (SEM) approach. *Resources Policy, 49*, 61-73.

Adetoba, L. A. (2012). The Nigerian marginal field initiative: recent developments. In Nigeria Annual International Conference and Exhibition. OnePetro.

Adeogun, O. (2015). Developing a Framework for Maximizing Marginal Oil and Gas Field Economics. In SPE Nigeria Annual International Conference and Exhibition. OnePetro.

Akinpelu, L. O., & Omole, O. A. (2009). Economics of Nigerian marginal oil fields-identifying high impact variables. In Nigeria Annual International Conference and Exhibition. OnePetro.

Akinwale, Y. O., & Akinbami, J. F. K. (2016). Economic evaluation of Nigerian marginal oil and gas field using financial simulation analysis. *International Journal of Energy Economics and Policy, 2016, 6*(3), 563-574.

Anderson, S. (2020). Nigeria's Marginal Field bid round – opportunities for investors? | Wood Mackenzie. Retrieved from https://www.woodmac.com/news/opinion/Nigerias-Marginal-Field-Bid-Roundopportunities-for-investors/

Ayomike, C. S., & Okeke, B. C. (2015). The Nigerian local content act and its implication on technical and vocational education and training (TVET) and the nation's economy. *International Journal of Education Learning and Development, 3(1)*, 26-35.

Atsegbua, L. (2005). Issues in the development of marginal oilfields in Nigeria. *Journal of Energy & Natural Resources Law, 23(3), 323-336.*

Balouga, J. (2012). Nigerian local content: challenges and prospects. *International Association for Energy Economics, 4*, 23-26.

Bello, T. (2017). Local Content in the Nigerian Oil and Gas Sector: A Classical Model for Indigenization. Available at SSRN 2971001.

Dagogo, O., Iledare, W., & Humphrey, O. (2018). Economic Viability of Infill Drilling Program for Marginal Oil Field Development: A Case Study of Sango Field in Nigeria. In *SPE Nigeria Annual International Conference and Exhibition*. OnePetro.

De Montclos, M. A. P. (2014). The politics and crisis of the Petroleum Industry Bill in Nigeria. *The Journal of Modern African Studies*, 403-424.

Ekeh, C., & Asekomeh, A. (2015). Optimality Test of Marginal Field Development Financing Arrangements in Nigeria. In *SPE Nigeria Annual International Conference and Exhibition*. Society of Petroleum Engineers.

Ezeani, E. C., & Nwuke, C. (2017). Local content and the marginal fields programme: challenges for indigenous participation in the Nigerian oil industry. Oil, gas and energy law (OGEL), 15(1), 1-20.

Ezekiel, M. P., & Okwuchukwu, O. B. (2020). A critique of the legal framework for the development of marginal oil fields in Nigeria. *Nnamdi Azikiwe University Journal of International Law and Jurisprudence, 11*(2), 135-149.

Gbegi, D. O., Adebisi, J. F., & Tosin, B. O. D. U. N. D. E. (2017). The effect of petroleum profit tax on the profitability of listed oil and gas companies in Nigeria. *American International Journal of Social Science, 6*(2), 40-46.

Humphrey, O., & Dosunmu, A. (2016). Strategies for economic development of marginal oil field in Nigeria. *Journal of Emerging Trends in Economics and Management Sciences, 7*(5), 322-327.

Hui, W., Guangyi, H., Hongjun, F., Lianyong, Y. U., & Zhenkun, L. (2012). Key technologies for the fluvial reservoir characterization of marginal oil fields. *Petroleum Exploration and Development, 39*(5), 667-673.

Iledare, O. O. (2010). Evaluating the Impact of Fiscal Provisions in the Draft Petroleum Industry Bill on Offshore E & P Economics and Take Statistics in Nigeria. In *Nigeria Annual International Conference and Exhibition. OnePetro.*

KPMG (2021). Petroleum Industry Bill (PIB) 2021 - A Game Changer. Retrieved from https://assets.kpmg/content/dam/kpmg/ng/pdf/tax/petroleum-industry-bill-(pib)-2021-a-game-changer.pdf.

Kue, Y. N., & Orodu, O. D. (2006). Economic analysis of innovative approaches to marginal field development. In *Nigeria Annual International Conference and Exhibition*. Society of Petroleum Engineers.

Lawal, K. T. (2013). Taxation of Petroleum Profit under the Nigeria's Petroleum Profit Tax Act. *Int'l J. Advanced Legal Stud. & Governance, 4*, 1.

Ogolo, O. (2021). Modification of the unit technical cost equation for the accurate determination of the cost of producing a barrel of oil in relation to the Contractor's revenue. *Journal of Petroleum Science and Engineering, 198*, 108-122.

Okon, T. E. (2006). Nigerian Fiscal Regime and Profitability Analysis. *Presentation made to NNPC Management.*

Onolemhemhen, R. U., Isehunwa, S. O., Iwayemi, A. P., & Adenikinju, A. F. (2017). The Technical and Economic Viability of Producing Marginal Oil Fields in the Niger-Delta Using Water Injection. In *IAEE Energy Forum Singapore Issue.*

Otombosoba, O. H., & Dosunmu, A. (2018). Examining the Legal and Regulatory Framework on Marginal Oil Field Development in Nigeria. In *SPE Nigeria Annual International Conference and Exhibition*. OnePetro.

Okwurionu, C. (2020). Nigeria: Analyzing the 2020 Marginal Fields Bid Round [Blog Post]. Retrieved from https://www.mondaq.com/nigeria/oil-gaselectricity/952366/analysing-the-2020-marginal-fields-bid-round?login=true

PIA (2021). Retrieved from https://hrpib.org.ng/wpcontent/uploads/2021/09/Petroleum-Industry-Act-2021-pdf.pdf

Saidu, S. & Mohammed, A. R. (2014). The Nigerian petroleum industry bill: An evaluation of the effect of the proposed fiscal terms on investment in the upstream sector. *Journal of Business and Management Sciences*, *2*(2), 45-57.

Sayne, A., Mahdavi, P., Heller, P. R., & Schreuder, J. (2012). The Petroleum Industry Bill and the Future of NNPC. *Retrieved from the Revenue Watch Institute website: https://resourcegovernance. org/sites/default/files/rwi_bp_nnpc_synth_rev2.pdf.*

Thurber, M. C., Emelife, I., & Heller, P. R. (2010). NNPC and Nigeria's oil patronage ecosystem. *Oil and Governance: State-Owned Enterprises and the World Energy Supply; Victor, D., Hults, D., Thurber, M., Eds*, 701-752.

Uwaga, A. O. (2008). Developing marginal fields in the Niger Delta: recovery & economics can be better than currently estimated. In Nigeria Annual International Conference and Exhibition. OnePetro.

Zhao, Z., Tang, Y., Wu, Z., Li, Y., & Wang, Z. (2016). Towing analysis of multi-cylinder platform for offshore marginal oil field development. In International Conference on Offshore Mechanics and Arctic Engineering (Vol. 49989, p. V007T06A073). *American Society of Mechanical Engineers.*

Part II
Microbial and Enzymatic Processes

Purification and Characterization of Phytase from a Local Poultry Isolate of *Aspergillus flavus* MT899184

E. A. Onibokun, A. O. Eni, and S. U. Oranusi

1 Introduction

Phytates and its isomers (myo-inositol phosphate esters (IP5 to IP2), even at low levels, constitute anti-nutritional factors that can have adverse effects on protein and mineral digestibility in animals particularly monogastric animals (Lee et al., 2020). These monogastric animals are unable to utilize all the nutrients in their feed because of the deficiency of their digestive system, particularly as they lack the enzyme phytase (Asmare, 2014). Being the main form by which phosphorus in cereals, leguminous crops, oil seeds, and nuts is stored (Gupta et al., 2015), commercial production of monogastric animal feeds requires the addition of exogenous phytase. Phytase is the most valuable commercial enzyme in poultry feed production (Lamp et al., 2020) and constitutes the major cost in the commercial production of poultry feed (Kim et al., 2019). The enzyme is able to hydrolyze the phytic acid present in the feed, thus improving the nutritional value of feed and also decreasing the amount of phosphorus excreted by animals (Muslim et al., 2018). Also phytase can be derived from other sources; however, plants are unable to produce extracellular phytase (Richardson et al., 2004). Microorganisms remain the most significant sources for production of enzymes (Atolagbe et al., 2016). They are generally regarded as cheaper means of producing industrial enzymes, since they can utilize readily available waste materials. In fungi, phytase production has been reported in species like *Aspergillus niger* (Afinah et al., 2010), *Aspergillus flavus* (Gaind & Singh, 2015) and *Aspergillus fumigatus* (Mittal et al., 2013). Although in most cases solid-state fermentation technique is regarded most suitable for commercial phytase

E. A. Onibokun (✉) · A. O. Eni · S. U. Oranusi
Department of Biological Sciences, College of Science and Technology,
Covenant University, Ota, Ogun State, Nigeria
e-mail: elizabeth.onibokun1@covenantuniversity.edu.ng

© The Author(s), under exclusive license to Springer Nature Switzerland AG 2022

A. O. Ayeni et al. (eds.), *Bioenergy and Biochemical Processing Technologies*, Green Energy and Technology, https://doi.org/10.1007/978-3-030-96721-5_9

production using *Aspergillus* species (Saithi and Tongta et al., 2016). In this study, submerged fermentation is employed due to the ease of product recovery which will minimize the cost of downstreaming processing and ultimately the overall cost of enzyme production. The aim of this study was therefore to characterize the phytase produced by a local fungal isolate to determine its suitability for inclusion as an exogenous supplement in animal feeds.

2 Materials and Methods

2.1 Sample Collection

A total of 100 g of soil samples were randomly collected from ten different spots (at 10 g per spot) of a poultry site in Lagos state (6°27′14.65" N 3°23'40.81" E), Nigeria. Samples were collected from the areas on which the poultry droppings fell. They were collected in black polythene bags and transported same day to the Covenant University, Microbiology Laboratory for analyses.

2.2 Screening and Selection of Phytase-Producing Fungi from Poultry Soil

The selected fungi were quantitatively screened for the production of phytase following Lee et al. (2005). Soil suspension (1 mL) was plated out onto phytase-screening medium (D-glucose – 15.0 g; sodium phytate – 3.0 g; NH_4NO_3–5.0 g; $MgSO4.7H_2O$ – 0.5 g; KCl – 0.5 g; $FeSO4.7H_2O$ – 0.01 g; $MnSO_4.4H_2O$ – 0.01 g; and Agar – 15.0 g in 1 L) (Qasim et al., 2016). Soil suspension was prepared by mixing 0.5 g of soil sample in 10 mL of 9% saline solution (Gontia-Mishra et al., 2013). The plates were left to incubate at 30 °C for 5 days. Fungal isolate was selected based on zone of hydrolysis.

2.3 Morphological Identification

Selected fungal isolate was purified by subculturing into potato dextrose agar (PDA) plates and subjected to lactophenol cotton blue test and identified based on morphological and microscopic characteristics referenced to fungi compendium (Domsch et al., 1980).

2.4 Molecular Identification

The DNA isolation was carried out following the CTAB method described by Dellaporta et al. (1983). PCR amplification of fungal internal transcribed spacer region was performed using the primer set ITS 1 and ITS 4 with the following sequences: ITS 1 (5'-TCCGTAGGTGAACCTGCG G-3′ and ITS 4 (5'-TCCTCCGCTTATTGATATGC-3′) (Alves et al., 2018). A 50 μL reaction master mix containing 0.5 μL Taq DNA polymerase (Transtat China), 5 μL 10X dilution of the manufacturer's buffer (Transtat China), 4 μL deoxynucleoside triphosphates (dNTPs), 1 μL each of primers ITS 1 and ITS 4, and 36.5 μL sterile DNAse free water was employed for the PCR amplification. The reaction conditions were as follows: initial denaturation at 95 °C for 10 min, 30 amplification cycles of denaturation at 95 °C for 1 min, annealing at 55 °C for 1 min (determined after gradient PCR), and primer extension at 72 °C for 90 min 1:30 min, followed by a final extension at 72 °C for 10 min. Aliquots of amplicons (7 μL) were analysed in 1% (w/v) agarose gel (Sigma, USA) by horizontal gel electrophoresis. DNA bands were visualized by UV excitation after staining with ethidium bromide (0.5 μL). Amplicon was purified according to the protocol described on EasyPure® PCR purification kit and quantified using Nanodrop 2000 Spectrophotometer. Then, appropriate concentration was packaged and sent for Sanger sequencing at Eurofins Genomics, Ebersberg, Germany. Forward and reverse sequencing reads were quality checked and assembled using reference sequence NR_121481.1 obtained from the NCBI Genbank database. Assembled reads were then compared against the NCBI Fungal ITS database, using the NCBI BLAST software, to check similar isolates. Multiple alignments were performed using the Geneious aligner (Kearse et al., 2012) and phylogenetic tree computed following Tamura-Nei distance model (Tamura and Nei, 1993) and Neighbor-Joining method with no outgroup.

2.5 Phytase Production

The selected fungus was quantitatively screened for the production of phytase by submerged fermentation (SmF). One mL of spore suspension containing 2 x 10^8 spores was inoculated into 100 mL phytase screening broth (PSB) (PSM without agar) and incubated in a shaking water bath at 30 °C for 5 days at 150 rpm. The fungal biomass was separated from the medium using Whatman No. 4 filter paper, and cell-free filtrate was employed for phytase assay (Qasim et al., 2016). The phytase activity was investigated by incubating 1 mL of crude phytase at 37 °C for 30 min in 1 mL 0.2 M sodium acetate buffer (pH 5.5) with 0.5% sodium phytate. The reaction was stopped by adding 1 mL of trichloroacetic acid (15% [w/v]). After the addition of 2 mL of freshly prepared colouring reagent (3.66 g of $FeSO_4{\cdot}7H_2O$, 0.5 g of $(NH_4)_6Mo_7O_{24}{\cdot}4H_2O$, and 1.6 mL of concentrated H_2SO_4 in 50 mL of distilled water), the chilled sample on ice was incubated for 10 min at 30 °C. The released phosphate was determined at 750 nm with a UV-VIS spectrophotometer (Lee et al., 2005). Protein quantification was carried following Lowry et al. (1951).

2.6 *Phytase Purification*

Ammonium Sulphate Precipitation

Ammonium sulphate precipitation was employed to purify the crude phytase using a modification of the methods of Sanni et al. (2019) and Kalsi et al. (2016). Crude phytase was purified by precipitation using 70% ammonium sulphate. The ammonium sulphate-crude phytase mixture was incubated overnight at 4 °C with gentle stirring. The resulting solution was centrifuged at 21,000 g for 20 min to obtain precipitated proteins in pellet form. The pellet was then dissolved in 10 mL of 0.1 M acetate buffer pH 6.0 and dialyzed using a pre-acetylated cellophane tubing against three changes of the same buffer solution for 24 h at 4 °C, to desalt the precipitated enzyme. The precipitated phytase sample was assayed for phytase activity as previously described by Lee et al. (2005) and protein content determined following Lowry et al. (1951).

Gel Filtration Chromatography

The dialyzed crude phytase from the ammonium sulphate precipitation was further purified using the modifications methods of Thyagarajan et al. (2014) and Ajith et al. (2019). A chromatographic column of (50 × 1.5 cm) was packed with Sephadex G-75 gel and glass wool placed in the lower part of the column. The Sephadex G-75 (16.7 g) was dissolved in 250 mL of the phosphate buffer pH 7.0 (following manufacturer's instruction of 1 g in 15 mL) and then boiled in a water bath for 5 min for swelling and degassing. The gel was allowed to settle for 18 h. The column was washed with phosphate buffer pH 7. After column equilibrium, 5 mL of ammonium sulphate phytase fraction was loaded gently on the surface of the gel and eluted with Tris-HCL buffer pH 7.8 at a 20 mL/h flow rate. Aliquots of 2 mL fractions of the eluent were collected; then the optical density at 280 nm and phytase activity (U/mL) was determined for each fraction. The phytase activity was determined following Lee et al. (2005), and protein content was determined following Lowry et al. (1951).

Effect of Carbon and Nitrogen Sources on Phytase Production

The effect of nutrient sources on phytase production was assessed by addition of 1% sucrose, in place of glucose and 0.1% of a different nitrogenous salt, $(NH_4)_2SO_4$ instead of NH_4NO_3. Phytase activity was determined following Lee et al. (2005).

Effect of Temperature on Phytase Activity

Optimum temperature for phytase activity was determined by subjecting the enzyme to different temperatures (30–70 °C). This was carried out by varying the incubation temperature at a 10 °C interval during the activity determination. Phytase activity was determined following the descriptions (Lee et al., 2005).

Effect of pH on Phytase Activity

Optimum pH was established by subjecting the phytase to varying pH across a pH range of 3 to 8. Sodium acetate buffer was prepared at different pH during the activity determination, while incubation temperature was held at the optimum temperature determined earlier for the phytase. Phytase activity was determined following the descriptions by Lee et al. (2005).

Thermal Stability of Phytase

Thermal stability of the phytase was determined by subjecting the phytase to temperatures of 70 °C, 80 °C, and 100 °C for 5, 10, 20, and 30 min. The phytase activity was determined following the descriptions by Lee et al. (2005).

3 Results and Discussions

3.1 Identification of Phytase-Producing Isolate

Based on zone of hydrolysis, isolate code H was selected as phytase-producing and identified morphologically as *Aspergillus* sp. (Table 1). Further identification by molecular techniques recognized the isolate as *Aspergillus flavus* (Fig.1). The amplified internal transcribed spacer region sequence of the fungi submitted to NCBI was allocated accession number MT899184 (Fig. 2). The isolate showed genetic relatedness to other *Aspergillus* species isolated from agricultural samples. Previous studies reporting phytase production by fungi exist which is consistent with this study (Ajith et al., 2019; Sanni et al., 2019; Ribeiro Corrêa et al., 2015;

Table 1 Morphological properties of isolated phytase-producing fungal species

Isolate code	Morphology	Diameter of zone (cm)	Number of spores/mL	Identity of fungi
H	Green colour, septate hyphae with sporangium borne, round spore	2	2×10^8	*Aspergillus* sp.

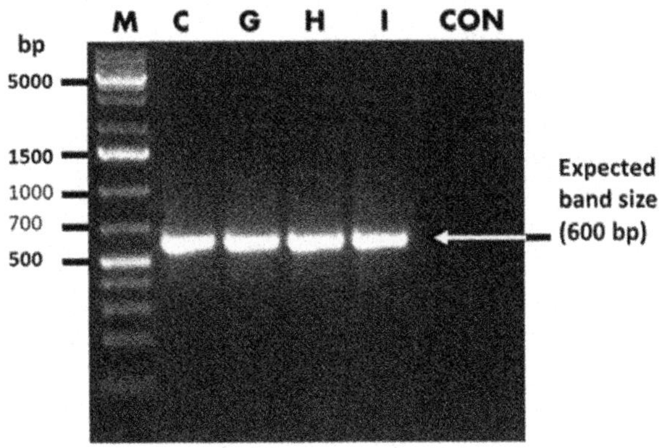

Fig. 1 Gel electrophoresis image of PCR amplification of the ITS region of isolated fungal sp. with *isolate* H on lane 4. (*KEY M* DNA molecular marker; *CON* Negative control without template DNA)

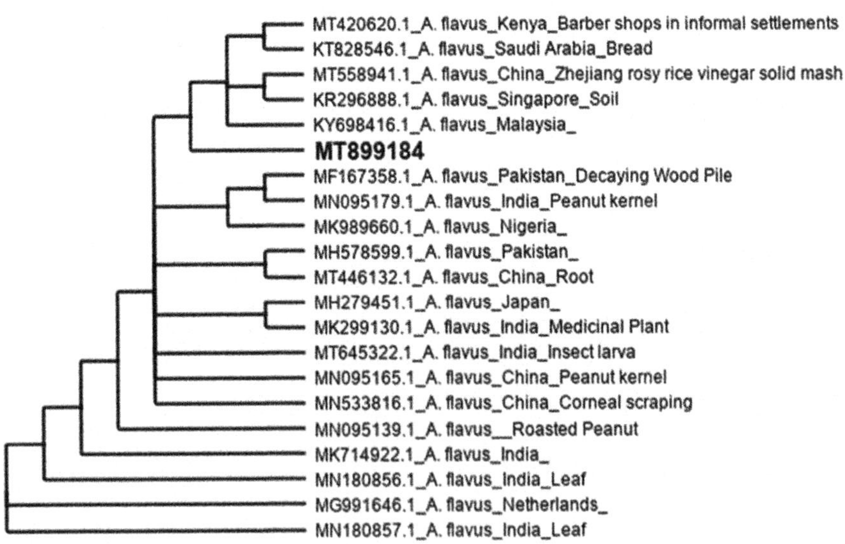

Fig. 2 Phylogenetic tree showing evolutionary relatedness of isolate H to other *Aspergillus flavus* from agricultural environments

Gaind and Singh, 2015; Mittal et al., 2013). Handa et al. (2020) noted that the genus *Aspergillus* is the best producers of phytase. The choice of members of the genus *Aspergillus* to produce phytase and indeed other industrially relevant enzyme may not be unconnected to their generally regarded as safe (GRAS) status, their ubiquitous nature, and being able to utilize wide variety of substrates and their minimal nutritional requirement (Obafemi et al., 2018; Saith et al., 2016). Phytase is inducible and produced in the presence of an anti-nutrient (Vasudevan et al., 2017). The high amount of anti-nutrient (phytate), in poultry environment owing to the inability of the birds to digest the phytate present in their feed grains, may have encouraged the growth of the phytate degrading fungi.

3.2 Phytase Production and Purification

Quantitative analysis of crude phytase showed a crude phytase with total activity of 609 U/mL and protein content of 314.6 mg/mL. Purification resulted in a consistent reduction in total activity but with a corresponding increase on phytase-specific activity from 1.9 U/mL to 18.4 U/mL in the crude enzyme fraction and the gel chromatography fraction, respectively (Table 2). The Sephadex G-75 chromatogram showed one very distinct elution peaks (Fig. 3).

Optimum phytase production was observed on day 5 as similarly reported by Neira-Vielma et al. (2018) and Qasim et al. (2016). Although differing report was presented by Sanni et al. (2019) where optimum phytase production was observed at 216 h (9 days) for *A. niger* and 144 h (6 days) for *A. flavus*. Shivanna and Govindarajulu (2014) also reported optimal activity on the sixth day. This prolonged fermentation time may be as a result of the inducible nature of the enzyme; hence a prolonged lag phase might be required by the fungi to produce the enzyme. Generally, fungi species have been reported to be slow growers requiring about 2–5 days for optimal growth depending on the nutrient medium in which they are grown (Meletiadis et al., 2001). The use of SmF in this study is corroborated by the report of Jain et al. (2016). SmF is associated with ease in the recovery of product (Sethi et al., 2016). As the aim of this study to maximize profit and reduce production cost to the barest minimum, a fermentation technique which will reduce the cost is highly encouraged. Additionally, SmF has extra advantage of easy control of

Table 2 Purification table for phytase produced by isolate H obtained from poultry soil in Abule-egba, Lagos State

Protein step	Total activity Total protein Phytase activity Yield Purification fold (U/mL) (mg/mL) (U/mL) (%)
Crude enzyme	609314.6 1.9100 1
$(NH_4)_2SO_4$ fraction (70%)	187.5 41.4 4.5 31 2.4
Sephadex G-75 fraction	7.6 4.2 18.4 12.8 9

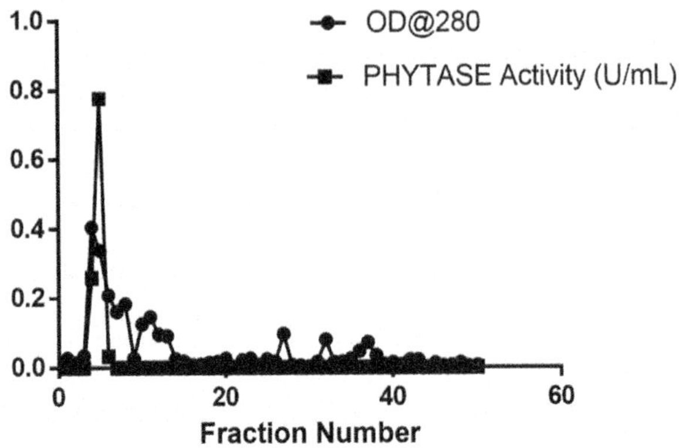

Fig. 3 Purification of phytase produced by *Aspergillus flavus* (isolate H) on Sephadex G-75 Gel filtration column (50 cm × 1.5 cm)

fermentation conditions such as aeration, pH, and temperature (Ornela and Guimaraes, 2019). The method is also the most commonly employed method for phytase production (Jain et al., 2016).

The quantitative assay of the cell-free supernatant showed an activity within the range of 609 U/mL. This result was higher than the report of Shivanna and Govindarajulu (2014) where they reported activity of 9.2 U/mL and 8.8 U/mL using a 6-day-old culture of *A. niger* and *A. ficuum*, respectively, under SmF. For total phytase activity, there was a continuous reduction from 609 U/mL in the crude phytase step to 77.6 U/mL in the Sephadex column fraction with each purification step. For specific phytase activity, there was a continuous increase from 1.9 U/mL in the crude phytase to 4.5 U/mL in the ammonium sulphate fraction and 18.4 U/mL in the Sephadex G-75 fraction. These values were higher when compared with the specific phytase activity of crude extract (1.13 U/mL), microfiltration fraction (1.15 U/mL), ultrafiltration fraction (1.93 U/mL), and DEAE Sepharose fraction (8.38 U/mL) reported by Neira-Vielma et al. (2018) when solid-state fermentation (SSF) was employed. Saithi and Tongta (2016) had earlier reported the limitation of SSF over SmF.

3.3 Phytase Characterization

Upon characterization, the phytase showed an optimum temperature and pH of 40 °C (Fig. 4) and 6 (Fig. 5), respectively. The thermal stability of the phytase over temperatures 70–80 °C showed that the phytase could withstand both temperatures for 5 minutes (Figs. 6 and 7), but had no activity at 100 °C (Fig. 8). For effect of carbon and nitrogen sources, results obtained showed that phytase production when glucose was employed as carbon source was 0.185 U/mL but no activity was observed using sucrose as carbon source (Fig. 9). However, the isolate could not

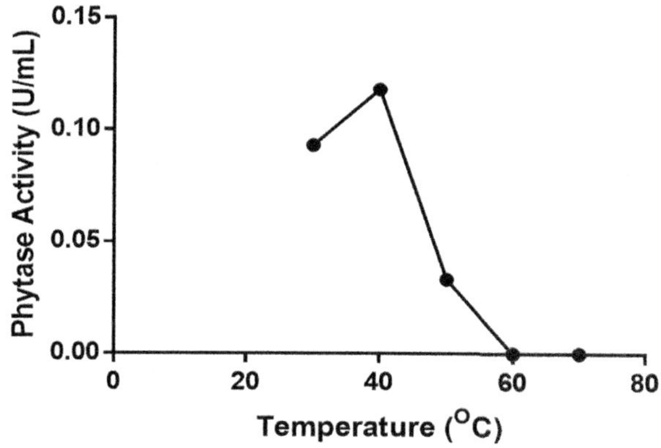

Fig. 4 Effect of temperature on the activity of phytase produced by isolate H

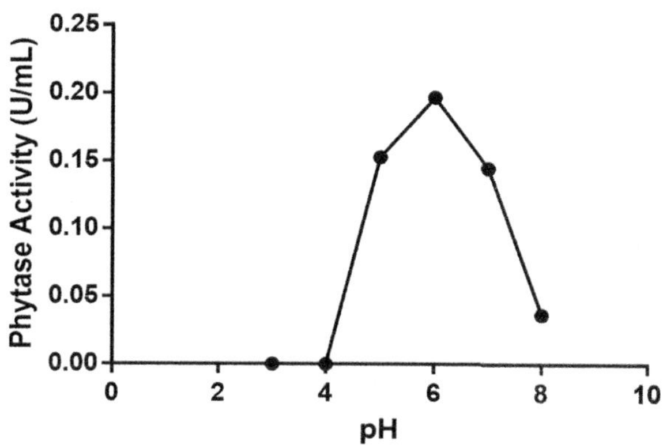

Fig. 5 Effect of pH on the activity of phytase produced by isolate H

utilize ammonium sulphate as an alternative nitrogen source to ammonium nitrate for phytase production (Fig. 9). Fungal phytases have a broad spectrum of optimal temperature ranging between 37 °C and 67 °C (Jatuwong et al., 2020). The optimal pH recorded in this study was at pH 6.0. Most fungal phytases are reported to be active at optimal pH within the acidic range between 2.0–6.0 (Pedri et al., 2015). However, there are studies reporting optimal activity around the neutral pH (Gaind and Singh, 2015), pH 6.5 (Thayagarajan et al., 2014), and pH 8–9 (Sanni et al., 2019). Different microorganisms exhibit varying optimum pH as pH greatly affects the transport of nutrients across the cell membrane of the organisms, thus contributing to cell growth and production of desired products (Gaind and Singh, 2015).

The isolate in this study showed optimal activity when glucose was employed as carbon source compared to sucrose, while ammonium sulphate showed better activity as nitrogen source as supported by the results of Pedri et al. (2015). Glucose is a simple sugar, hence may require less energy to metabolize. When exposed to temperature of 70 °C, the phytase in this study retained its activity after 5 min, similarly

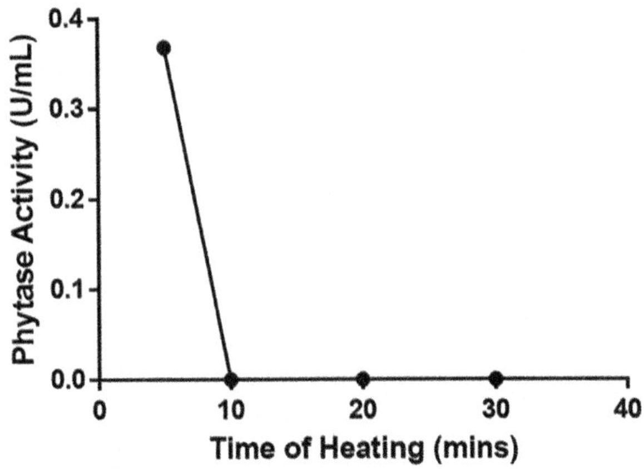

Fig. 6 Temperature stability at 70 °C on phytase produced by isolate H

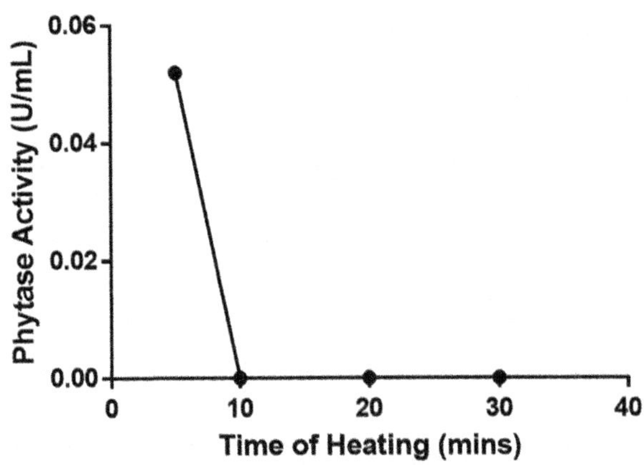

Fig. 7 Temperature stability at 80 °C on phytase produced by isolate H

at 80 °C. This result is of utmost importance since inclusion of this enzyme in animal feed processing means that it can withstand a temperature of 70 °C and above and still remain active. Although pelleting commonly occurs at temperatures of 80 °C, reducing the temperature to about 65 °C does not in any alter the overall effect of the process (Boroojeni et al., 2016); therefore phytase produced in this current study can find application as feed supplement if the feed is pelleted at temperatures 80 °C and below. Phytase is regarded as a thermo-tolerant enzyme; the results from the enzyme produced in this study are similar to the reports of Neira-vielma et al. (2018), where the enzyme retained more than 70% activity at 80 °C for 1 min, 40% activity for 2 min, and 9% after 5 min.

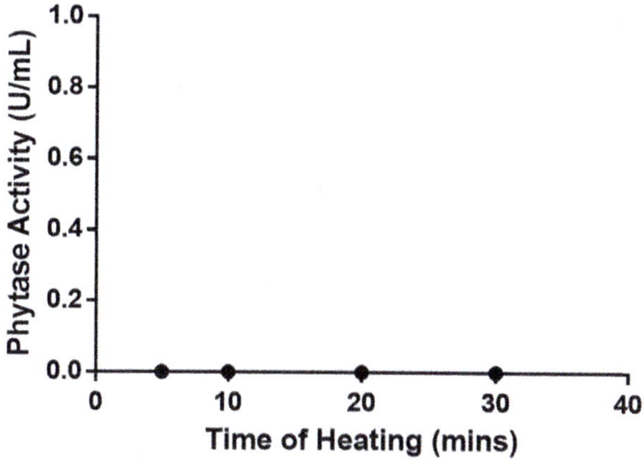

Fig. 8 Temperature stability at 100 °C on phytase produced by isolate H

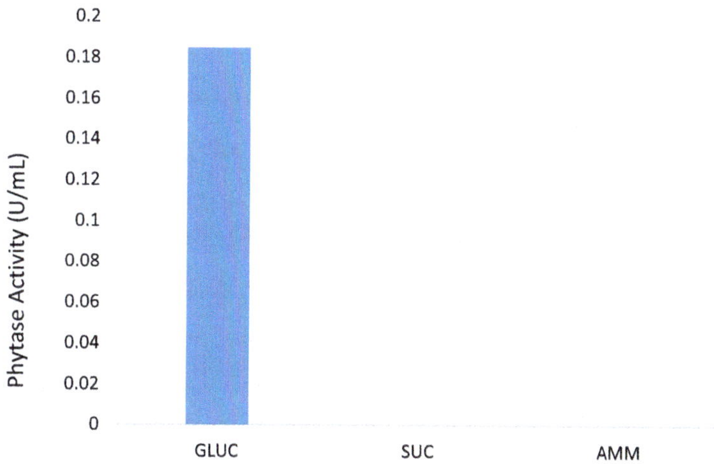

Fig. 9 Effect of carbon and nitrogen sources on the production of phytase produced by *A. flavus* (isolate H). (*GLUC* Medium containing glucose as carbon source and ammonium nitrate as nitrogen source, *SUC* Medium containing sucrose as carbon source and ammonium nitrate as nitrogen source, *AMM* Medium containing glucose as carbon source and ammonium sulphate as nitrogen source)

4 Conclusion

This present study has demonstrated the ability of *Aspergillus flavus* MT899184, local poultry isolate to degrade sodium phytate under submerged fermentation technique by producing phytase enzyme. There was continuous increase in specific activity of the phytase produced with each purification step. The phytase has optimal conditions of 40 °C and 6.0 of temperature and pH, respectively. The enzyme also demonstrated enzymatic activity after 5 minutes of heating at both 70 °C and

80 °C. These properties make it suitable as an exogenous additive in poultry feed. The properties of this enzyme also suggest that the feed formulation processes specifically the pelleting will not have adverse effects on the feed characteristics. The exogenous addition of this enzyme also makes its application easy without technical hitches, thus user friendly for easy utilization by the local farmers.

Acknowledgements We also acknowledge Covenant University for support in payment of publication costs.

Disclaimer None.

Conflict of Interest None.

Funding This study was funded by the International Foundation for Science (IFS) for its support through a grant awarded to Onibokun Adeola Elizabeth (Research grant agreement NO. I-3-C-6179-1).

References

Afinah S, Yazid AM, Anis Shobirin MH, Shuhaimi M (2010) Phytase: application in food industry. International Food Research Journal 17(1): 13-21

Ajith S, Ghosh J, Shet D, ShreeVidhya S, Punith BD, Elangova AV (2019) Partial purification and characterization of phytase from *Aspergillus foetidus* MTCC 11682. AMB Express 9(1): 3

Alves RC, Borges IL, Dutra V, Garino Junior F, Miranda Neto EGD, Dantas AFM, Riet-Correa F, de-Galiza GJN (2018) Systemic infection by *Aspergillus flavus* in a mare. Acta Sci. Vet 46(supl): 345

Asmare B (2014) Effect of common feed enzymes on nutrient utilization of monogastric animals. International Journal Molecular Biology Research 5(4): 27–34

Atolagbe OM, Ajayi AA, Edegbo O (2016) Characterization of α-Amylase from Soursop (*Annona muricata* Linn.) Fruits Degraded by *Rhizopus stolonifer*. Pakistan Journal of Biological Sciences 9:77–81

Boroojeni FG, Svihus B, von Reichenbach HG, Zentek J (2016) The effects of hydrothermal processing on feed hygiene, nutrient availability, intestinal microbiota and morphology in poultry—A review. Animal Feed Science and Technology. 220: 187-215

Dellaporta SL, Wood J, Hicks JB (1983) DNA Extraction. Plant Molecular Biology Reporter 1(4): 19–21

Domsch K, Gams W, Anderson T (1980) Compendium of soil fungi (Vol. 1)

Gaind S and Singh S, 2015. Production, purification and characterization of neutral phytase from thermotolerant *Aspergillus flavus* ITCC 6720. International Journal of Biodeterioration and Biodegradation 99: 15–22

Gontia-Mishra I, Deshmukh D, Tripathi N, Bardiya-Bhurat K, Tantwai K, Tiwari S (2013). Isolation, morphological and molecular characterization of phytate-hydrolysing fungi by 18S rDNA sequence analysis. Brazilian Journal of Microbiology 44(1): 317

Gupta RK, Gangoliya SS, Singh NK (2015) Reduction of phytic acid and enhancement of bioavailable micronutrients in food grains. Journal of Food Science and Technology 52(2): 676

Handa V, Sharma D, Kaur A, Arya SK (2020) Biotechnological applications of microbial phytase and phytic acid in food and feed industries. Biocatalysis and Agricultural Biotechnology 25: 101600.

Jain J, Sapna, Bijender S (2016) Characteristics and biotechnological applications of bacterial phytases. Process Biochemistry 51: 159–169.

Jatuwong K, Suwannarach N, Kumla J, Penkhrue W, Kakumyan P, Lumyong S (2020) Bioprocess for Production, Characteristics, and Biotechnological Applications of Fungal Phytases. Frontiers in Microbiology 11: 188.

Kalsi HK, Singh R, Dhaliwal HS, Kumar V (2016) Phytases from Enterobacter and Serratia species with desirable characteristics for food and feed applications. 3 Biotech 6(1):64. doi: https://doi.org/10.1007/s13205-016-0378-x. Epub 2016 Feb 13

Kearse M, Moir R, Wilson A, Stones-Havas S, Cheung M, Sturrock S, Buxton S, Cooper A, Markowitz S, Duran C Drummond A (2012) Geneious Basic: An integrated and extendable desktop software platform for the organization and analysis of sequence data. Bioinformatics. 28(12): 647–1649

Kim JW, Sanjayan N, Leterme P, Nyachoti CM (2019) Relative bioavailability of phosphorus in high-protein sunflower meal for broiler chickens and effects of dietary phytase supplementation on bone traits, growth performance, and apparent ileal digestibility of nutrients. Poultry Science 98(1): 298–305

Lamp AE, Ward NE, Wilson JW, Moritz JS (2020) In vitro phytase activity after steam conditioning and pelleting may not consistently correlate with in vivo measurements among commercial phytases. Journal of Applied Poultry Research 29(2): 420–434

Lee DH, Choi SU, Hwang YI (2005) Culture Conditions and Characterizations of a New Phytase-Producing Fungal Isolate, *Aspergillus sp.* L117. Mycobiology 33(4): 223–229

Lee SA, Lupatsch I, Gomes GA, Bedford MR (2020) An advanced *Escherichia coli* phytase improves performance and retention of phosphorus and nitrogen in rainbow trout (*Oncorhynchus mykiss*) fed low phosphorus plant-based diets, at 11 °C and 15 °C. Aquaculture 516: 734549

Lowry O, Rosebrough N, Farr A, Randal R (1951) Protein measurement with the Folin phenol reagent. Journal of Biological Chemistry 193: 265-275

Meletiadis J, Mouton JW, Meis JF, Bouman BA, Donnelly JP, Verweij PE, Eurofung Network (2001) Colorimetric Assay for Antifungal Susceptibility Testing of Aspergillus Species. Journal of Clinical Microbiology 39(9): 3402–3408

Mittal A, Gupta V, Singh G, Yadav A, Aggarwal NK (2013) Phytase: A Boom in Food Industry. Acta Journal of Biosciences. 1(2): 158–169

Muslim SN, Ali AN, AL-Kadmy IM, Khazaal SS, Ibrahim SA, Al-Saryi NA, Al-Saadi LG, Muslim SN, Salman BK, Aziz SN (2018) Screening, nutritional optimization and purification for phytase produced by *Enterobacter aerogenes* and its role in enhancement of hydrocarbons degradation and biofilm inhibition. Microbial Pathogenesis 115: 159-67

Neira-Vielma AA, Aguilar CN, Ilyina A, Contreras-Esquivel JC, das Graça Carneiro-da-Cunha M, Michelena-Álvarez G, Martínez-Hernández JL (2018) Purification and biochemical characterization of an *Aspergillus niger* phytase produced by solid-state fermentation using triticale residues as substrate. Biotechnology Reports 17: 49–54.

Obafemi YD, Ajayi AA, Olasehinde GI, Atolagbe OM, Onibokun EA (2018) Screening and partial purification of amylase from *Aspergillus niger* isolated from deteriorated tomato (*Lycopersicon esculentum* mill.) Fruits. African Journal of Clinical and Experimental Microbiology 19: 47–57.

Ornelaa PH, de O, Guimarães LHS (2019) Purification and characterization of an alkalistable phytase produced by *Rhizopus microsporus* var. microsporus in submerged fermentation. Process Biochemistry 81: 70–76.

Pedri ZC, Lozano LMS, Hermann KL, Helm CV, Peralta RM, Tavares LBB (2015) Influence of nitrogen sources on the enzymatic activity and grown by *Lentinula edodes* in biomass *Eucalyptus benthamii.* Brazilian Journal of Biology 75(4): 0-0

Qasim SS, Shakir KA and Al-Shaibani AB (2016) Purification of Phytase produced from a Local fungal isolate and its Application in Food Systems. Iraqi Journal of Agricultural Sciences 47: 112–120

Ribeiro Corrêa T, de Queiroz MV, de Araújo EF (2015) Cloning, recombinant expression and characterization of a new phytase from *Penicillium chrysogenum.* Microbiology Research 170: 205–212.

Saithi S, Tongta A (2016) Phytase Production of *Aspergillus niger* on Soybean Meal by Solid-State Fermentation Using a Rotating Drum Bioreactor. Agricultural Science 11: 25–30.

Sanni DM, Lawal OT, Enujiugha VN (2019) Purification and characterization of phytase from Aspergillus fumigatus Isolated from African Giant Snail (*Achatina fulica*). Biocatalysis and Agricultural Biotechnology 17: 225–232

Sethi BK, Jana A, Nanda PK, Dasmohapatra PK, Sahoo SL, Kumar J (2016) Production of α-Amylase by *Aspergillus terreus* NCFT 4269.10 Using Pearl Millet and Its Structural Characterization. Frontiers in Plant Science 1: 639

Shivanna GB, Govindarajulu V (2014) Phytase Production by *Aspergillus niger* CFR 335 and *Aspergillus ficuum* SGA 01 through Submerged and Solid-State Fermentation. Science World Journal 2014: 1–6

Tamura K, Nei M (1993) Estimation of the number of nucleotide substitutions in the control region of mitochondrial DNA in humans and chimpanzees. Molecular Biology and Evolution 10(3): 512–526

Thyagarajan R, Namasivayamm KS, Narendrakumar G (2014) Partial purification of phytase from *Hypocrea lixii* SURT01, a poultry isolate. International Journal of Pharmacy and Biological Sciences 5(4): 680–687

Vasudevan UM, Krishna S, Jalaja V, Pandey A (2017) Microbial phytase: Impact of advances in genetic engineering in revolutionizing its properties and applications. Bioresource Technology 245: 1790–1799.

Optimisation of Soursop Juice Recovery by Alpha Amylase Produced by *Aspergillus niger* Using Statistical Tool

O. M. Atolagbe, A. A. Ajayi, and G. I. Olasehinde

1 Introduction

Annona muricata L. is generally called the soursop fruit. It is found commonly in many of the southern states of Nigeria, and it is mostly eaten as fresh fruits (Abbo et al., 2006). The soursop fruit originated from Central America and the Caribbean and spread to many regions with tropical climates, and this includes the rainforest part of Nigeria and other tropical belts of Africa. It is regarded as edible Annona and called by other local and common names such as 'soursop', 'corossol', 'graviola', 'guanabana' and 'guyabano'. These different names largely depend on the geographic locations of the plant (Ajayi and Adedeji, 2015). Soursop fruits are large and may be consumed fresh (raw), fermented or cooked. They may have various shapes and are known for their distinctive flavour and pleasant aromatic juicy flesh. It has medicinal properties such as antibacterial, anti-ageing, diuretic, antihaematuria, anticancerous, sedative and astringent (Asif et al., 2018). They may also be turned into other different products which are valuable economically, such as juice.

Soursop juice has a pleasant aroma and taste and also contains a large number of vitamins such as vitamins E and C and a large amount of antioxidants (Abbo et al., 2006). Its fruit is known for its many medicinal values such as containing certain natural compounds with therapeutic abilities that confer on it the benefit of being able to treat many ailments. Soursop is also known to prevent osteoporosis and beneficial for strong bones because the fruit contains phosphorus and calcium. The

O. M. Atolagbe (✉) · G. I. Olasehinde
Department of Biological Sciences, College of Science & Technology, Covenant University, Ota, Ogun State, Nigeria
e-mail: bunmi.atolagbe@covenantuniversity.edu.ng

A. A. Ajayi
Faculty of Science, Augustine University, Epe, Lagos State, Nigeria

© The Author(s), under exclusive license to Springer Nature Switzerland AG 2022

A. O. Ayeni et al. (eds.), *Bioenergy and Biochemical Processing Technologies*, Green Energy and Technology, https://doi.org/10.1007/978-3-030-96721-5_10

fruit flesh contains about 84% water which helps to hydrate the body. It can work as a natural diuretic for the treatment of oedema or retention of water (Keam et al., 2007). Because it has high moisture content, soursop can help to prevent and treat urethritis, haematuria and urinary tract infections (UTI) (Rachmani et al., 2012). Soursop fruit juice can be administered orally in case of haematuria, liver ailments and urethritis (Amusa et al., 2003).

Enzymes serve as catalysts in biological systems by helping to speed up reactions in living cells. They can be obtained from plant, animal and microbial cells and are used to speed up reactions for a large spectrum of processes that are of commercial importance (Rani et al., 2015). Amylases are important, and they are a notable category of enzymes which are used in the breaking down of starch and glycogen into simple sugars (Krishna et al., 2011). They break down the α 1–4 glycosidic linkages of amylopectin, glycogen and other associated compounds (Rani et al., 2015).

2 Methodology

2.1 Production and Clarification of Soursop Juice

Freshly bought ripe soursop fruits were washed carefully under flowing tap water, peeled, cut into halves, deseeded and then juiced with a mechanical juice extractor. The effect of different concentrations of alpha amylase produced by *Aspergillus niger* OMA1 strain was checked on 50 ml of the soursop juice. The concentrations 2 ml, 5 ml, 10 ml, 15 ml and 20 ml were used to clarify 50 ml of the extracted soursop juice at varying temperatures and times. The temperatures employed were 20 °C, 25 °C, 30 °C and 35 °C. Clarification intervals were 10 min, 15 min, 20 min, 25 min and 30 min.

3 Optimisation of Enzyme Concentration Using Response Surface Methodology (RSM)

Optimisation refers to a process of accelerating a system's performance, process or product with the view of obtaining the highest possible benefits from it (Bezerra et al., 2008). Response surface methodology (RSM) was employed to develop a mathematical model to relate the amount of soursop juice recovery which is the response to the different constituent factors (volume of enzyme used, temperature of recovery and length of time of recovery) and to explore and predict for optimum conditions.

4 Experimental Design

The central composite design (CCD) technique was employed in the experimental design and optimisation of soursop juice recovery with the use of alpha amylase produced by *A. niger* OMA1 (Table 1). This tool was used because it has been recorded to be an efficient tool in the improvement of bioprocessing (Betiku and Ajala, 2014). A class of three level complete factorial designs (central composite design) was used, which generated a total of 20 runs. The important variables selected for the modelling and optimisation are enzyme volume (ml), temperature (°C) and time (min) separately designated as A, B and C, respectively (Table 1).

The experimental data were analysed by fitting to a second-order polynomial model, which was statistically validated by performing analysis of variance (ANOVA) and lack-of-fit test to evaluate the significance of the model. Desirability function was used to find optimum conditions where the maximum soursop juice recovery was obtained with the use of enzymes.

An empirical model was developed to correlate the response of soursop juice recovery to the effect of temperature, time and volume of enzyme used, and this was based on the second-order quadratic model as given by Eqn. 1:

$$Y = \beta 0 + \Sigma I = 1k\beta ixi + \Sigma I = 1k\,\Sigma\, j = 1k\beta ijxixj + \Sigma\, i = 1k\beta iixii2 + \varepsilon \qquad (1)$$

where Y = the response variable; β_0 = the intercept; β_i, β_{ij} and β_{ii} = coefficients of the linear effect, double interactions; x_i, x_j = range of the independent variables or factors; and ε = random error.

5 Results and Discussion

5.1 *Development of Regression Model for Clarified Soursop Juice Recovery Using Purified α-Amylase from* **a. niger** *OMA1*

The statistical significance of the model is shown in Table 2. This represent the analysis of variance of the variables and their levels for the central composite design. The *P*-value of ($p < 0.0007$) seen in the table shows that the model is highly significant and can therefore be used to predict and optimise the response for soursop juice clarification with different volumes of enzyme treatment and varying temperature and time.

The value of coefficient of determination R^2, adjusted R^2 and predicted R^2 for the developed RSM central composite design model are 89.83%, 80.67% and 39.14%, respectively, as shown in Table 3. The relatively high value of R^2 (89.83%) implies that the second-order polynomial prediction model is adequate.

Table 1 Design summary of optimisation of soursop juice recovery with α amylase

Type of study	Response surface	Subtype	Randomised
Type of design	Central composite	Runs	20
Model design	Quadratic	Blocks	No blocks

Factor	Name	Units	Type	Subtype	Minimum	Maximum	Coded	Values	Mean	Std. Dev.
A	Enzyme (volume)	Ml	Numeric	Continuous	2	20	−1.000 = 2	1.000 = 20	11	6.52929
B	Temperature	Celsius	Numeric	Continuous	20	35	−1.000 = 20	1.000 = 35	27.5	5.44107
C	Time	Mins	Numeric	Continuous	10	30	−1.000 = 10	1.000 = 30	20	7.25476

From *Aspergillus niger* OMA1

Table 2 Analysis of variance (ANOVA) on response surface quadratic model for the volume of soursop juice recovery with α-amylase produced by *A. niger* OMA1

Source	Sum of squares	DF	Mean square	F value	p-value Prob > F	
Model	180.98	9	20.11	9.81	0.0007	Significant
A-amylase (*A. niger* – Volume)	96.72	1	96.72	47.19	<0.0001	
B-temperature	29.24	1	29.24	14.27	0.0036	
C-time	17.42	1	17.42	8.50	0.0154	
AB	1.53	1	1.53	0.75	0.4077	
AC	3.25	1	3.25	1.59	0.2365	
BC	1.20	1	1.20	0.59	0.4616	
A^2	7.82	1	7.82	3.82	0.0793	
B^2	2.83	1	2.83	1.38	0.2676	
C^2	18.07	1	18.07	8.82	0.0141	
Residual	20.50	10	2.05			
Lack of fit	20.50	5	4.10			
Pure error	0.000	5	0.000			
Cor. Total	201.48	19				

Table 3 Coefficients of determination for the developed model for soursop juice recovery for α-amylase

R-Squared	0.8983
Adj R-squared	0.8067
Pred R-squared	0.3914
Adeq precision	12.130

From *A. niger* OMA1

The result shown in Figs. 1 and 2 represents the adequacy of the model as carried out and validated by experimental observations. The normal plot of residuals and the plot of the residuals against the predicted responses are seen in Figs. 1 and 2, respectively. In Fig. 3, the plot of residual and run number is shown. It is assumed that the residuals are independent and normally distributed.

Figure 4 depicts the plot of the predicted response and the actual response of soursop juice recovery as seen in experimental results. The clustering of the response variables around the 45° line indicates a high confidence limit. Table 4 presents the central composite design run order and the observed and predicted soursop juice recovery with different volumes of enzyme treatment at varying temperatures and time.

Figures 5, 6, and 7 show the contour plots for the response of soursop juice recovery with the three factors considered. The blue colour indicates the least effects and the red colour indicates the maximum effector response of the dependent variable (temperature, time and volume of enzyme used) in the juice clarification.

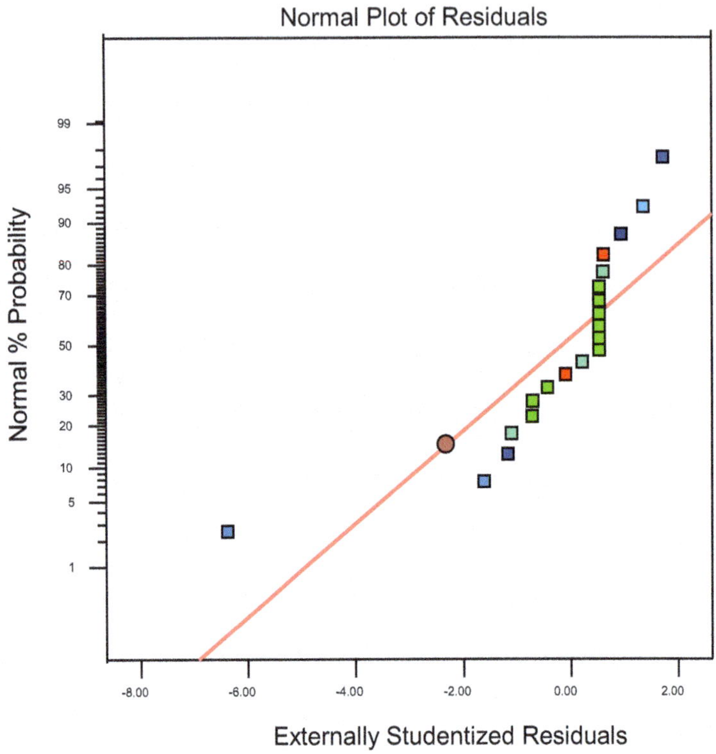

Fig. 1 Normal plot of residuals for soursop juice recovery with α-amylase. (From *A. niger* OMA1 from mathematical model checking)

In addition, Figs. 8, 9, and 10 show the 3D-surface plot. The curves are a clear indication that the response is a curvature as defined by the quadratic equation model. The response surface plot indicates clearly an increasing volume of juice recovery as the time, temperature and volume of enzyme treatment increase.

6 Predictive Optimisation of Soursop Juice Recovery with α-Amylase from *a. niger* OMA1

The model indicates that the volume of α-amylase from *A. niger* OMA1 to be used for clarification of a volume of 50 ml of soursop juice to get the optimum juice recovery is 20 ml. The model suggests in Fig. 11 that this experiment should be carried out at 27.5 °C for 30 min. This agrees with the results obtained from the experiments carried out in the laboratory, where an increase in time, temperature and volume of enzyme treatment used brought about a corresponding increase in soursop juice recovered.

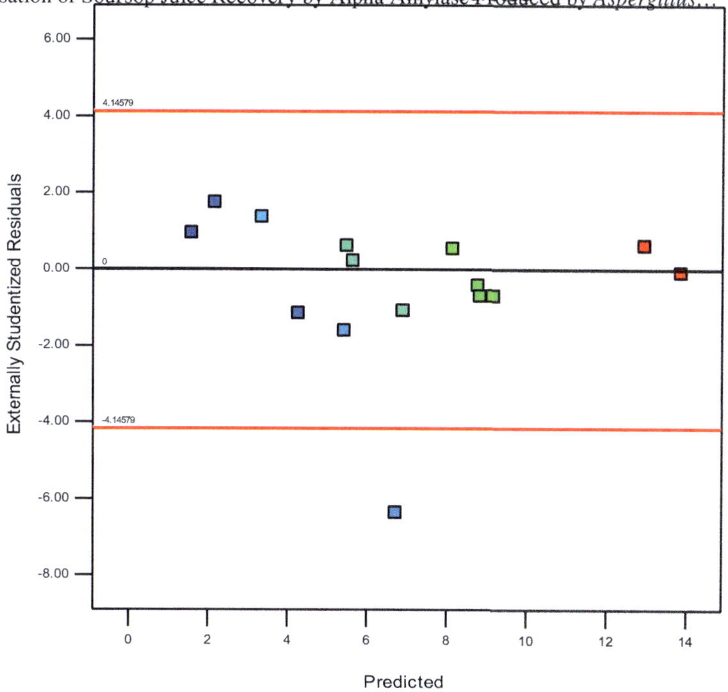

Fig. 2 The plot of the residuals against the predicted responses for soursop juice recovery with α- amylase. (From *A. niger* OMA1 from mathematical model checking)

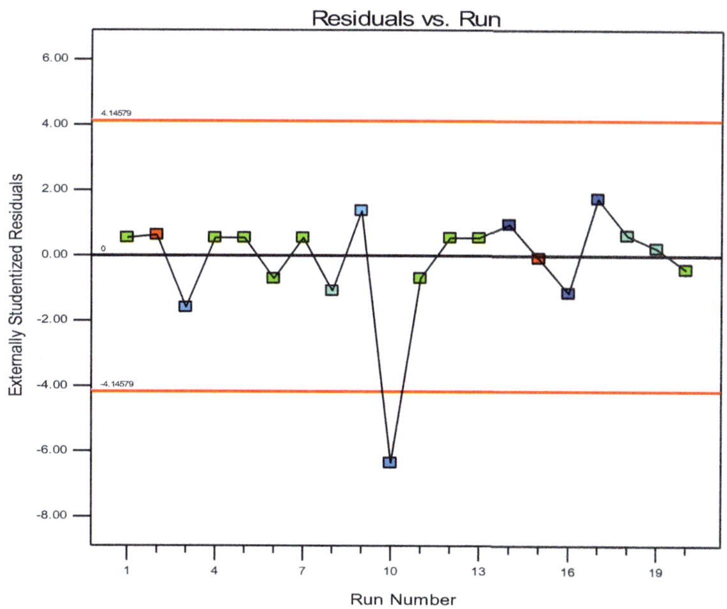

Fig. 3 The plot of the residual and run number for soursop juice recovery with α-amylase. (From *A. niger* OMA1 from mathematical model checking)

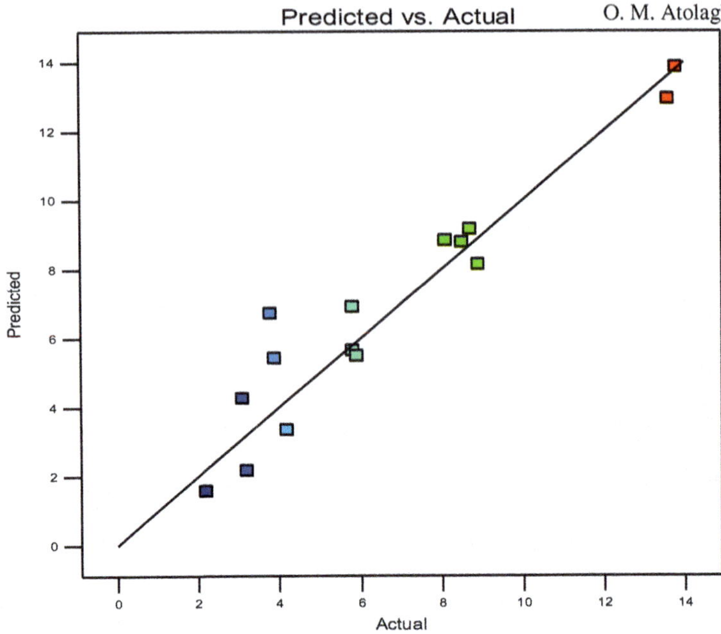

Fig. 4 Actual value versus predicted value of soursop juice recovery with α-amylase. (From *A. niger* OMA1)

Table 4 Values of the predicted response from the model and observed values from the experiment for soursop juice recovery for α-amylase

Run order	Actual value	Predicted value	Residual	Leverage
1	8.90	8.15	0.75	0.118
2	13.60	12.94	0.66	0.491
3	3.90	5.42	−1.52	0.491
4	8.90	8.15	0.75	0.118
5	8.90	8.15	0.75	0.118
6	8.10	8.84	−0.74	0.491
7	8.90	8.15	0.75	0.118
8	5.80	6.90	−1.10	0.491
9	4.20	3.35	0.85	0.793
10	3.80	6.72	−2.92	0.491
11	8.70	9.17	−0.47	0.793
12	8.90	8.15	0.75	0.118
13	8.90	8.15	0.75	0.118
14	2.20	1.58	0.62	0.793
15	13.80	13.86	−0.057	0.793
16	3.10	4.26	−1.16	0.491
17	3.20	2.17	1.03	0.793
18	5.90	5.49	0.41	0.793
19	5.80	5.65	0.15	0.793
20	8.50	8.79	−0.29	0.793

From *A. niger* OMA1

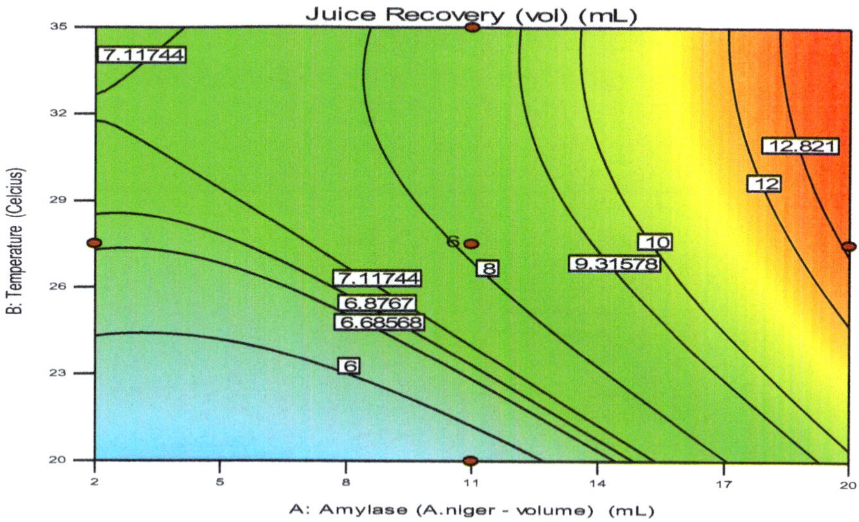

Fig. 5 Contour plots for the effect of factors (temperature and volume of enzyme) on the amount of juice recovery with α-amylase. (From *A. niger* OMA1)

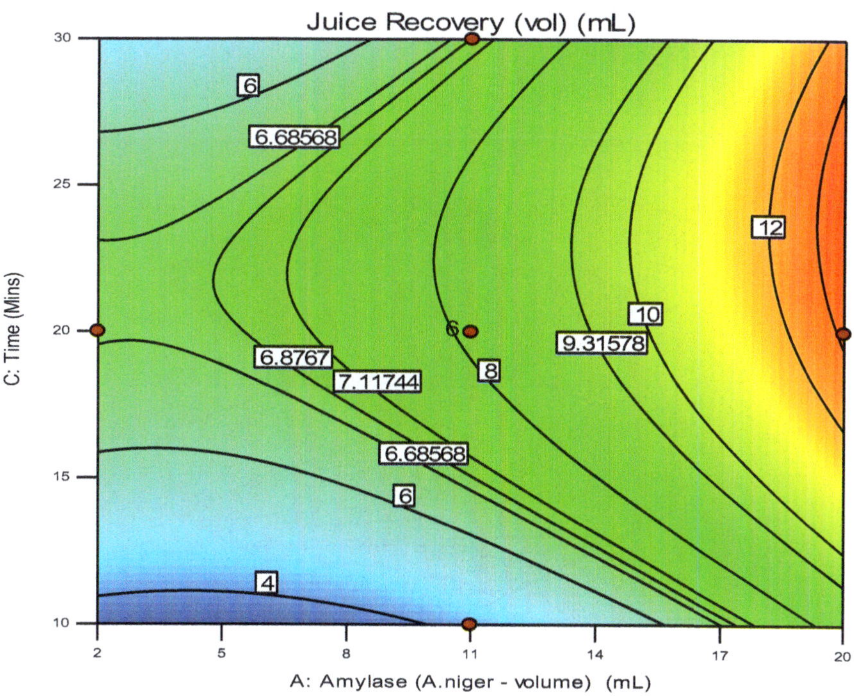

Fig. 6 Contour plots for the effect of factors (time of incubation and volume of enzyme) on the amount of juice recovery with α-amylase. (From *A. niger* OMA1)

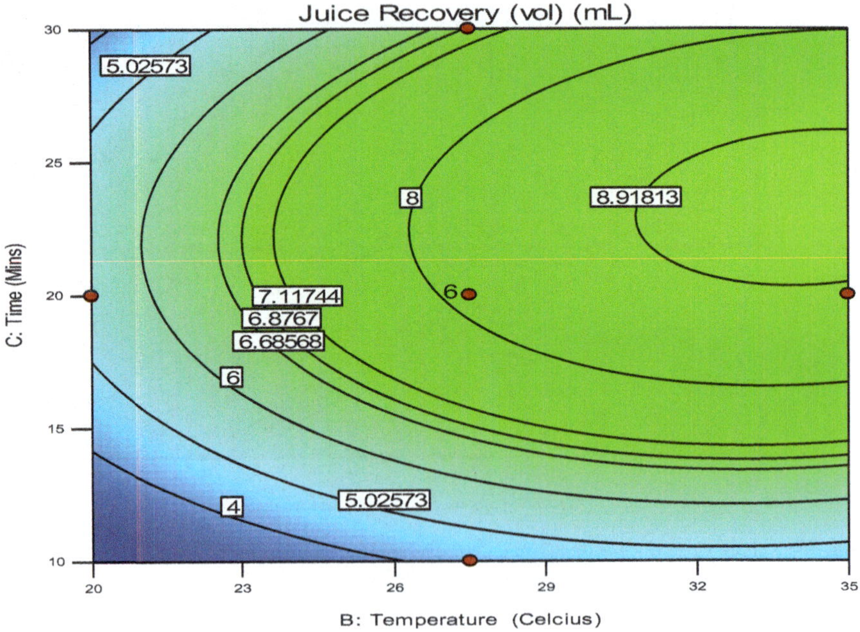

Fig. 7 Contour plots for the effect of factors (temperature and time of incubation) on the amount of juice recovery with α-amylase. (From *A. niger* OMA1)

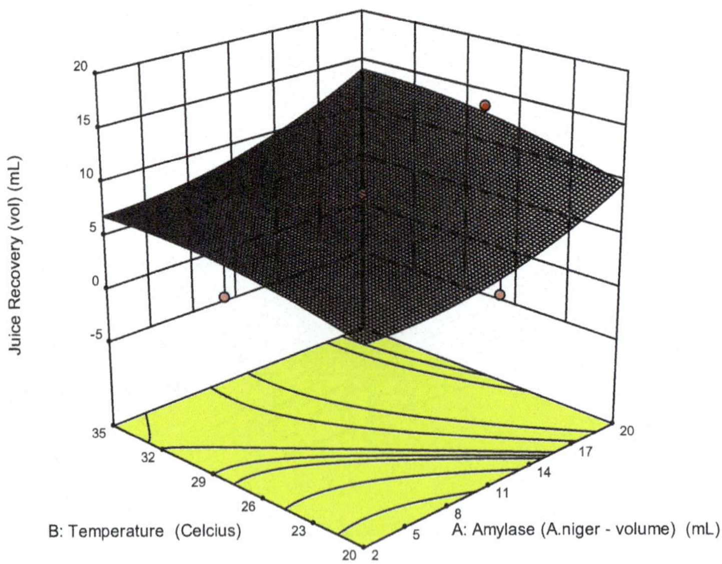

Fig. 8 Response surface in 3-D for the effect of factors (temperature and volume of enzyme) on the amount of juice recovery with α-amylase. (From *A. niger* OMA1)

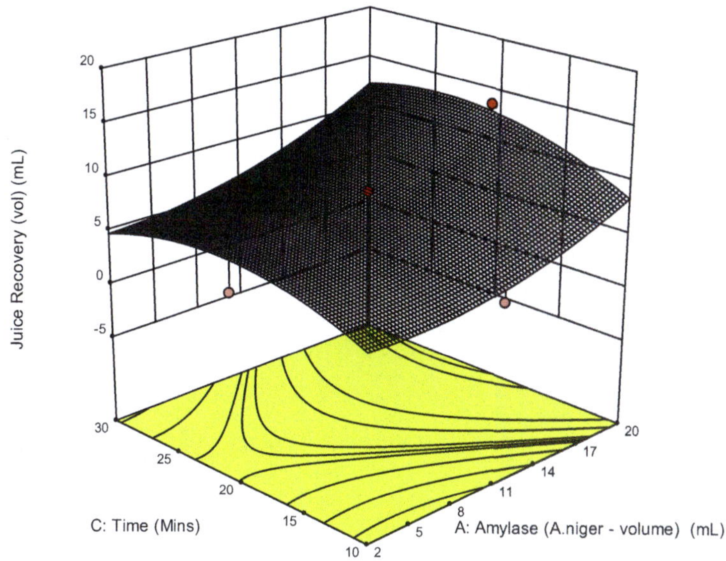

Fig. 9 Response surface in 3-D for the effect of factors (time of incubation and volume of enzyme) on the amount of juice recovery with α-amylase. (From *A. niger* OMA1)

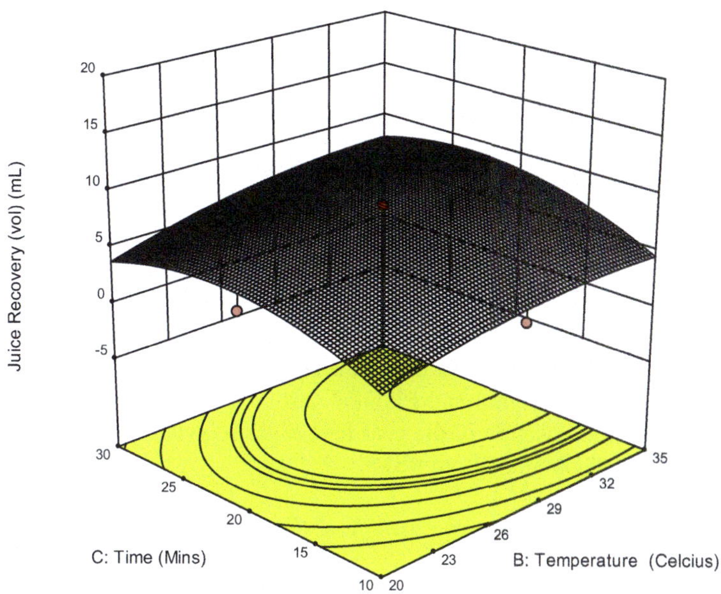

Fig. 10 Response surface in 3-D for the effect of factors (time of incubation and temperature) on the amount of juice recovery with α-amylase. (From *A. niger* OMA1)

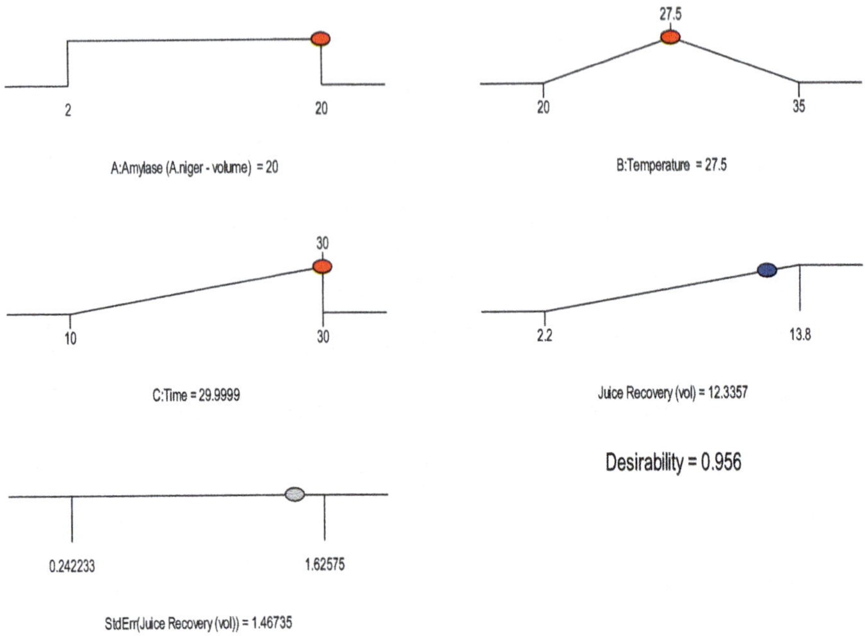

Fig. 11 Predictive optimisation plot of soursop juice recovery with α-amylase. (From *A. niger* OMA1)

The response surface plot clearly indicates an increasing volume of juice recovery as the time, temperature and volume of enzyme treatment increase. The response surface methodology gave a model for optimisation of juice recovery which was significant and had a high percentage of desirability at 95.6%. The predicted values of the model were also close to the actual experimental values. Ajayi et al. (2018) reported that when RSM (CCD) was applied to optimise the parameters (pH, temperature, substrate concentration and time of heating), a total of 30 experiments were successful during the production of pectinase by *A. niger*. In 2011, Ibrahim and Elkhidir reported that RSM is a collection of statistical techniques for designing experiments, building models, evaluating the effects of factors and searching for the optimum conditions (Ibrahim and Elkhidir, 2011). The maximisation of enzyme production and minimisation of the cost for cellulase, xylanase and phytase were achieved using data obtained from RSM.

Acknowledgements The authors would like to thank Covenant University for partial sponsorship of this work.

Disclosure Authors declare no conflict of interest.

References

Abbo ES, Olurin TO, Odeyemi G. Studies on the storage stability of soursop (*Annona muricata* L.). *African Journal of Biotechnology*. 2006; 5:1808–1812.

Ajayi AA, Salubi AE, Lawal B, Onibokun AE, Ajayi OM, Ogunleye TA. Optimization of pectinase production by *Aspergillus niger* using central composite design. *African Journal of Clinical and Experimental Microbiology*. 2018; 19(4): 314-319.

Ajayi AA, Adedeji OM. Characterisation of partially purified cell wall-degrading enzymes: polygalacturonase and cellulase from tomato fruits degraded by *Aspergillus niger*. *Canadian Journal of Pure and Applied Sciences*. 2015; 9(2): 3383-3391.

Amusa NA, Ashaye OA, Oladapo MO, Kafaru OO. Pre-harvest deterioration of soursop (*Annona muricata*) at Ibadan Southwestern Nigeria and its effect on nutrient composition. *African Journal of Biotechnology*. 2003; 2: 23–25.

Asif KQ, Jawed AS, Rahat J, Sanjib C, Larry AW, Zafar S, Dwight TJ, Surinder KB, Muzafar AM. Emerging therapeutic potential of graviola and its constituents in cancers *Carcinogenesis*. 2018; 39(4): 522–533.

Betiku E, Ajala SO. Modeling and optimization of *Thevetia peruviana* (yellow oleander) oil biodiesel synthesis via *Musa paradisiacal* (plantain) peels as heterogeneous base catalyst: A case of artificial neural network vs. response surface methodology. *Industrial Crops and Products*. 2014; 53:314-322.

Bezerra MA, Santelli ER, Oliveira EP, Villar LS, Escaleira LA. Response surface methodology (RSM) as a tool for optimization in analytical chemistry. *Talanta*. 2008; 76(5): 965-977.

Ibrahim HM, Elkhidir EE. Response Surface Method as an Efficient Tool for Medium Optimisation. *Trends in Applied Sciences Research*. 2011; 6: 121-129.

Keam B, Im SA, Kim HJ, Oh DY, Kim JH, Lee SH, Bang YJ, et al. Prognostic impact of clinicopathologic parameters in stage II/III breast cancer treated with neoadjuvant docetaxel and doxorubicin chemotherapy: paradoxical features of the triple negative breast cancer. *BMC Cancer.* 2007; **7**: 1- 11.

Krishna PR, Srivastava AK, Ramaswamy NK, Suprasanna P, D Souza SF. Banana peel as substrate for alpha-amylase production using *Aspergillus niger* NCIM 616 and process optimization. *Indian Journal of Biotechnology.* 2011;11: 314-319.

Rachmani EPN, Suhesti TS, Widiastuti R, Aditiyono A. The breast of anticancer of anticancer from leaf extract of *Annona muricata* against cell line in T47D. *International Journal of Applied Science and Technology*. 2012; 2(1): 198-2003.

Rani K, Rana R, Datt S. Review on characteristics and application of amylases. *International Journal of Microbiology and Bioinformatics*. 2015; 5(1): 1-5

Molecular Detection of ESBLs, TEM, SHV, and CTX-M in Clinical *Pseudomonas aeruginosa* Isolates in Ogun State

H. U. Ohore, P. A. Akinduti, E. F. Ahuekwe, A. S. Ajayi, and G. I. Olasehinde

1 Introduction

Immunocompromised individuals are particularly susceptible to *Pseudomonas aeruginosa* infections in the hospital setting as it is an opportunistic pathogen that could colonize patient implants and cause myriads of diseases including sepsis, pneumonia, and cystic fibrosis (Ayepola et al. 2018; Agbo et al. 2019). There has been a growing incidence in *Pseudomonas* infections (Ahmad et al. 2016) that the mortality rate is as high as 61% (Kang et al. 2003). *Pseudomonads* are challenging organisms as their infections have limited therapeutic options due to their natural resistance to antibiotics. Most antibiotics are ineffective against *Pseudomonas* spp. because of poor permeability of their outer membrane and other fortifications like efflux pumps (Pérez et al. 2019). Beyond their natural resistance, they can acquire artificial resistance to antibiotics from their environment, increasingly reducing the efficacy of treatment options.

Extended-spectrum beta-lactamases have shown to be causes of resistance in Gram-negative bacteria. This negates the actions of beta-lactam antibiotics, the most commonly prescribed antimicrobials (Shaikh et al. 2015). The ESBL enzymes are plasmid-mediated and operate by hydrolyzing the oxyimino beta-lactams and the monobactams (aztreonam) but do not affect the cephamycins (cefoxitin, cefotetan) and the carbapenems (imipenem) (Kaur and Singh 2018).

There is a paucity of data on the ESBL genes that are present in clinical settings in Nigeria (Adesina et al. 2019). The primary aim of this study is to characterize the common resistant gene types harbored by *Pseudomonas aeruginosa* strains in

H. U. Ohore (✉) · P. A. Akinduti · E. F. Ahuekwe · A. S. Ajayi · G. I. Olasehinde (✉)
Department of Biological Sciences, College of Science and Technology, Covenant University, Ota, Ogun State, Nigeria
e-mail: grace.olasehinde@covenantuniversity.edu.ng

© The Author(s), under exclusive license to Springer Nature Switzerland AG 2022

A. O. Ayeni et al. (eds.), *Bioenergy and Biochemical Processing Technologies*, Green Energy and Technology, https://doi.org/10.1007/978-3-030-96721-5_11

selected hospitals in Ogun State as a means to generate local data for planning and advocacy for empiric therapy, antibiotic stewardship, and infection control. Because these resistance genes are easily transferred among the *Pseudomonads*, this current study screened 150 clinical samples for the presence of ESBL genes, TEM, SHV, and CTX-M. Ethical approval was obtained, and samples were subjected to molecular methods for the characterization of these genes.

2 Methodology

A total of one hundred and fifty (150) extra-intestinal samples were obtained at three hospitals across Ogun State from September 2020 to December 2020. The samples were recovered from various sites including ear, blood, high vaginal swabs, urine, and surface wound. The samples were characterized and only 27 samples were positive for *P. aeruginosa* isolates.

2.1 Molecular Technique

Genotypic identification of *P. aeruginosa* isolates was carried out by amplifying B-lactamase genes CTX-M, SHV, and TEM using polymerase chain reaction. DNA was extracted from the identified *P. aeruginosa* using Zymo DNA extraction kit following manufacturer's instructions and quantified using the NanoDrop spectrophotometer.

2.2 Polymerase Chain Reaction (PCR)

Polymerase chain reaction (PCR) was used to determine the presence of the genes using the One Taq Hot Start 2X Mix with standard Buffer Kit (Biolabs, UK). Then the procedures from the PCR kit were strictly followed, and all work was done on ice. The thermal cycles were carried out in the thermocycler (SIGMA Companies).

2.3 Detection of CTX-M*, SHV, and TEM Genes*

For detection of *CTX-M*, *SHV*, and *TEM* gene which codes for resistance *in P. aeruginosa*, the presence of this gene was determined using the PCR technique described by Murakami et al. (1991). PCR was performed in 12.5 μl volume amplification mix containing 1 μl of primer, 1.25 μl of RNA free water, 6.25 μl master mix, and 4 μl of genomic DNA. DNA amplification for 30 cycles was carried out. The amplified

Table 1 PCR primers used and their sequence

Gene	Primer sequence 5′ → 3′	Amplicon size	PCR			No of cycles	References
			Denaturation	Annealing	Extension		
CTX-M	*Forward*: ACCGCCGATAATTCCAGAT *reverse*: GATATCGTTGGTGGTGCCATA	583 bp	94 °C	60 °C	72 °C	30	Naziri et al. (2020)
SHV	*Forward*: AAGATCCACTATCGCCAGCAG *reverse*: ATTCAGTTCCGTTTCCCAGCG	231 bp	94 °C	64 °C	72 °C	30	Rezai et al. (2018)
TEM	*Forward*: GAGTATTCAACATTTCCGTGT *reverse*: TAATCAGTGAGGCACCTATCT	848 bp	94 °C	53 °C	72 °C	30	Esohe et al. (2017)

product was resolved on 1.0% w/v agarose gel stained with 4 μl ethidium bromide and visualized under UV light (Table 1).

2.4 *Gel Electrophoresis*

Using agarose gel electrophoresis, the PCR amplified products were resolved and visualized under the UV transilluminator. A 100 bp DNA molecular weight marker was used as a reference to estimate the size of the gene-specific DNA band in the amplicon. 100 ml of 1 × TBE buffer (40 mmol/I Tris-acetate, 1 mmol/I EDTA) was used for the analysis of PCR products. For visualization, ethidium bromide was used. The DNA bands were viewed with UV light illumination and images recorded by photography (Akinduti et al. 2021).

3 Results and Discussion

The prevalence of *Pseudomonas aeruginosa* infection in this study was 18%. The results presented here are from a subset of 27 (18%) participants from whose samples *Pseudomonas aeruginosa* was isolated. Study participants were mostly female (74.07%) with an average age of 29 years. The majority of the samples were obtained from the ear (37.04%) and wound (25.93%) as shown in Fig. 1.

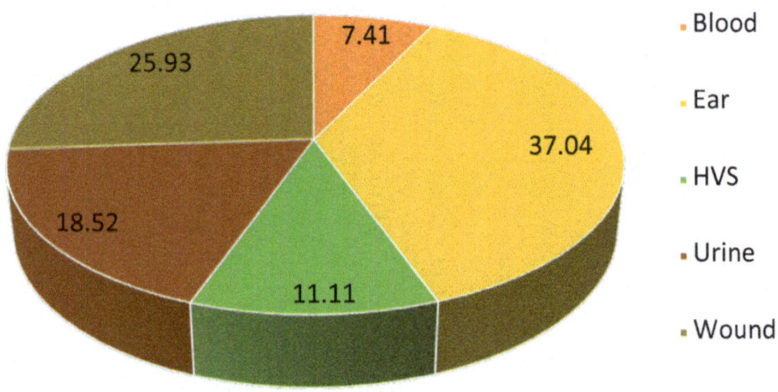

Fig. 1 Prevalence of *Pseudomonas aeruginosa* according to demographic data of the patients recruited for the study

Table 2 Association between β-lactamase production and source of isolation

	β-Lactamase production				
	Absent		Present		*P* value
Sample	*n*	%	*n*	%	
Blood	1	50.0	1	50.0	0.194
Ear	0	0.0	10	100.0	
HVS	1	33.3	2	66.7	
Urine	0	0.0	5	100.0	
Wound	0	0.0	7	100.0	

N number or sample
% percentage of b-lactamase production

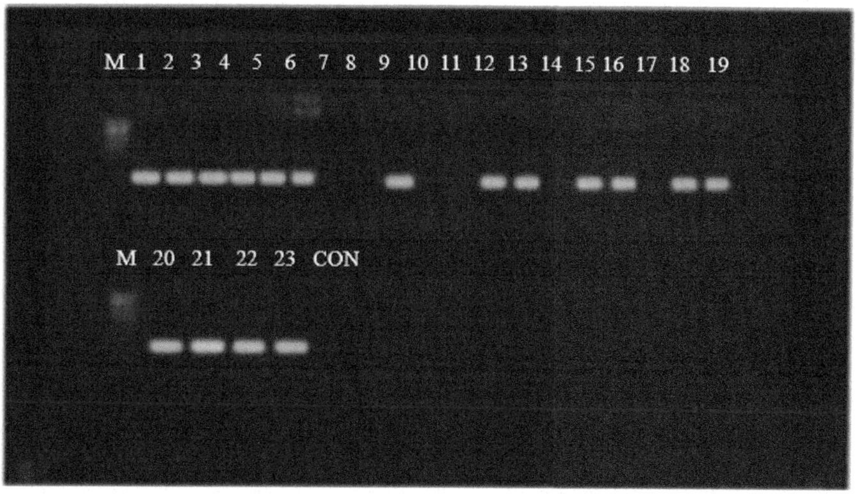

Fig. 2 Amplicons of TEM gene; Lane 1: 1 kb molecular weight marker

3.1 Phenotypic Detection of β-Lactamase

A total of 25 (92.6%) of the isolates in this study tested positive in the β-lactamase phenotypic detection. All isolates from the ear, urine, and wound tested positive, while isolates from HVS and blood were the least likely to test positive in the assay (Table 2).

3.2 Detection of CTXM, SHV, and TEM Genes

Of the 25 isolates that screened positive in the phenotypic β-lactamase assay, 4 (16%) were negative for all three genes screened, 7 (28%) positive for at least one gene, and 14 (56%) positive for two or three genes (Fig. 2). TEM was the most

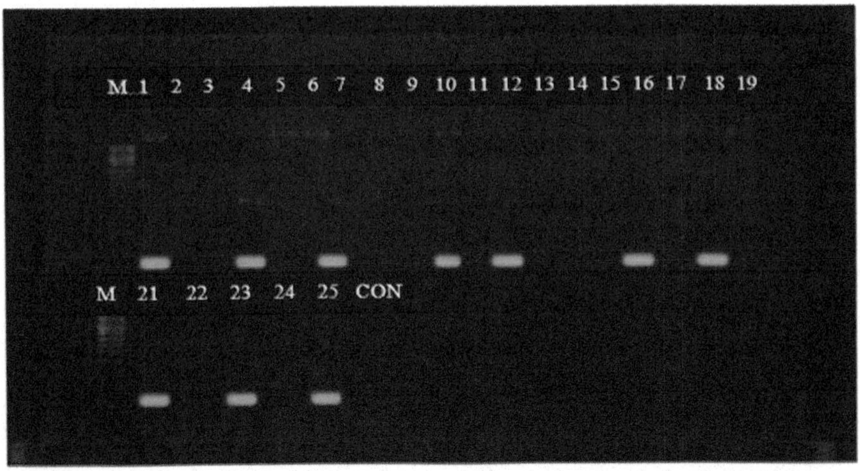

Fig. 3 Amplicons of SHV gene; Lane 1: 1 kb molecular weight marker

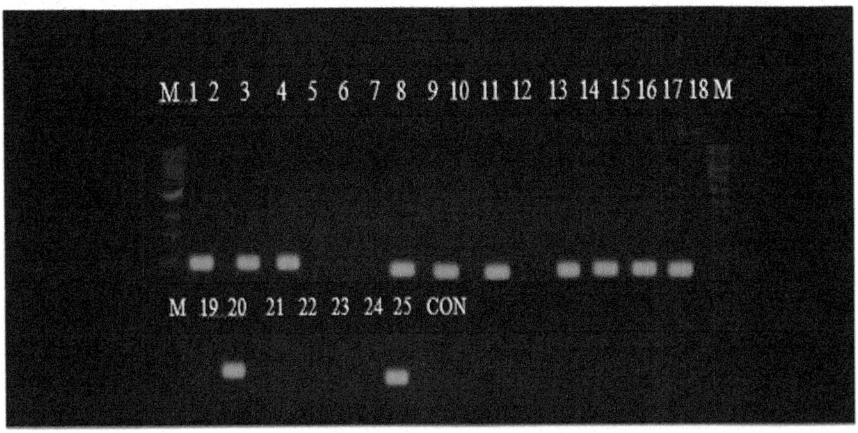

Fig. 4 Amplicons of CTX-M gene; Lane 1: 1 kb molecular weight marker

dominant gene with 16 positives (64%); SHV had 12 positive isolates (52%), and CTX-M had the least dominance with 10 positives (44%). The correlation between the three genes assayed was weak. The highest correlation was found between SHV and CTX-M (R = 0.368), while there was very little correlation between TEM and CTX-M (Figs. 3 and 4).

ESBLs have been more frequently reported in members of the *Enterobacteriaceae* (Elhariri et al. 2017; Oli et al. 2017; Kpoda et al. 2018; Jesumirhewe et al. 2020) and even then, from urine samples, in the *Pseudomonads* from other clinical samples. This has caused these organisms to be named priority by the World Health Organization (Kpoda et al. 2018). Research in Nigeria has focused on the presence of ESBLs in enteric isolates, with little attention given to the *Pseudomonadaceae*.

The recorded *P. aeruginosa* prevalence in clinical samples in this study is consistent with what has been reported in other studies in Nigeria (Ajibade et al. 2019; Jombo et al. 2010; Ogbolu et al. 2008) and around the world (Rezai et al. 2018; Pachori et al. 2019; Shi et al. 2019). Some other studies have found significantly lower or higher rates of *P. aeruginosa* infections than is reported in this study. For example, a study conducted in Afghanistan reported a prevalence rate of 6.67% (Khan et al. 2008), while another conducted in Karachi reported a prevalence rate of 30% (Khan et al. 2014). The differences in prevalence is probably as a result of sampling duration.

The most frequent ESBL genotype found in this study of *P. aeruginosa* isolates was TEM with 64%. Most of the isolates harbored two to three of the genes. The clinical isolates screened by Ogbolu et al. (2018) represented Ibadan, Ogbomoso, and Osogbo, in Southwest Nigeria, with CTX-M emerging dominant at 83.3%. This opposes our study that found CTX-M to be the least dominant at 44%. This may be accounted for by the difference in antibiotic consumption in the various locations, though within the same geopolitical zones in Nigeria. The dominance of the SHV gene is found to be at par, with Ogbolu et al. (2018) recording 58.3% and ours 52%.

In Egypt, ESBL-producing *P. aeruginosa* was retrieved from Camel meat (Elhariri et al. 2017). CTX-M, SHV, and TEM genes were recorded among the highest. ESBL producers found in food samples and consumed by humans could pose dangers to clinical treatment. Kpoda et al. (2018) also found CTX-M to be the most prevalent gene assessed. In contrast, Agbo et al. (2019) did not find any of the isolates screened to harbor the CTX-M, TEM, and SHV genes. It is likely that these genes are not ensuing resistance within Enugu, Nigeria.

Our study found TEM as the most dominant, contradicting most of the figures of research within Nigeria and Africa. The analysis of Hosu et al. (2021) in South Africa confirmed *bla*TEM (79.3%) and *bla*SHV (69.5%) to be most detected, in an order similar to our findings. CTX-M was also their least prevalent gene. Clinical isolates of *P. aeruginosa* with blaTEM and *bla*CTX-M were also found by Madaha et al. (2020) in Cameroon. This could signify that the *bla*TEM is spreading within Ogun State, Nigeria, and *bla*CTX-M has not gained as much ground.

The variations in results could be as a result of the use of products from different pharmaceutical brands, the availability of those brands among consumers, the lifestyle of consumers within a certain geographical location, the kind of microorganism's endemic in those locations, and the means by which the resistance elements are transferred. A correlation has been found between the level of antibiotic prescription and the rates of antibiotic resistance (Hosu et al. 2021).

Moreover, the ESBLs produced by *Pseudomonas* spp. are yet to be researched in depth so it is hard to say what genotypes pose significant challenges in Nigeria. If uncurbed, the rate of fatalities from *Pseudomonas* infections will continue to surge. According to available data we researched, this is one of the first research to focus exclusively on the detection of ESBL genes in *P. aeruginosa* in Ogun State, and it is worth watching closely by health professionals. The prescription of antibiotics and its ease of accessibility have been abused by individuals which has consequently led to the quick development of resistance to antibiotics by the pathogens, which in turn has led to the rise of resistance genes in the pathogen. The dispense of antibiotics should be reviewed critically in other to mitigate the antibiotic resistance by the *P. aeruginosa* pathogen.

4 Conclusion

ESBL-producing *Pseudomonas aeruginosa* were increasingly found in clinical samples. Their distribution in the hospital and environment further threatens the measures currently available to treat infections they cause. In this study, although *P. aeruginosa* infections were not highly prevalent, the 27 isolates already possess ESBLs which could spread and lead to more resistance in individuals in this area. Several studies have been carried out in different states, and this seems to be a growing problem. Poor antibiotic usage and a failing health system in Nigeria has increased purchase of over-the-counter drugs and seeking of medical attention from unskilled individuals. Clinical professionals require urgent control measures to monitor and reduce the spread of ESBLs and produce effective antipseudomonal agents to return infected patients to a state of health.

Acknowledgments The authors would like to acknowledge the support of Covenant University Center for Research, Innovation, and Development (CUCRID).

References

Adesina T, Nwinyi O, De N, Akinnola O, Omonigbehin E (2019) First detection of carbapenem-resistant escherichia fergusonii strains harbouring beta-lactamase genes from clinical samples. Pathogens 8(4):164

Agbo MC, Ezeonu IM, Odo MN et al (2019) Phenotypic and molecular characterization of extended spectrum β-lactamase producing *Pseudomonas aeruginosa* in Nigeria. African J Biotechnol 18(32):1083–1090

Ahmad M, Hassan M, Khalid A et al (2016) 'Prevalence of Extended Spectrum β-Lactamase and Antimicrobial Susceptibility Pattern of Clinical Isolates of *Pseudomonas* from Patients of Khyber Pakhtunkhwa, Pakistan'. BioMed Res Int. Hindawi Limited

Ajibade O, Oladipo EK, Aina KT et al (2019) 'Incidence of *Pseudomonas aeruginosa* Resistance in Clinical Isolates from Selected Hospitals in Oyo State, Nigeria'. Appl Microbiol 5(2)

Akinduti P, Obafemi YD, Isibor PO et al (2021) Antibacterial kinetics and phylogenetic analysis of Aloe vera plants. Open Access Maced J Med Sci 9(A): 946-954

Ayepola OO, Olasupo LA, Egwari LO, Schaumburg F (2018) Characterization of panton-valentine leukocidin-positive staphyloccoccus aureus from skin and soft tissue infections and wounds in Nigeria: A cross-sectional study. F1000Res 7:1155

Elhariri M, Hamza D, Elhelw R et al (2017) 'Extended-spectrum beta-lactamase-producing *Pseudomonas aeruginosa* in camel in Egypt: Potential human hazard'. Ann Clin Microbiol Antimicrob 16(1):21.

Esohe A, Oronsaye F, Omorodion N et al (2017) 'Presence of bla TEM and bla SHV Genes in ESBLs Producing *Klebsiella Pneumonia* among Female Patients in University of Benin Teaching Hospital'. 724.

Hosu MC, Vasaikar SD, Okuthe GE et al (2021) 'Detection of extended spectrum beta-lactamase genes in *Pseudomonas aeruginosa* isolated from patients in rural Eastern Cape Province, South Africa'. Sci Rep 11(1):7110

Jesumirhewe C, Springer B, Allerberger F et al (2020) 'Whole genome sequencing of extended-spectrum β-lactamase genes in Enterobacteriaceae isolates from Nigeria'. *PLoS ONE* 15(4)

Jombo G, Springer B, Allerberger F et al (2010) 'Multidrug resistant *Pseudomonas aeruginosa* infections complicating surgical wounds and the potential challenges in managing post–operative wound infections: University of Calabar Teaching Hospital experience'. Asian Pac J Trop Med 3(6):479–482

Kang CI, Kim SH, Kim HB et al (2003) '*Pseudomonas aeruginosa* bacteremia: Risk factors for mortality and influence of delayed receipt of effective antimicrobial therapy on clinical outcome'. Clin Infect Dis 37(6):745–751

Kaur, A. and Singh, S. (2018) ' Prevalence of Extended Spectrum Betalactamase (ESBL) and Metallobetalactamase (MBL) Producing *Pseudomonas aeruginosa* and *Acinetobacter baumannii* Isolated from Various Clinical Samples '. J Pathog 1–7.

Khan, F., Khan, A. and Kazmi, SU. (2014) 'Prevalence and susceptibility pattern of multi drug resistant clinical isolates of *Pseudomonas aeruginosa* in Karachi'. Pakistan J Med Sci 30(5):951

Khan JA, Iqbal Z, Rahman SUR (2008) 'Report: prevalence and resistance pattern of *Pseudomonas aeruginosa* against various antibiotics.'. Pak J Pharm Sci 21(3):311–5.

Kpoda DS, Ajayi A, Somda M et al (2018) 'Distribution of resistance genes encoding ESBLs in Enterobacteriaceae isolated from biological samples in health centers in Ouagadougou, Burkina Faso'. BMC Res Notes 11(1):471

Madaha EL, Gonsu HK, Bughe RN et al (2020) 'Occurrence of blaTEM and blaCTXM Genes and Biofilm-Forming Ability among Clinical Isolates of *Pseudomonas aeruginosa* and Acinetobacter baumannii in Yaoundé, Cameroon'. Microorganisms 8(5):708

Murakami K, Minamide W, Wada, K et al (1991). Identification of methicillin-resistant strains of staphylococci by polymerase chain reaction. J clin microbiol 29(10):2240-2244.

Naziri Z, Derakhshandeh A, Soltani BA, (2020) 'Treatment Failure in Urinary Tract Infections: A Warning Witness for Virulent Multi-Drug Resistant ESBL- Producing *Escherichia coli*'. Infect Drug Resist 13:1839–1850

Ogbolu DO, Ogunledun A, Adebiyi OE et al (2008) 'Antibiotic susceptibility patterns of *Pseudomonas aeruginosa* to available antipseudomonal drugs in Ibadan, Nigeria.'. Afr J Med Med Sci 37(4):339–44

Ogbolu DO, Terry AOA, Webber MA et al (2018) 'CTX-M-15 is established in most multidrug-resistant uropathogenic Enterobacteriaceae and Pseudomonaceae from hospitals in Nigeria'. Eur J Microbiol Immunol 8(1):20–24

Oli AN, Eze DE, Gugu TH et al (2017) 'Multi-antibiotic resistant extended-spectrum beta-lactamase producing bacteria pose a challenge to the effective treatment of wound and skin infections', Pan Afr Med J 27:1–12

Pachori P, Gothalwal R and Gandhi P. (2019) 'Emergence of antibiotic resistance *Pseudomonas aeruginosa* in intensive care unit; a critical review'. Genes Dis 6(2):109–119

Pérez A, Gato E, Pérez-Llarena J et al (2019) 'High incidence of MDR and XDR *Pseudomonas aeruginosa* isolates obtained from patients with ventilator-associated pneumonia in Greece, Italy and Spain as part of the MagicBullet clinical trial'. J Antimicrob Chemother 74(5):1244–1252

Rezai MS, Ahangarkani F, Rafiei A (2018) 'Extended-Spectrum Beta-Lactamases Producing *Pseudomonas aeruginosa* Isolated From Patients With Ventilator Associated Nosocomial Infection'. Arch Clin Infect Dis 13(4)

Shaikh S, Fatima J, Shakil S et al (2015) 'Prevalence of multidrug resistant and extended spectrum beta-lactamase producing *Pseudomonas aeruginosa* in a tertiary care hospital', Saudi J Biol Sci 22(1):62–64

Shi Q, Huang C, Xiao T et al (2019). A retrospective analysis of *Pseudomonas aeruginosa* bloodstream infections: Prevalence, risk factors, and outcome in carbapenem-susceptible and -non-susceptible infections. Anti Resist Infect Control 8(1):68

Pectinase Production and Application in the Last Decade: A Systemic Review

G. D. Ametefe, A. O. Lemo, H. U. Ugboko, E. E. J. Iweala, and S. N. Chinedu

1 Background

The degradation of the polymers of pectin located in the plant cell wall can be undertaken with a class of enzymes called pectinases. The increase in demand for pectinases accounts for its importance as one of the enzymes with a high market value among the commercial enzymes employed in industries (El Enshasy et al. 2018; Garg et al. 2016). The variety of areas for its applications are probably the main reason for the increase in demand for the enzyme (Zhu et al. 2019; Ametefe et al. 2017; Liu et al. 2012).

Pectinases are mainly produced by fermentation (Cherekara and Pathak 2020; Chinedu et al. 2017; Subramaniyam and Vimala 2012; Ward 1992). Biodegradation of complex compounds into simpler ones may be regarded as fermentation, with the production of the desired products along with carbon dioxide and alcohols

G. D. Ametefe (✉) · A. O. Lemo
Department of Biochemistry, Covenant University, College of Science and Technology, Ota, Ogun State, Nigeria
e-mail: george.ametefe@stu.cu.edu.ng

H. U. Ugboko
Department of Biological Sciences, Covenant University, College of Science and Technology, Ota, Ogun State, Nigeria

Covenant University Biotechnology Research Cluster, Covenant University, Ota, Ogun State, Nigeria

E. E. J. Iweala · S. N. Chinedu
Department of Biochemistry, Covenant University, College of Science and Technology, Ota, Ogun State, Nigeria

Covenant University Public Health and Wellbeing Research Cluster, Covenant University, Ota, Ogun State, Nigeria

© The Author(s), under exclusive license to Springer Nature Switzerland AG 2022

A. O. Ayeni et al. (eds.), *Bioenergy and Biochemical Processing Technologies*, Green Energy and Technology, https://doi.org/10.1007/978-3-030-96721-5_12

(Subramaniyam and Vimala 2012; Ward 1992). Both solid-state fermentation (SSF) and submerged fermentation (SMF) methods have been used for the production of this enzyme (Cherekara and Pathak 2020; Ametefe et al. 2017; Chinedu et al. 2017). The SSF is a fermentation method that utilizes substrates that are not soluble in water, such that the moisture is only absorbed inside the substrates; however, the SmF takes place with the submersion of the nutrients and substrates in water (Balakrishnan et al. 2021; Kumar et al. 2021; Catalan and Sanchez 2020). Corn cobs, orange peels, and banana peels are some examples of the substrates employed for pectinase production (Ametefe et al. 2017; Widowati et al. 2017; Okafor et al. 2010).

Moreover, the raw materials for pectinase production (as earlier captured) for local production of the enzyme are cheaply available. The microorganisms are also cheaply available in our immediate environment. Besides, the increase in pollution caused by agro-wastes is a source of concern, especially in developing and underdeveloped countries (Govindraraji and Vuppu 2020). Hence, the pollution from agro-wastes pollution can be reduced to a great extent with the conscious employment of these wastes for pectinase production.

The application of pectinases (pectin enzymes) is not only limited to the food and environment sectors. Pectin enzymes are also beneficial in the health sector, as in the degradation of tablets coated with pectin to release the content (drug), as reported for the treatment of colon, and in studies concerning pulsatile drug delivery (Zhu et al. 2019; Liu et al. 2012). The many application areas of pectinases are probably the reason researchers harness ways to optimally produce the enzyme. The main tools used for the design of the studies for pectinase production, optimization, and applications are the "one-factor-at-a-time" (OFAT) method and the response surface methodology (RSM) (Sudeep et al. 2020; Ametefe et al. 2017; Ghazala et al. 2015). The OFAT is designed such that one factor is measured at a time, as the name suggests (Ametefe et al. 2017). The RSM is a tool for the prediction of the behavior of a system (and it's optimum), using mathematical and statistical techniques, as interactions among the variables are more adequately considered in this design of experiment (Chollom et al. 2019; Najib et al. 2017). The RSM method has proven to produce enzymes with lesser experimental runs and more reliable outcomes (regarding obtaining the best enzyme activity) than the OFAT method (Jasni et al. 2020; Said and Amin 2015). The two main RSM designs employed for pectinase production are the Box-Behnken and the central composite designs (Chollom et al. 2019).

Systematic reviews have been seen as a method for reviewing or summarizing data on a specific question in a standardized and systematic or organized way to derive more beneficial outputs (results) that are comprehensive in any required area of study (Tawfik et al. 2019). The principal steps used for systematic reviews begin with outlining the question, searching for relevant (criteria) materials (articles) in the area, evaluation of the quality of the materials for inclusion in the study, condensing of the pieces of evidence obtained, and explanation of the outcomes (results) (Khan et al. 2003; Rys et al. 2009).

It is important to note that there is a dearth of information on systematic review regarding the production and application of pectinases in the last decade as undertaken in this study. It is for these reasons that a systematic review of literature concerning the production and application of pectinase articles on the SCOPUS indexed journals in the last decade (2011–2020) was investigated. Furthermore, the findings and some recognizable gaps have been uncovered for further research.

2 Methods

The Preferred Reporting Items for Systematic Reviews and Meta-Analyses (PRISMA) guidelines were followed for the study (Page et al. 2021). The SCOPUS database (website) was searched using *pectinase* as the keyword. Exclusion criteria employed were the articles that did not produce and apply the enzyme (in the laboratory). Therefore, the criteria for inclusion of articles was in preference for articles reported in the English language on the laboratory application of the (local) production of pectinase. The results were obtained after a manual inspection of the titles, abstracts, methods, results, and conclusion of each of the articles obtained (Fig. 1). The searches were limited to articles from 2011 to 2020, which were open access, covering the various subject areas: biochemistry, genetics, and molecular biology, agriculture, and biological sciences. The study used percentages for analyses of the frequency in the outcomes obtained (Table 1). The rationale for the choice of percentage for analysis was due to the qualitative evaluation of literature in this study.

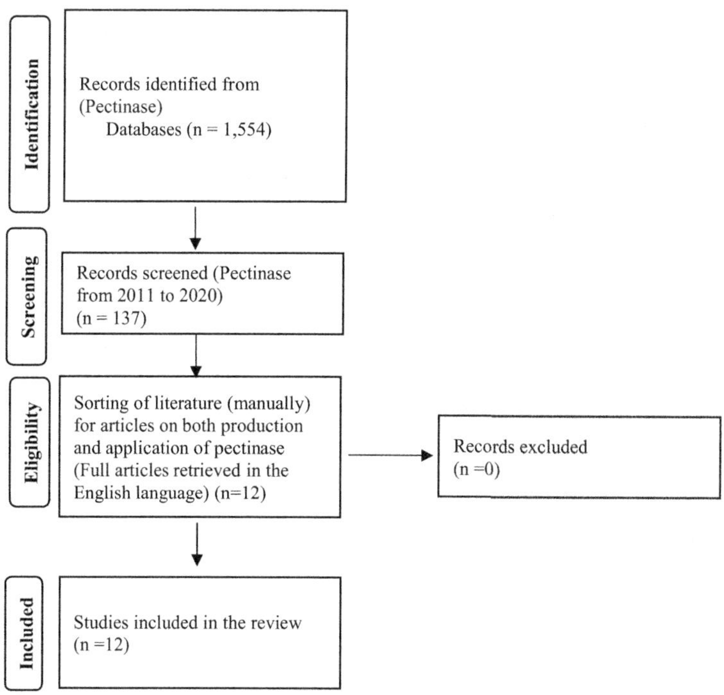

Fig. 1 Systematic review flow chart

Table 1 Systematic review of literature in the last 10 years (2011–2020)

S/No	Author	Year	Microbe	Fermentation type/substrates	Application	RSM/OFAT and crude/purified	Findings	Gap
1	Rajendra et al.	2011	*Fusarium* sp.	0.5% substrate concentration	Bioscouring of cotton fabric	OFAT/partially purified	Scouring of cotton with considerably higher tensile strength than the conventionally treated sample	RSM be used for the optimization
2.	Poletto et al.	2015a	*Aspergillus niger* LB-02-SF	SSF/wheat bran and citric pectin	Fruit juice clarification	OFAT/ultrafiltration for pectinase concentration	Total pectinase activity increased by 73.6-fold compared to the crude Findings were statistically comparable to commercial enzymes	RSM be used for the study
3.	Ghazala et al.	2015	*Bacillus mojavensis* 14	Liquid state fermentation/ carrot (*Daucus carota*) peels	Extraction of oil sesame	RSM	Isolation of a new pectinase-producing microorganism and thermostable in alkaline condition and identification of pectinase production from carrot peels	Other relevant conditions such as particle size of the substrate be included in the design
4.	Barman et al.	2015	*Aspergillus niger*	SSF/banana (*Musa balbisiana*) peel	Banana juice clarification	OFAT/ethanol precipitation (partial purification)	Authors obtained. Three times polygalacturonase activity obtained during partial purification in relation to the crude	RSM be used for the study
5.	Ismail et al.	2016	*Trichoderma viride* EF-8	SMF/Egyptian onion (*Allium cepa* L.) skin wastes	Fruit juices clarification	OFAT/partial purification (ethanol precipitation)	Identification of pectinase-producing microorganism	RSM be used for the study

6.	Mahmoodi et al.	2017	*Aspergillus niger*	SSF/orange pomace	Apple juice clarification	OFAT/crude enzyme used	Pectinase improved the clarity, viscosity, and yield of the juice	RSM be used for the study
7.	Sharma et al.	2017	*Bacillus pumilus* AJK strain (MTCC Accession No. 10414)	SMF/ 2% wheat bran and 2% citrus peel	Bio- bleaching of plywood	OFAT/ultrafiltration	A combination of xylanase and pectinase was used for bio-bleaching of plywood	RSM be used for the study
8.	Abdulrachman et al.	2017	*Aspergillus aculeatus* in *Pichia pastoris*	*Escherichia coli* DH5α as host for the propagation of plasmid	Bio-scouring of cotton fabrics	Optimization using Gene Designer Software/ recombinant enzyme	Functions of the *A. aculeatus* EndoPG in the processing of textile	The use of the coding sequence of other pectinase-producing microbes be investigated
9.	Sudeep et al.	2020	*Aspergillus* spp. Gm	SMF/1% substrate concentration	Orange juice clarification	OFAT/partial purification with ice-cold acetone	The thermal stability of the enzyme decreased by 50% in 10 minutes of heating at 60 °C. Increase in clarity of the juice with an increase in pectinase concentration. Identification of a strain of *Aspergillus* species with potency for pectinase production	RSM be used for the study

(continued)

Table 1 (continued)

S/No	Author	Year	Microbe	Fermentation type/substrates	Application	RSM/OFAT and crude/purified	Findings	Gap
10.	Cherekar and Pathak	2020	*Halomonas pantellerinsis* strain SSL8	SMF/pectin	Extraction and clarification of apple juice	OFAT/partial purification with chilled acetone and ammonium sulfate precipitation.	Production of halo alkaline pectinase with stability at high pH	RSM be used for the study
11	Govindaraiji and Vuppu	2020	*Streptomyces fumigatiscleroticus* VIT-SP4	Culinary fruit peels	Drug delivery	Taguchi method and RSM (central composite design)/ purified pectinases and confirmed by High-Performance Liquid Chromatography (HPLC)	Isolation of pectinolytic microorganisms for optimal production of the enzyme for drug delivery at pH of 6. Characterization of pectin from some fruit peels was also carried out	Inclusion of other relevant parameters for optimization of pectinase production not investigated
12.	Azzaz et al.	2020	*Aspergillus terreus*	SMF/sugar beet pulp	Improvement in the diet degradation of dairy animals	OFAT/crude enzyme	The use of an electron microscope to study the impact of the enzyme on the diet of dairy animals	RSM be used further to include other relevant parameters for (a more) comprehensive study

3 Results and Discussion

Search for “pectinases” on the SCOPUS database on the 4th of May, 2021, revealed 1554 results (Fig. 1). Exploration of the keyword “pectinase” from articles indexed in the SCOPUS website resulted in 137 outcomes (2011 to 2020). On further manual screening of the results (i.e., the articles), using their various titles, abstracts, methods, results, and conclusions for each of the 137 documents, 12 articles reported on pectinase production and its actual application (Table 1). Hence, the 12 articles met the benchmark for inclusion in this review (Table 1).

3.1 Microorganisms and Fermentation Method Used for Pectinase Production

A variety of microorganisms comprising about 66.7% fungi to 33.3% bacteria were employed in the production of pectinase, with about 50% of the studies applying the enzyme in fruit juice treatment (Table 1). The use of the genus *Aspergillus* for production of pectinase and the application in the treatment of fruit juices (Sudeep et al. 2020; Mahmoodi et al. 2017; Barman et al. 2015; Poletto et al. 2015a), in bio-scouring of cotton fabrics (Abdulrachman et al. 2017), and in the improvement of animals diet (Azzaz et al. 2020) indicated its importance in pectinase studies. Hence, it is no surprise the utilization of the genus, *Aspergillus*, for the production of pectinase with optimal activity (Sudeep et al. 2020; Ezike et al. 2014). The utilization of the genus *Aspergillus* for pectinase production is predominantly the solid-state fermentation, which suggests its inability to produce the enzyme in saturated moisture content as obtained in submerged fermentation (Table 1). However, bacteria employed were principally shown to produce the enzyme in relatively more moisture content than obtained with the largely studied fungi (Cherekar and Pathak 2020; Govindaraiji and Vuppu 2020; Sharma et al. 2017; Ghazala et al. 2015). Furthermore, the advantages of SSF over SmF like in less effluents generated after fermentation, simulation of the natural habitat of microorganisms, and the easier recovery of the enzyme produced could suggest the rationale for the utilization of fungi in SSF for the production of pectinases (Sharma and Rishishwar 2015).

Chart 1 Pectinase search on the SCOPUS database limited 2011 to 2020

Chart 2 Pectinase search on the SCOPUS database limited to 2011 to 2020 and further filtered

It is also important to state that the SSF also has some disadvantages, as the aeration in this method of fermentation poses a challenge due to the relatively higher concentration of solid and the increased period of fermentation compared to the submerged fermentation (Perez-Guerra et al. 2003). The microorganisms used for the production of pectinases in this study were mostly isolated from soil samples (Sudeep et al. 2020; Ismail et al. 2016; Ghazala et al. 2015; Rajendran et al. 2011) and rotten fruits like orange (Mahmoodi et al. 2017). Whereas microbes not isolated in the studies reviewed and used for pectinase production, also showed promise for pectinase production and application (Azzaz et al. 2020; Sharma et al. 2017; Barman et al. 2015; Poletto et al. 2015a). The use of microorganisms for the production of pectinases makes the product easier to predict since, in the process of utilization of the substrate (pectin), pectinases are released and left in the medium (Raveendran et al. 2018). Hence, the phenomenon exploits the feeding behavior of microbes for the production of pectin enzymes.

3.2 *Optimization and Application*

Optimization of conditions for pectinase production is known to enhance the activity of pectinases (Cherekar and Pathak 2020; Ametefe et al. 2017). The application of the enzyme has also been proven to be enhanced after optimization of the relevant conditions for pectinase activity (Ametefe et al. 2017; Rajendran et al. 2011) and in the production of the enzyme (Azzaz et al. 2020). However, pectinase optimization could be undertaken using predominantly one-factor-at-a-time (OFAT) and the response surface methodology (RSM). The OFAT design was significantly employed ($\approx$75%) for pectinase production (Table 1). The RSM is a tool for the prediction of the behavior of a system and its optimum, using mathematical and statistical techniques (Chollom et al. 2019; Najib et al. 2017). The RSM (system) designs experimental runs and analyzes the data to fit the experimental data obtained to a model (quadratic, cubic, linear, etc.) for adequate prediction of the optimum responses. Despite the importance and the expected reliability of the RSM owing to the application of statistics and mathematical techniques for pectinase production, this method of optimization was found to be underutilized compared to the OFAT tool in the last decade for the production of pectinases and applications (Table 1). It is important to state that the OFAT design does not adequately account for the interactions as in the RSM where all the prevailing conditions run concurrently in a statistically organized manner. This could further explain the reliability with RSM with a lesser number of experimental runs, thereby making the method less expensive relative to the OFAT method in the long run (Chollom et al. 2019; Tetteh et al. 2017; Bezerra et al. 2008). Additionally, the RSM and artificial neural network design (another design not considered in this study) employed in studies have proven to produce enzymes and other products with lesser experimental runs and a more reliable outcome (regarding obtaining the best enzyme activity) than the OFAT method which has been widely used (Jasni et al. 2020; Sonawane et al. 2020; Said and Amin 2015). Studies from 2000 to 2013 spanning different fields reported that out of the 3190 articles on the SCOPUS database, 352 documents were shown to optimize systems using the RSM approach (Chollom et al. 2019; Bashir et al. 2015). The RSM approach was predominantly utilized in the fields of engineering (15.6%), biochemistry (11.2%), and agricultural and biological sciences (10.6%) relative to chemical engineering (9.5%) and environmental science (6%) (Bashir et al. 2012, 2015). The RSM also generates models for the prediction of responses (Chollom et al. 2019; Bashir et al. 2015). Hence, further justifying the need for increased utilization of the response surface methodolody (RSM) in optimization studies.

Furthermore, the extraction solvents in this study were predominantly used for obtaining the laboratory-produced pectinase after the expected fermentation duration (Mahmoodi et al. 2017; Ismail et al. 2016; Barman et al. 2015). Acetate buffer, distilled water, 0.1 M Na_2SO_4, and NaCl have also been utilized in literature for pectinase extraction (Sudeep et al. 2020; Ametefe et al. 2017; Ibrahim et al. 2015; Patil and Chaudhari 2010; Silva et al. 2005; Singh et al. 1999). Hence, the solvents can be said to ease the release of the trapped enzyme in the fermented substrates (most especially, in SSF) for recovery and subsequent utilization.

Application of the enzyme in the studies reviewed (Table 1) showed the employment of the enzyme for the improvement of dairy diets, drug delivery, treatment of cotton fabrics, plywood bleaching, and the extraction of oil. It was interesting to note that crude pectinases extracted were applied without purification (Azzaz et al. 2020; Mahmoodi et al. 2017). Nevertheless, purification of the pectinases and recombinant enzyme studies were also investigated and applied (Govindaraiji and Vuppu 2020; Abdulrachman et al. 2017). The rationale for the non-purification of the enzyme (crude pectinases) could be attributed to the presence of other substrates in the medium required for degradation, such as in fruit juice treatment (Ametefe et al. 2017). In a study, a combination of pectin enzyme and xylanase was reported to be concurrently applied in the bio-bleaching of plywood, hence further justifying the application of crude enzymes or a mixture of enzymes (Sharma et al. 2017). Furthermore, the technicalities associated with the purification of enzymes could account for the non-purification of the laboratory-produced pectin enzymes for application purposes. Swain and Ray (2010) reported an increase in crude pectinase application (13.3%) on carrot juice yield relative to two percent Pectinex (Novozyme, Denmark) for the same purpose. Additionally, the absorbance of apple juice decreased (showing an increase in the juice clarity) with an increase in the enzyme concentration on application with crude pectinase (Cocok et al. 2017). Hence, the study by Swain and Ray (2010) further showed the 'relative efficacy' in the utilization of crude pectinase for application purposes. The following purification methods have been employed for the enzyme, including ethanol precipitation (Ismail et al. 2016; Trentini et al. 2015), dialysis of precipitate and column chromatography (Mahesh et al. 2016), ultrafiltration with activated charcoal (Poletto et al. 2015b), and ammonium sulfate precipitation before ion-exchange chromatography (Ahmed et al. 2016). Novel pectinolytic microorganisms and application areas were among the additional findings obtained (Table 1). The research gaps were mainly found to be in the use of one-factor-at-a-time (OFAT), while in studies where the RSM were used, relevant conditions such as particle size of the substrates, and inoculum volume, among others, should have been included in the design and the actual laboratory experimentation, for an all-inclusive study (Govindaraiji and Vuppu 2020; Ghazala et al. 2015). Optimization of substrate particle size and inoculum size has also been reported in the literature for maximum production of pectinase (Jacob and Prema 2008; Maria et al. 2006; Patil and Dayanand 2006; Kuhad et al. 1998).

4 Conclusion

The study has explored pectinase-related research articles indexed in the SCOPUS database in the last decade (2011–2020), with a view of identifying research findings and areas for further study in the application of locally produced pectin enzymes. One hundred and thirty-seven documents were obtained from the literature search in the last decade, suggesting the importance of this class of enzymes. Hence, the application of laboratory-produced pectin enzymes should be encouraged in studies. Additionally, despite the benefits of the response surface methodology (RSM) for process optimization of pectinase and its application in the number of fields investigated, the study has further exposed the underutilization of this methodology in publications related to the production and application of pectinases. In areas where the design were used, there were significant conditions not considered during the optimization process for a more comprehensive study. Hence, pectinase still has research areas to be explored, with an advantage in the cheaply available raw materials for its production. Encouragingly, a variety of microorganisms have been isolated from the environment for the local production of pectinase in the last decade. Thus, there is need for further research into both production and industrial applications of the enzyme using the response surface methodology. This is because most of the software packages available are updated periodically. Thereby, increasing the reliability of the findings or results generated from the optimization of pectinases using the available RSM soft ware packages compared to the one-factor at a time method. The study has, therefore, catered for the dearth of literature regarding systematic reviews involving the production and application of pectinases in the last decade.

Acknowledgments The authors are thankful to the Covenant University Centre for Research, Innovation, and Discovery (CUCRID) for financing this manuscript.

Funding There was no aid from any funding agency for this research.

References

Abdulrachman D, Thongkred P, Kocharin K et al (2017) Heterologous expression of *Aspergillus aculeatus* endo-polygalacturonase in *Pichia pastoris* by high cell density in textile scouring. BMC Biotech 17:15.

Ahmed I, Zia MA, Hissain MA et al (2016) Bioprocessing of citrus waste peel for induced pectinase production by *Aspergillus niger;* its purification and characterization. J Radiat Res Appl Sci 9: 148-154.

Ametefe GD, Dzogbefia VP, Apprey C et al (2017) Optimal conditions for pectinase production by *Saccharomyces cerevisiae* (ATCC 52712) in solid-state fermentation and its efficacy in orange juice extraction. IOSR J Biotech and Biochem 3(6):78-86.

Azzaz HH, Murad HA, Hassan NA et al (2020) Pectinase production optimization for improving dairy animal's diets degradation. Int. J. Dairy Sci 15(1):54-61.

Balakrishnan M, Jeevarathinam G, Kumar SKS et al (2021) Optimization and scale-up of α-amylase production by *Aspergillus oryzae* using solid-state fermentation of edible oil cakes. BMC Biotechnol 21:33.

Barman S, Sit N, Badwaik LS et al (2015) Pectinase production by *Aspergillus niger* using banana (*Musa balbisiana*) peel as substrate and its effect on clarification of banana juice. J Food Sci Technol 52(6):3579-3589.

Bashir M et al (2012) An overview of wastewater treatment processes optimization using response surface methodology (RSM). In: The 4th International Engineering Conference–Towards engineering of 21st century, Gaza, 15-16 January 2012.

Bashir MJ, Amr SSA, Aziz SQ et al (2015) Wastewater treatment processes optimization using response surface methodology (RSM) compared with conventional methods: Review and comparative study. Middle-East J. Sci. Res 3:244-252.

Bezerra MA, Santelli RE, Oliveira EP et al (2008) Response surface methodology (RSM) as a tool for optimization in analytical chemistry. Talanta 76:965-977.

Catalan E, Sanchez A (2020) Solid-state fermentation (SSF) versus submerged fermentation (SmF) for the recovery of cellulases from coffee husks: A life cycle assessment (LCA) based comparison. Energies MDPI Open Access J 13(11):1-20.

Cherekar MN, Pathak AP (2020) Production and characterization of a haloalkaline pectinase from *Halomonas pantellerinsis* strain SSL8 isolated from Sambhar Lake, Rajasthan. Curr Trends Biotechnol Pharm 14(3):319-326.

Chinedu SN, Dayo-Odukoya OP, Iheagwam FN (2017) Partial purification and kinetic properties of polygalacturonase from *Solanum macrocarpum* L. fruit. Biotech 16(1):27-33.

Chollom MN, Rathilal S, Swalaha FM et al (2019) Comparison of response surface methods for the optimization of an up-flow anaerobic sludge blanket for the treatment of slaughterhouse wastewater. Environ Eng Res 25(1):114-122.

Cocok AMB, Fahrurrozi B, Anja M (2017) Pectinase production and clarification treatments of apple (*Malus domestica)* juice. Ann Bogor 21(2):63-68.

El Enshasy HA, Elsayed EA, Suhaimi N et al (2018) Bioprocess optimization for pectinase production using *Aspergillus niger* in a submerged cultivation system. BMC Biotech 18:1–13.

Ezike TC, Eze SOO, Nsude CA et al (2014) Production of pectinases from *Aspergillus niger* using submerged fermentation with orange peels as carbon source. Sylwan 158(8): 434-440

Garg G, Singh A, Kaur A et al (2016) Microbial pectinases: an eco-friendly tool of nature for industries. 3 Biotech. 6(1):1–13.

Ghazala I, Sayari N, Romdhane MB et al (2015) Assessment of pectinase production by *Bacillus mojavensis* 14 using an economical substrate and its potential application in oil sesame extraction. J Food Sci Technol 52(12):7710-7722.

Govindaraiji PK, Vuppu S (2020) Characterization of pectin and optimization of pectinase enzyme from novel *Streptomyces fumigatiscleroticus* VIT-SP4 for drug delivery and concrete crack-healing applications: An eco-friendly approach. Saudi J Biol Sci 27:3529-3540.

Ibrahim D, Weloosamy H, Lim SH (2015) Effect of agitation speed on the morphology of *Aspergillus niger* HFD5A-1 hyphae and its pectinase production in submerged fermentation. World J Biol Chem 6(3):265-271.

Ismail AS, Abo-Elmagd HI, Housseiny MM (2016) A safe potential juice clarifying pectinase from *Trichoderma viride* EF-8 utilizing Egyptian onion skins. J Genet Eng Biotechnol 14:153-159.

Jacob N, Prema P (2008) Novel process for the simultaneous extraction and degumming of banana fibers under solid-state cultivation. Braz. J. Microbiol 39:115-121.

Jasni AB, Kamyab H, Chelliapan S et al (2020) Treatment of wastewater using response surface methodology: a brief review. Chem Eng Trans 78:535-540.

Khan KS, Kunz R, Kleijnen J et al (2003) Five steps to conducting a systematic review. J R Soc Med 96(3):118–21.

Kuhad R, Manchanda M, Singh A (1998) Optimization of xylanase production by a hyperxylanolitic mutant strain of *Fusarium oxysporum*. Proc Biochem 33:641.

Kumar V, Ahluwalia V, Sara S et al (2021) Recent developments on solid-state fermentation for production of microbial secondary metabolites: challenges and solutions. Bioresour Technol 323:124566.

Liu J, Zhang L, Jia Y et al (2012) Preparation and evaluation of pectin-based colon-specific pulsatile capsule *in vitro* and *in vivo*. Arch Pharm Res 35:1927–1934.

Mahesh M, Arivizhivendhan KV, Maharaja P et al (2016) Production, purification and immobilization of pectinase from *Aspergillus ibericus* onto functionalized nanoporous activated carbon (FNAC) and its application on treatment of pectin containing wastewater. J Mol Catal B Enzym 133:43-54.

Mahmoodi M, Najafpour GD, Mohammadi M (2017) Production of pectinases for quality apple juice through fermentation of orange pomace. J Food Sci Technol 54(12):4123-4128.

Maria L, Garcia S, Samia MT (2006) Optimization of xylanase biosynthesis by *Aspergillus japonicas* isolated from a Caatinga area in the Brazilian state of Bahia. Afr J Biotechnol 5(11):1135-1141.

Najib T, Solgi M, Farazmand A et al (2017) Optimization of sulfate removal by sulfate reducing bacteria using response surface methodology and heavy metal removal in a sulfidogenic UASB reactor. J Environ Chem Eng 5:3256-3265.

Okafor UA, Okochi VI, Chinedu SN et al (2010) Pectinolytic activity of wild-type filamentous fungi fermented on agro-wastes. Afr J Microbiol Res 4(24):2729-2734.

Raveendran S, Parameswaran B, Ummalyma SB et al (2018) Applications of microbial enzymes in food industry. Food Technol Biotechnol 56(1):16-30.

Page MJ, Moher D, Bossuyt PM et al (2021) PRISMA 2020 explanation and elaboration: updated guidance and exemplars for reporting systematic reviews. BMJ (OPEN ACCESS); 372:n160. DOI: https://doi.org/10.1136/BMJ.n160.

Patil SR, Dayanand A (2006) Optimization of process for the production of fungal pectinases from deseeded sunflower head in submerged and solid-state conditions. Bioresour Technol 97(18):2340-2344.

Patil NP, Chaudhari BL (2010) Production and purification of pectinase by soil isolate *Pencillium* sp. and search for better agro-residue for its SSF. Recent Res. Sci. Technol 2(7):36-42.

Pérez-Guerra N, Torrado-Agrasar A, Lopéz-Macais C et al (2003) Agricultural and food chemistry. Elec J Env Agricult Food Chem 2:343-360.

Poletto P, Borsoi C, Zeni M et al (2015a) Downstream processing of pectinase produced by *Aspergillus niger* in solid-state cultivation and its application to fruit juices clarification. Food Sci Technol 35(2):391-397.

Poletto P, Renosto DR, Baldasso C et al (2015b) Activated charcoal and microfiltration as pretreatment before ultrafiltration of pectinases produced by *Aspergillus niger* in solid-state cultivation. Sep Purif Technol 151:102-107.

Rajendra R, Sundraram SK, Radhai R et al (2011) Bioscouring of cotton fabrics using pectinase enzyme its optimization and comparison with conventional scouring process. Pak J Biol Sci 14(9):519-525.

Rys P, Wladysiuk M, Skrzekowska-Baran I et al (2009) Review articles, systematic reviews and meta-analyses: which can be trusted? Polskie Archiwum Medycyny Wewnetrznej 119(3):148–56.

Said KAM, Amin MAM (2015) Overview on the response surface methodology (RSM) in extraction processes. J Appl Sci Eng 2(1):8-17.

Sharma V, Rishishwar P (2015) Effect of pH on commercially applicable pectinase enzyme derived from strain improved *Aspergillus niger* using agro-industrial waste by solid-state fermentation process. Res J Sci IT Management 5 (1):11-14.

Sharma D, Agrawal S, Yadav RD et al (2017) Improved efficacy of ultra-filtered xylanase-pectinase concoction in biobleaching of plywood waste soda pulp. 3 Biotech 7:2.

Silva D, Tokuioshi K, Martins ES et al (2005) Production of pectinase by solid-state fermentation with *Penicillium viridicatum* RFC3. Proc Biochem 40:2885-2889.

Singh SA, Platter H, Diekmann H (1999) Exopolygalacturonase-lyase from thermophilic *Bacillus* sp. Enzyme Microb Technol 25:420-424.

Sonawane AS, Pathak SS, Pradhan RC (2020) Optimization of a process for the enzymatic extraction of nutrient-enriched bael fruit using artificial neural network (ANN) and response surface methodology (RSM). Int J Fruit Sci 20(S3):S1845-S1861.

Subramaniyam R., Vimala R (2012) Solid-state and submerged fermentation for the production of bioactive substances: A comparative study. Review article. Int J Sci Nat (IJSN) 3(3):480-486.

Sudeep KC, Upadhyaya J, Joshi DR et al (2020) Production, characterization, and industrial application of pectinase enzyme isolated from fungal strains. Ferment 6:59.

Swain MR, Ray RC (2010) Production, Characterization and application of a thermostable exo-polygalacturonase by *Bacillus subtilis* CM5. Food Biotechnol 24:37-50.

Tawfik GM, Dila KAS, Mohamed MYF et al (2019). A step-by-step guide for conducting a systematic review and meta-analysis with simulation data. Tropical Medicine and Health 47: 46

Tetteh EK, Rathilal S, Chollom MN (2017) Treatment of industrial mineral oil wastewater-optimization of coagulation flotation process using response surface methodology (RSM). Int J Appl Eng Res 12:13084-13091.

Trentini MM, Toniazzo G, Zeni J et al (2015) Purification of pectinases from *Aspergillus niger* ATCC 9642 by ethanol precipitation. Biocatal Agric Biotechnol 4:315-320.

Ward O P (1992) Fermentation Biotechnology: Principles, processes, and products, 2nd edition. John Wiley and Sons, New York, USA. Pp 52-53.

Widowati E, Utami R, Mahadjoeno E et al (2017) Effect of temperature and pH on polygalacturonase production by pectinolytic bacteria *Bacillus licheniformis* strain GD2a in submerged medium from Raja Nangka (*Musa paradisiaca var. formatypica*) banana peel waste. IOP Conf Ser: Mater Sci Eng 193:012018.

Zhu W, Han C, Dong Y et al (2019) Enzyme-responsive mechanism based on multi-walled carbon nanotubes and pectin complex tablets for oral colon-specific drug delivery system. J Radioanal Nucl Chem 320:503–512.

Biogas Production From Thermo–Alkaline Pretreated Corn Stover Co-digested with Rumen Content

D. Adebowale, O. Oziegbe, Y. D. Obafemi, E. F. Ahuekwe, and S. U. Oranusi

1 Introduction

Developing nations are faced with several environmental challenges like pollution due to their substantial reliance on fossil fuels, pollution from solid wastes, and low or no accessibility to adequate energy resources (Jain et al. 2015; Kamp and Forn 2016). This has informed various studies on the search for alternative energy sources (Su et al. 2015; Lynd et al. 2015; Chirambo 2016). Currently, biogas energy using anaerobic digestion (AD) has been proven due mainly to the ready availability of substrates, low budget, and uncomplicated mechanic requirements (Kwietniewska and Tys 2014). Anaerobic digestion is a biological process that involves the breakdown of organic materials in the absence of oxygen to produce biogas (Marcell and Jorg 2020).

As a biotechnological process, AD takes place in a bioreactor; as such, the technological advancement had been slow and sparse before the twentieth century (Adnan et al. 2019). AD remains an eco-friendly and economical approach for biogas production from agricultural and animal wastes, forest residues, energy crops, and sewage boasting a high net-energy yield than expenditure (Naran et al. 2016).

The biochemical stages involved in AD include hydrolysis, acidogenesis, acetogenesis, and methanogenesis stages (Roopnarain and Adeleke 2017). Several species of acid- and methane-producing microbes reportedly involved in these various stages include *Clostridium*, *Bacteroides*, *Salmonella*, *Escherichia*, *Citrobacter*, *Bacillus*, *Pseudomonas*, *Klebsiella*, *Aspergillus*, *Mucor*, *Rhizopus*, and *Penicillium*, *Methanococcus*, and *Methanosarcina* (Alfa et al. 2014; Dahunsi et al. 2017a, b, c).

D. Adebowale · O. Oziegbe · Y. D. Obafemi · E. F. Ahuekwe · S. U. Oranusi (✉)
Department of Biological Sciences, College of Science and Technology, Covenant University, Ota, Ogun State, Nigeria
e-mail: solomon.oranusi@covenantuniversity.edu.ng

© The Author(s), under exclusive license to Springer Nature Switzerland AG 2022

A. O. Ayeni et al. (eds.), *Bioenergy and Biochemical Processing Technologies*, Green Energy and Technology, https://doi.org/10.1007/978-3-030-96721-5_13

Biogas generation from lignocellulosic biomass (LCB) such as wood, municipal wastes, and agricultural wastes does not require the use of arable lands. Also, it is an ample source of unexpended biomass in the world that cannot be neglected in the search for human sustenance and economic growth (Khoufi et al. 2015; Giwa et al. 2017). The utilization of this biomass as an option has, however, stopped the contest within feed, food, and energy production (Ezeonu and Ezeonu 2016). Animal wastes such as dung and poultry droppings have been used in anaerobic digestion process. Rumen content from slaughtered animals has found use in biogas production. It comprises the digested and undigested feed in the rumen of ruminant animals, serving as a good source of microbial proteins, minerals, vitamins, and energy, although varying in nutrient from one ruminant animal to the other (Elfaki and Abdelatti 2016). However, the major challenge in the use of these materials is the complex structure of the lignocelluloses (lignin, cellulose, and hemicelluloses). Several pretreatment methods have been tested on some lignocellulosic materials. These include physical, thermal, biological, and chemical processes, as well as a combination of the pretreatment methods Su et al. (2015) to fasten hydrolysis of feedstock. Studies show that pretreatment has proven to enhance physical and chemical changes in structural composition of biomass. As such, it is imperative to consider the type of pretreatment suitable for any candidate biomass for use in biogas generation.

The successful production of biogas in an anaerobic condition involves several microorganisms under a suitable environment. The biogas digester can be controlled by putting into consideration important parameters like pH, temperature, and retention time. Certain factors like substrate type, feedstock mixture, type of inoculum, and loading rate can also affect biogas production (Adnan et al. 2019). In this study, thermo-alkaline pretreated corn stover biomass was mixed together with cattle's rumen content as slurry for biogas production in a batch system bioreactor. The corn stover was pretreated prior to its usage as inoculum in order to breakdown the lignocellulose structure. Temperature, pH, and microbial load were assessed as the anaerobic co-digestion progressed for the 20-day period.

2 Materials and Methods

2.1 Sample Collection

The samples for the anaerobic digestion include corn stover (CS) and rumen content (RC) which serves as the inoculum. Corn stover was obtained from Covenant University farm, while the cattle rumen content was collected from Alubarika Abattoir Market, Igboloye, Ota, Ogun State. The cattle's rumen content was sampled in an airtight container right after the cattle was slaughtered to avoid air exposure. Both CS and RC were carefully transported to the Microbiology Laboratory of the Department of Biological Sciences, Covenant University for processing, storage, and further work.

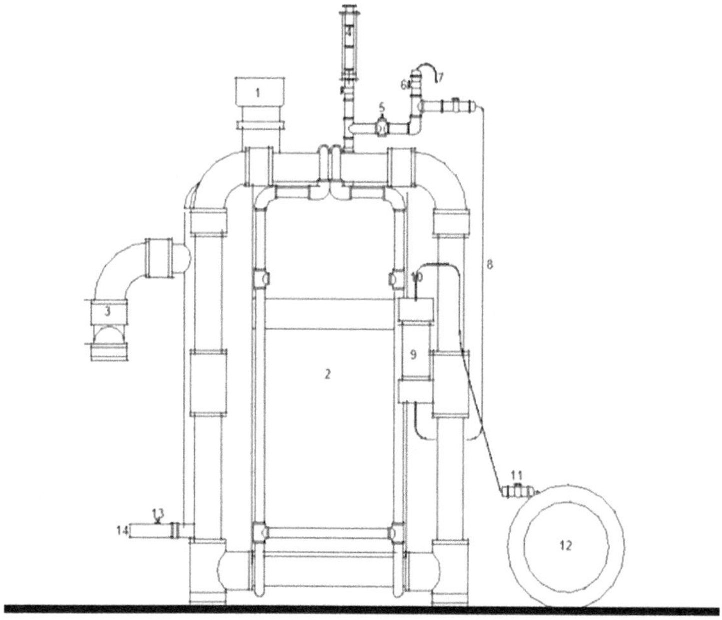

Fig. 1 Schematic representation of the fabricated anaerobic digester. 1. Feed inlet. 2. Digestion tank. 3. Slurry overload discharge. 4. Gas measuring syringe. 5. Lock and key. 6. Gas sampling lock. 7. Gas sampling syringe. 8. Connecting hose. 9. Filtering chamber. 10. Hose. 11. Lock and key. 12. Gas storage tube. 13. Lock and key. 14. Tap for sample collection

2.2 Fabrication of Anaerobic Digester

The anaerobic digester used for this study is a single-stage batch anaerobic digestion system with vertical orientation. It was fabricated using polyvinyl chloride (PVC) and plastic materials as seen in Fig. 1. The digester is an 18 L-capacity AD system with a working system of 15 L and 3 L as head space. There is an inlet for substrate loading, an outlet for flowing out excess slurry, a 200 ml-capacity gas measuring syringe for measurement of the volume of produced biogas, and a biogas filter for de-watering and de-sulfuring the biogas generated. The digester is fitted with a sampling tap and a mini gas storage bag for filtered biogas storage. The AD system is modeled, in that, the substrate is fed into the anaerobic digester, and batch anaerobic degradation occurs in it. Afterwards, the produced biogas builds up pressure and volume, with the gas volume indicated by the gas measuring syringe; then the biogas flows into the storage bag.

2.3 Pretreatment of Corn Stover

CS samples were air-dried for 14 days before pretreatment. This was done by first chopping the corn stover with a sterile knife and grinding using an electric grinder (Bajaj twister mixer grinder-410025) to obtain a powdered form. The thermal treatment was performed using a CLIFTON water bath (Nickel-Electro Ltd., 57599) at 80 °C for 3 hours (Liu et al. 2015; Dahunsi et al. 2017c, 2018) and was further treated chemically using the alkaline sodium hydroxide (NaOH). The alkaline pretreatment was based on ratio of dry weight of the corn stover:NaOH:H_2O at 10:0.2:60 (w/w/v) at 25 °C for 3 days. After pretreatment, the biomass was used directly with the rumen content as the biodigester inoculum. The use of NaOH for pretreatment has been documented to be cost-effective and highly efficient for biological conversion (Zheng et al. 2009; Li et al. 2015).

2.4 Lignocellulose Content Analysis

Compositional analysis of the pretreated biomass for the lignin, hemicellulose, cellulose, and extractive was performed in threefolds. The mean values with standard error of means were recorded. Gravimetric method was used as it exhibits minimal instrumental error and does not require the use of sophisticated instruments (Ayeni et al. 2014, 2015). Also, the extractive was broken down to stop obstruction of biomass downstream analysis.

Determination of Extractives

Two and a half grams (2.5 g) of the pretreated dry sample was introduced into a cellulose thimble with a Soxhlet extractor construct using 150 ml acetone as extraction solvent. The heating mantle (KDM-500, USA) was used for the boiling and rising stages of the extraction with the residence time adjusted to 25 mins at 70 °C for 4 hours. After successful excerption, the samples were then air-dried at room temperature for 15 minutes. The constant weight of the excerpted material was achieved in a conventional oven from Genlab at 105 °C. The percentage weight per weight of the extractive content of each sample was estimated as the difference between the weight of raw extractive-laden and extractive free biomass (Li et al. 2004; Lin et al. 2010).

Hemicellulose Determination

One gram (1 g) of treated sample was measured and conveyed into a 250 ml conical flask, and 150 ml of 500 mol/m^3 NaOH was added, before heating the blend for 3½ hours in a water bath. It was filtered after cooling using a filter paper and wiped until a neutral pH was accomplished. The resultant residue was dried to a constant weight in a conventional oven at 105 °C. The hemicellulose content (%w/w) was taken as sample weight obtained before and after (Li et al. 2004; Lin et al. 2010).

Lignin Content Determination

Following standard methods, a portion of the sample (0.3 g) was measured into test tubes with addition of 3 ml of 72% H_2SO_4 into each tube. The samples were held for 2 hours at room temperature while being carefully shaken at 30 minutes intervals for initial hydrolysis to take place. Thereafter, 84 ml of distilled water was introduced into each test tube. The subsequent step of hydrolysis was done in an autoclave at 121 °C for 1 hour. The slurry was left at room temperature to cool and hydrolyzates filtered using an 11 μm filtering crucible in a vacuum. The acid insoluble lignin fraction of the biomass was evaluated by drying of residuals at 105 °C in an oven and documenting ash by incinerating the hydrolyzed samples at 575 °C in a muffle furnace (Chesterfield, UK-EEF3).

The acid soluble lignin component evaluation is by measurement of the acid hydrolyzed samples absorbance with a spectrophotometer (Thermoscientific; GENESYS, IOS UV-VIS) at 320 nanometers. Then the composition of lignin was recorded as the total value of acid soluble lignin and acid insoluble lignin (Sluiter et al. 2008).

Determination of Cellulose

The cellulose composition (%w/w) was recorded as the differences in other components. On assumption, samples comprise lignin, hemicellulose, cellulose, ash, and extractives (Lin et al. 2010; Li et al. 2004).

2.5 *Microbial Analysis*

The microbiological analysis of substrate and inoculum involved the use of MacConkey agar (Oxoid CM1169, United Kingdom), eosin methylene blue agar (Rapid Labs, Essex, United Kingdom), nutrient agar (Himedia Laboratories, India), thioglycolate broth (Rapid Labs, Essex, United Kingdom), compounded basal medium for isolation of anaerobes, and potato dextrose agar for fungi isolation. All reagents used were of analytical grade and measured using an analytical weigh balance (Model EK-1200A). All media used were prepared following manufacturer's guide and then sterilized at 121 °C for 15 minutes at 15 psi in an autoclave (Surgifriend medicals, England- SM280E).

The basal medium was compounded following the methods of Balch et al. (1977). At a final pH of 6.7, the medium composition of 1000 ml was 0.4 g potassium di-hydrogen phosphate, 1.0 g ammonium chloride, 0.4 g dipotassium hydrogen phosphate, 0.1 g magnesium chloride, 0.0001 g resazurin, 0.5 g sodium sulfide, 0.5 g cysteine HCl, 7 g sodium bicarbonate, 2 g yeast extract, 10 g calcium carbonate, 10 ml mineral solution, 10 ml vitamin solution, and 20 g agar.

The vitamins solution included 5 mg cyanocobalamin, 4 mg p-aminobenzoic acid, 1 mg biotin, 10 mg nicotinic acid, 5 mg calcium pantothenate, 15 mg pyridoxamine–2HCl, and 10 mg thiamine–HCl. The trace elements composition was 1.6 mM HCl, 7 mg $ZnCl_2$, 100 mg FeC12.7H_2O, 10 mg $MnCl_2.4H_2O$, 13 mg $CoCl_2.6H_2O$, 0.6 mg H_3BO_3, 2.4 mg $NiCl_2.6H_2O$, 0.2 mg $CuCl_2.2H_2O$, 0.26 mg $Na_2SeO_3.5H_2O$, 0.66 mg Na_2WO_4, and 3.6 mg $Na_2MoO_4.2H_2O$, all of which were dissolved in double distilled water.

Thioglycolate broth was used for the preservation of anaerobes. One gram of samples was aseptically pulled out and diluted in sterile distilled water (9 ml) to dilutions of 10^4 and 10^6 before plating on the different media. Plates were incubated for 24 hours at 37 °C for aerobes, while anaerobic organisms were incubated at 37 °C for 5–7 days. Fungal isolates were placed at room temperature for 3–5 days.

Identification of Isolates

Presumptive identification of isolates was based on morphological and phenotypic identification involving various biochemical tests. These included Gram staining technique, starch hydrolysis, indole, methyl red, Voges-Proskauer, citrate, oxidase, and sugar fermentation. The API kits were employed for further verification of the presumptive isolates.

3 Results and Discussion

3.1 Lignocellulosic Composition of Corn Stover

The results of the lignocellulosic composition as shown in Table 1 reveal that the thermo-alkaline pretreatment has percentage reduction in lignin, hemicelluloses, ash, and extractive components. However, the cellulose component increased from untreated to treated. This is an indication of the effectiveness of the thermo-alkaline pretreatment applied on the biomass, and it helped in reducing the recalcitrant components and thus aided in effective biogas production. Results of lignocellulosic component study corroborate the reports of Ayeni et al. (2015) and Liu et al. (2015),

Table 1 Lignocellulosic Component of Corn stover

Composition of LCB*	Untreated	Treated	% Lignocellulose reduction
Extractives	17.50 ± 0.05	15.92 ± 0.04	9.0
Cellulose	35.19 ± 0.09	48.42 ± 0.19	37.60
Lignin	11.25 ± 0.14	6.11 ± 0.11	45.69
Ash	4.45 ± 0.12	2.4 ± 0.10	46.10
Hemicellulose	31.07 ± 0.12	26.59 ± 0.01	14.42

**LCB* lignocellulosic biomass

who documented extensive lignin degradation and successive depletion when alkaline-thermal pretreatment was employed. The reduction of hemicelluloses observed in this study after thermo-alkaline pretreatment is related to those highlighted in prior studies that employed alkaline thermo-mechanical, diluted acids, thermo-alkaline, ionic liquids, steam explosion, and ammonia fiber expansion on various materials (Espinoza-Acosta et al. 2014; Dahunsi et al. 2016; Zou et al. 2016).

3.2 *Operational Parameters and Biogas Yield*

The pH of the co-digestion regimes was alkaline all through the digestion process. This ranged from 7.05 to 7.84 while the temperature was maintained at a mesophilic temperature range of between 29.8 to 31.1 °C as represented in Table 2. The biogas production started on the 5th day with little fluctuations till a constant volume was produced from 18th to 20th day. The total biogas volume of the anaerobic digestion of corn stover and rumen content was 54.97 m^3. This was estimated by summation of all the biogas produced and recorded per day with the digester's inbuilt gas-measuring syringe, for the 20 days co-digestion period. The operational pH values obtained in this study were between 7.05 to 7.84 and agree with the reports of Miikue-Yobe and Sunday (2019), Dahunsi et al. (2017a,b), and Dahunsi and Oranusi (2013). More so, different studies from Dahunsi et al. (2016) and Dahunsi et al. (2017c) submit that an effective AD system works best at a pH between 6.5 and 8.0.

The temperature of the digestion has been reported to influence and improve bioconversion stability as well as promote richness in microbial population (Kwietniewska and Tys 2014; Zou et al. 2016). Miikue-Yobe and Sunday (2019) reported a temperature range of 22–32 °C in a study on biogas production from sawdust co-digested with African dwarf goat excreta. The thermal and alkaline pretreatment employed in several studies helps in increasing the biogas yield, unlike those that did not either employ any pretreatment or used a single pretreatment method.

The highest biogas produced recorded between the 16th to 20th day could be attributed to the operational parameters (pH and temperature) and also the diversity of anaerobic organisms within the period. This report suggests that the optimum pH range for proper augmentation of bioreactor organisms may range between 7.05 to

Table 2 Biogas yield and operational parameters of digestion

Days	Biogas yield (m^3)	Temperature (°C)	pH
0	0	30.4	7.05
1	0	30.6	7.07
2	0	31.1	7.42
3	0	30.9	7.55
4	0	30.8	7.76
5	3.01	30.9	7.78
6	3.01	30.7	7.73
7	3.01	31.0	7.80
8	3.05	29.9	7.84
9	3.02	29.8	7.84
10	3.20	30.0	7.84
11	3.15	30.1	7.80
12	3.50	30.5	7.80
13	3.50	29.8	7.75
14	3.60	29.5	7.67
15	3.40	30.1	7.69
16	3.70	30.0	7.69
17	3.82	29.8	7.70
18	4.00	29.9	7.80
19	4.00	29.9	7.84
20	4.00	30.1	7.84

7.2, although opine that some anaerobic systems can tolerate may range from pH 6.5 to 8.0 (Cioabla et al. 2012; Dahunsi et al. 2016). Similarly, Ennouri et al. (2016) suggest that AD pH range of less than 6.5 or greater than 8.0 may give rise to the AD methanogenesis stage failure.

Reports of Priebe et al. (2016) and Suksong et al. (2016) on AD suggest that the optimum temperature for effective production of biogas is a mesophilic range of 30 °C–37 °C which supports bacterial growth. A decrease or increase above this temperature range may lead to reduced production of biogas. To this end, it is imperative to ensure a suitable temperature and pH for proper AD process, adequate microbial growth, balance, and early biogas generation (Dahunsi et al. 2017b). Previous studies on corn stover pretreated with biological treatment by liquid fraction at different concentrations showed a biogas production between 20.69 to 33.91 litres (Li et al. 2016; Ohimain and Izah 2017); which, in comparison to the results obtained in this study, is reduced. More so, corn stover pretreated with NaoH and steam explosion reported a lower biogas yield compared to the volume obtained in this study (Liu et al. 2015; Miikue-Yobe and Sunday 2019). The difference in the biogas yield could be attributed to a myriad of factors which may include the class of LCB, type of inoculum, environmental conditions of the digester, the pretreatment method employed, and the retention time of digestion (Jain et al. 2015; Kim et al. 2015). In light of these, Li et al. (2015) opine that utilization of adequate pretreatment methods may amplify biogas production from LCB.

Table 3 Microbial composition of digestion process

Substrate	Aerobic bacteria	Anaerobic bacteria	Fungi
Rumen content	*Bacillus* spp. *Enterococcus* sp. *Pseudomonas aeruginosa*	*Clostridium* spp. *Bacteroides* sp. *Fusobacterium* sp.	*Aspergillus niger* *Aspergillus flavus* *Penicillium* sp. *Mucor* sp.
Corn stover	*Bacillus* spp. *Proteus* spp.	*Clostridium* spp.	*Penicillium* sp.
RC* + CS* before digestion	*Citrobacter* sp. *Bacillus* spp. *Enterococcus* sp. *Pseudomonas aeruginosa*	*Clostridium* sp. *Fusobacterium* sp. *Bacteroides* sp.	*Aspergillus niger* *Rhizopus* sp. *Penicillium* sp.
RC* + CS* after digestion	*Bacillus* spp.	*Clostridium* sp., Methanococcus sp.	*Aspergillus niger*

**RC* rumen content, *CS* corn stover

3.3 Microbial Composition of the Feedstocks and Slurry Used for Digestion

Table 3 shows the consortium of microorganisms isolated with progression of the co-digestion in this study. This included species of *Bacillus*, *Staphylococcus*, *Proteus*, *Enterococcus*, *Pseudomonas*, *Bacteroides*, *Clostridium*, *Fusobacterium*, *Aspergillus*, *Penicillium*, *Mucor*, and *Rhizopus*. Most of the organisms associated with the digestion process are comparable with previous studies on anaerobic digestion using some other substrates (Kim et al. 2015; Ennouri et al. 2016; Dahunsi et al. 2017a, b, c). An investigation of microorganisms associated with biogas generation from *Telfairia occidentalis* (vegetable), banana peel, and pig dung includes species of *Pseudomonas*, *Bacillus*, *Staphylococcus*, *Escherichia coli*, and fungi (Asikong et al. 2014). The co-digestion of sawdust and African dwarf Goat excreta as conducted by Miikue-Yobe and Sunday (2019) identified microbial species of *Escherichia coli*, *Staphylococcus* spp., *Bacillus* spp., *Clostridium* spp., *Pseudomonas* spp., *and Enterococcus* spp. for biogas production.

4 Conclusion

The result of this study has shown that biogas can be produced from organic wastes such as corn stover and rumen content with the help of microorganisms under an alkaline pH and a mesophilic temperature range. The study also shows that the thermal and alkaline pretreatment is effective in reducing non-digestible parts of the lignocellulosic biomass, making them accessible to microbial action, and thus, resulting in increased biogas recovery. The ready availability of the waste substrates used and the high biogas yield obtained show promise as an alternative solution to the global energy need. In light of these, the study recommends the use of thermo-alkaline pretreated corn stover and cattle rumen fluid for biogas production as an alternate energy source, as it also presents an effective waste management strategy capable of leading to an improved bioeconomy.

Acknowledgments We appreciate the support of Covenant University Centre for Research Innovation and Discovery (CUCRID).

Conflicts of Interest The authors declare no conflict of interest.

References

Adnan AI, Ong MY, Nomanbhay S et al (2019) Technologies for Biogas Upgrading to Biomethane: A Review. Bioengineering 6:92

Alfa IM, Dahunsi, SO, Iorhemen, OT et al (2014) Comparative evaluation of biogas production from poultry droppings, Cow dung and Lemon grass. Bioresour Technol 157:270–277

Asikong BE, Idire SO, Tiku DR (2014) Microorganisms Associated with Biogas production using vegetable *Telfairia occidentalis* wastes, Banana peel and pig Dung as substrate. Br Microbiol Res J, 16(3):1-12

Ayeni AO, Omoleye JA, Mudliar S et al (2014) Utilization of lignocellulosic waste for ethanol production: Enzymatic digestibility and fermentation of pretreated shea tree sawdust. Korean J Chem Eng 31(7):1180-1186

Ayeni AO, Adeeyo OA, Oresegun OM et al (2015) Compositional analysis of lignocellulosic materials: Evaluation of an economically viable method suitable for woody and non-woody biomass. Am J Eng Res 4:14-19

Balch WE, Schoberth S, Tanner RS et al (1977) Acetobacterium, a new genus of hydrogen oxidizing, carbon dioxide-reducing, anaerobic bacteria. Int J of Syst Bacter 27(4) :355-361

Chirambo D (2016) Addressing the renewable energy financing gap in Africa to promote universal energy access: Integrated renewable energy financing in Malawi. Renew Sustain Energy Reviews 62:793–803

Cioabla AE, Lonel L, Dumitrel GA et al (2012) Comparative study on factors affecting anaerobic digestion of agricultural vegetal residues. Biotechnol Biofuels 5:39

Dahunsi SO, Oranusi SU (2013) Co-digestion of food waste and human excreta for biogas production. Br Biotechnol J, 3(4):485-499

Dahunsi SO, Oranusi S, Efeovbokhan VE (2017a) Anaerobic mono-digestion of *Tithonia diversifolia* (Wild Mexican sunflower). Energy Convers Manag 148:128–145

Dahunsi SO, Oranusi S, Efeovbokhan VE (2017b) Bioconversion of *Tithonia diversifolia* (Mexican Sunflower) and poultry droppings for energy generation: optimization, mass and energy balances, and economic benefits. Energy Fuels 31:5145–5157

Dahunsi SO, Oranusi S, Efeovbokhan VE (2017c) Cleaner energy for cleaner production: Modeling and optimization of biogas generation from *Carica papayas* (Pawpaw) fruit peels. J Clean Prod 156:19–29

Dahunsi SO, Oranusi S, Owolabi JB et al (2016) Comparative biogas generation from fruit peels of Fluted Pumpkin (*Telfairia occidentalis*) and its optimization. Bioresour Technol 221:517–525

Dahunsi SO, Oranusi S, Efeovbokhan VE et al (2018) Biochemical conversion of fruit of *Telfairia occidentalis* (fluted pumpkin) and poultry manure. Energy sources 40:2799-2811.

Elfaki OA, Abdelatti KA (2016) Rumen Content as Animal Feed: A Review. University of Khartoum Journal of Veterinary Med and Animal Prod 7(2):80-88

Ennouri H, Miladi B, Diaz SZ, et al (2016) Effect of thermal pretreatment on the biogas production and microbial communities balance during anaerobic digestion of urban and industrial waste activated sludge. Bioresour. Technol 214: 184-191

Espinoza-Acosta JL, Torres-Chávez PI, Carvajal-Millán E et al (2014) Ionic liquids and organic solvents for recovering lignin from lignocellulosic biomass. Bioresour 9:3660–3687

Ezeonu CS, Ezeonu NC (2016) Alternative Sources of Petrochemicals from Readily Available Biomass and Agro-Products in Africa: A Review. J Petr Environ Biotech 7(5):12-23

Giwa A, Alabi A, Yusuf, A et al (2017) A comprehensive review on biomass and solar energy for sustainable energy generation in Nigeria Renew Sustain Energy Reviews 69, 620–641

Jain S, Wolf IT, Lee J et al (2015) A comprehensive review on operating parameters and different pretreatment methodologies for anaerobic digestion of municipal solid waste. Renew Sustain Energy Reviews 52: 142–154

Kamp LM, Forn EB (2016) Ethiopia's emerging domestic biogas sector: Current status, bottlenecks and drivers. Renew Sustain Energy Reviews 60: 475–488

Kim D, Lee K, Park LY (2015) Enhancement of biogas production from anaerobic digestion of waste activated sludge by hydrothermal pre-treatment. Int Biodegr Biodeter 101:42-46

Khoufi S, Louhichi A, Sayadi S (2015) Optimization of anaerobic co-digestion of olive mill wastewater and liquid poultry manure in batch condition and semi-continuous jet-loop reactor. Bioresour Technol 182:67–74

Kwietniewska E, Tys J (2014) Process characteristics, inhibition factors and methane yields of anaerobic digestion process, with particular focus on microalgal biomass fermentation. Renew Sustain Energy Reviews 34:491–500

Li C, Zhang G, Zhang Z et al (2016) Alkaline thermal pretreatment at mild temperatures for biogas production from anaerobic digestion of antibiotic mycelial residue. Bioresour Technol 208:49-57

Li L, He Q, Ma Y et al (2015) Dynamics of microbial community in a mesophilic anaerobic digester treating food waste: Relationship between community structure and process stability. Bioresour Technol 189:113–120

Li S, Xu S, Liu S et al (2004) Fast pyrolysis of biomass in free-fall reactor for hydrogen-rich gas. Fuel Process Technol 85(8–10):1201-1211

Lin L, Yan R, Liu Y et al (2010) In-depth investigation of enzymatic hydrolysis of biomass wastes based on three major components: cellulose, hemicellulose, and lignin. Bioresour Technol 101(21):8217-8223

Liu L, Zhang T, Wan H et al (2015) Anaerobic codigestion of animal manure and wheat straw for optimized biogas production by the addition of magnetite and zeolite. Energy Convers Manag 97:132–139

Lynd LR, Sow M, Chimphango AFA., et al (2015) Bioenergy and African transformation. Biotechno for Biofuels 8(18):1-18

Marcell N, Jorg K (2020) Anaerobic Digestion in the 21st Century. Bioengineering 7:157. doi:https://doi.org/10.3390/bioengineering7040157

Miikue-Yobe TF, Sunday EM (2019) Production of Biogas from Saw-Dust and African Dwarf Goat Excreta. The International Organization of Scientific Research Journal of Applied Physics 11:2278-4861

Naran E, Toor, UA, Kim D, (2016) Effect of pretreatment and anaerobic codigestion of food waste and waste activated sludge on stabilization and methane production. Int Biodegrad Biodeter 113:17–21

Ohimain EI, Izah S. (2017) A review of biogas production from palm oil mill effluents using different configurations of bioreactors. Renew Sustain Energy Reviews 70:242–253

Priebe GPS, Kipper E, Gusmao AL, et al (2016) Anaerobic digestion of chrome-tanned leather waste for biogas production. J Clean Pro. 129:410-416

Roopnarain A, Adeleke R (2017) Current status, hurdles and future prospects of biogas digestion technology in Africa. Renew Sustain Energy Reviews 67:1162–1179

Sluiter, A., Hames, B., Ruiz, R., Scarlata, C., Sluiter, J. & Templeton, D. (2008). Determination of structural carbohydrates and lignin in biomass: laboratory analytical procedure (LAP). Golden, CO: National Renewable Energy Laboratory; NREL Report No.: TP-510-42618. Contract No.: DE-AC36-99-G010337. Sponsored by the U.S. Department of Energy. Accessed 15 January 2020

Su H, Liu L, Wang S et al (2015) Semi continuous anaerobic digestion for biogas production: influence of ammonium acetate supplement and structure of the microbial community. Biotechno Biofuels 8(13):1-13

Suksong W, Kongjan P, Prasertsan P et al (2016). Optimization and microbial community analysis for production of biogas from solid waste residues of palm oil mill industry by solid-state anaerobic digestion. Bioresour. Technol:214, 166-174

Zheng M, Li X, Li L et al (2009) Enhancing anaerobic biogasification of corn stover through wet state NaOH pretreatment. Bioresoure Technology 100:5140-5145

Zou S, Wang X, Chen Y, et al (2016). Enhancement of biogas production in anaerobic co-digestion by ultrasonic pretreatment. Energy Convers Manag 112:226–235

Assessing the Safety of Tiger Nut Drinks Produced from *Cyperus esculentus* L. Seeds Sold in Ota

M. B. Alade, E. F. Ahuekwe, A. O. Adeyemi, and O. C. Nwinyi

1 Introduction

Aflatoxins (AF) are difuranocoumarins (Peterson et al. 2001) synthesized by *Aspergillus parasiticus* and *Aspergillus flavus in* relatively warm and humid conditions, most commonly in food or feeds. Among all mycotoxins, aflatoxins are best known for their huge impacts on human and animal health (IARC 2015). Primary forms of aflatoxins include AF (B_1, B_2, G_1, and G_2) are named after the fluorescence they display in ultraviolet (UV) light (B for blue and G for green). These have recently gained attention among researchers due to their public health concerns. Aflatoxin AFM_1 and AFM_2 hydroxylated metabolites occur in the animal products such as milk and other dairy products (D'Mello and MacDonald 1997). The increased infection with aflatoxins produced by *A. parasiticus* and *A. flavus* in pistachio, cotton, nut, and maize has been attributed to preharvest weather conditions associated with periods of drought and stress during fruit growth and flowering (Kebede et al. 2012). Aflatoxins are stable in foods during baking, extrusion, and cooking, where it induces great problem in processed food (Marin et al. 2013). Aflatoxin in humans and animals leads to disease condition called aflatoxicosis (Adeyeye 2016). They are considered to be harmful due to their teratogenic, carcinogenic, immunosuppressive, and mutagenic (DNA damaging) effects (JECFA 2017). Tiger nut and its products are highly susceptible to aflatoxin contamination

M. B. Alade (✉) · E. F. Ahuekwe · O. C. Nwinyi (✉)
Department of Biological Sciences, College of Science and Technology Covenant University, Ota, Nigeria
e-mail: mazeedat.aladepgs@stu.cu.edu.ng; obinna.nwinyi@covenantuniversity.edu.ng

A. O. Adeyemi
Department of Biochemistry, College of Science and Technology Covenant University, Ota, Nigeria

© The Author(s), under exclusive license to Springer Nature Switzerland AG 2022

A. O. Ayeni et al. (eds.), *Bioenergy and Biochemical Processing Technologies*, Green Energy and Technology, https://doi.org/10.1007/978-3-030-96721-5_14

either in the field or during storage, particularly in the hot and humid climatic conditions (Shamsuddeen and Aminu 2016; Akomolafe and Awe 2017). Aflatoxin production is also influenced by high moisture, temperature, relative humidity, and insect damage (Margherita et al. 2012). Due to the warm and humid climatic condition and poor practices during harvesting, processing, and storage, agricultural commodities in tropical and subtropical are prone to aflatoxin contamination (Olopade et al. 2020; Shamsuddeen and Aminu 2016).

The scientific name of tiger nut is *Cyperus esculentus*. It is a perennial monocotyledon grass-like plant with spheroid tubers and pale-yellow cream kernel surrounded by a fibrous sheath. Interestingly, tiger nut is a small tuber discovered 4000 years ago (Asante et al. 2014; Maduka and Ire 2018). It has been recognized as one of the best nutritional crops used to augment diets, since a substantial intake decreased reported cases of various health-related conditions such as diabetes, cardiovascular disease, obesity, and cancer. It is also ideal for older people, athletes, and children, as well as an excellent source of calcium and iron for body growth and development (Shamsuddeen and Aminu 2016). From studies, tiger nuts have been reported to lower low-density lipoprotein (LDL) while increasing high-density lipoprotein (HDL) cholesterol (Gambo and Da'u 2014). According to Chukwuma et al. (2010), tiger nut drinks are beneficial in preventing heart problems, thrombosis, arteriosclerosis, and activating blood circulation. Tiger nut is also suitable for lactose-intolerant and gluten patients because it provides lipase, catalase, amylase, and stimulant that aids digestion (Ogodo et al. 2018). Unfortunately, despite these potentials in tiger nuts, it has been a neglected crop in Nigeria. This is due to lack of knowledge on its production, nutritional value, and utilization (Belewu and Abodunrin 2006). Tiger nut can be consumed raw, dried, baked roasted, or processed into beverage (Belewu and Abodunrin 2006; Sa'id et al. 2017).

Tiger nuts are important source of nourishment and essential component of healthy and balanced diets (Nwinyi and Umane 2019; Oladele and Aina 2007); however, it may harbor various microbial flora which can be a potential source of acute or chronic foodborne illnesses when contaminated with pathogenic microorganisms or microbial metabolites such as toxins (Negedu et al. 2011). This is because the tuber grows in the soil and may contain toxigenic molds and their mycotoxins. Thus, this study is aimed at assessing the presence of fungi mycoflora and their metabolites in *Cyperus esculentus* L. seeds or milk products sold in Ota, Ogun state.

2 Materials and Methods

2.1 Sample Area

The study site was Sango Ota, Ogun state (see Fig. 1). The samples of tiger nut seeds were purchased from major dealers within Ota (Fig. 2). Sango Ota coordinates are 6° 42′ 0″ North, 3° 14′ 0″ East.

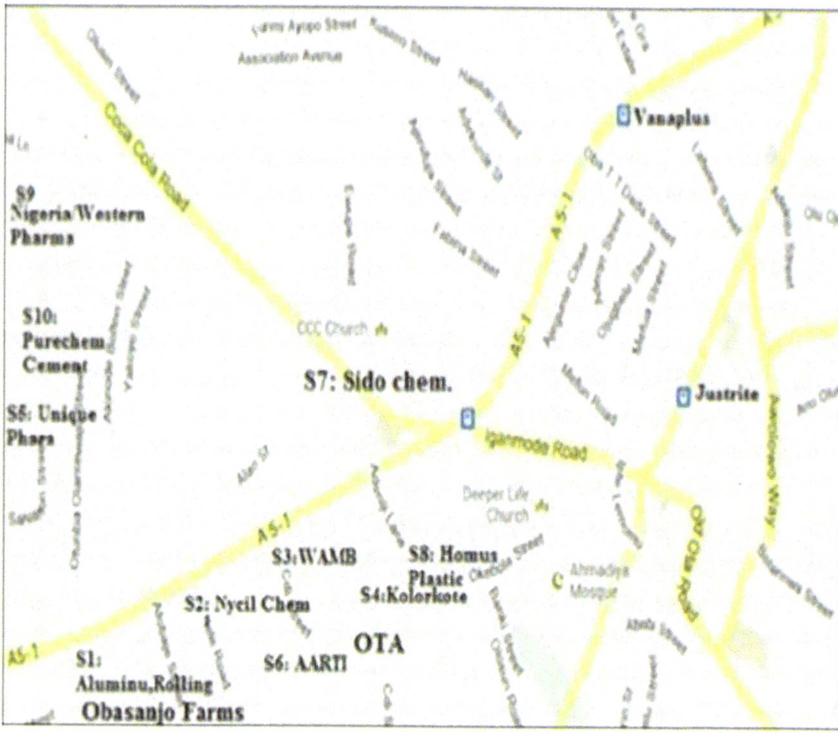

Fig. 1 Map of the study area, Sango Ota. (Reproduced from Daramola et al. 2018)

Fig. 2 Tiger nut sampled from selected markets. (Reproduced from Maduka and Ire 2018)

2.2 Collection of Samples

Snowball sampling technique was used to obtain dry tiger nut seeds in Sango Ota, Ogun state. A total of nine samples with an equivalent weight of two hundred and fifty gram (250 g) were purchased from three major dealers in Ota, Ogun state. Samples were placed in ziplock bag and transported to the Microbiology Laboratory of Covenant University where they were analyzed after milling with a sterile mechanical blender (Kenwood high power).

2.3 Preparation of Tiger Nut Drink

The tiger nut drink was prepared from the purchased seeds in the laboratory using a modified technique. The seeds were processed by removing foreign particles and unwanted materials. About 120 g of the tiger nuts were manually sorted. For the tiger nut drinks, the nuts were soaked in 250 ml of potable water at room temperature for 3 days. The soaked nuts were milled into slurry using electrical blender and 500 ml of potable water. The homogenized slurry was pressed with clean muslin cloth to extract the milk and then stored at 4 °C for 15 hours in a white plastic container.

3 Methodology

3.1 Serial Dilution and Isolation of Fungi Microflora

In nine milliliter (9 ml) of sterile Ringer solution, one gram of each blended sample was diluted, vortexed and serially diluted to 10^5 after which one milliliter of the sample was transferred into duplicate petri dishes that were appropriately labeled. Thereafter, using pour plate methods, about 15–20 ml of sterile Rose Bengal Chloramphenicol was poured into the petri dishes. The samples were mixed by swirling the plates, and then allowed to solidify before being incubated at 25 °C for 7 days. Also the raw seeds samples (unblended) were washed in ringer solution and plated out using the pour plate method. The number of colonies on the agar plates was expressed as number of CFU per 1 ml of tiger nut milk and 1 g of tiger nut seed. Following the colony count, isolates from tiger nut and tiger nut milk were subcultured successively on potato dextrose agar to obtain a pure culture and incubated at 25 °C for 7 days. Using the standards of Klich (2002) and Pitt and Hocking (2009), pure isolates were identified based on the macroscopic and microscopic characteristics. For the morphological identification, lactophenol blue-stained fungal isolates were mounted on slides, overlaid with cover slides, and placed on the stage of a microscope to observe the micromorphological attributes for identification.

3.2 Qualitative Screening for Toxigenic Potential of the Isolates

The isolates toxigenic strains were screened using ammonium vapor assay. *Aspergillus* isolates were grown as single colonies in the center of the plate on yeast extract sucrose agar (2% yeast extract, 15% sucrose, and 1.5% agar) and incubated at 25 °C for 2 weeks according to Saito and Machida (1999) and Kumar et al. (2007). After 14 days incubation, duplicate plates were inverted after ethereal sprayed 2 ml of ammonium hydroxide for 10–15 minutes. The toxigenic isolates produced a color change of the culture medium producing pink or red coloration. However, no color change was observed in the non-toxic isolates. Further confirmation of the toxigenic potentials shall be carried out quantitatively via HPLC chromatography.

4 Results and Discussion

The fungal count isolated from tiger nut seeds and tiger nut drink samples are as shown in Tables 1 and 2, respectively. The fungal count ranged between 1.1×10^4 cfu/g - 3.0×10^6 cfu/g and 1.0×10^4 cfu/ml - 2.0×10^4 cfu/ml for the tiger nut and tiger nut drink samples, respectively.

Table 3 shows the morphological characteristics, i.e., the macroscopic features and the macroscopic features of fungal isolates in the tiger nut and tiger nut drink samples. A further microscopic structure is shown in Plate 1.

Table 4 shows the result distinguished between the toxigenic and atoxigenic isolates using ammonia vapor assay. Out of 15 investigated isolates, 3 presented as toxigenic potentials based on the presence of color change (See Plate 2).

Table 1 Fungal load of tiger nut seed

Sample	Fungal count
AOJU1	2.0×10^4 cfu/g
AOJU2	1.0×10^6 cfu/g
AOJU3	2.7×10^4 cfu/g
BIYA1	3.5×10^4 cfu/g
BIYA2	2.1×10^6 cfu/g
BIYA3	1.9×10^4 cfu/g
CSAN1	3.0×10^4 cfu/g
CSAN2	1.1×10^4 cfu/g
CSAN3	3.0×10^6 cfu/g

AOJU1-3 three dealers in Oju-Ore, *BIYA1-3* three dealers in Iyana, *CSAN1-3* three dealers in Sango Ota

Table 2 Fungal load of tiger nut drink samples

Sample	Fungal count
AOJU(D)	1.0×10^4 cfu/ml
BIYA(D)	1.7×10^4 cfu/ml
CSAN(D)	2.0×10^4 cfu/ml

AOJU(D) drink made from Oju-Ore sample, *BIYA(D)* drink made from Iyana sample, *CSAN(D)* drink made from Sango-Ota sample

GA02 group A code 02, *GB04* group B code 04, *GC06* group C code 06, *GD08* group D code 08, *GE10* group E code 10, *GF12* group F code 12, *GG14* group G code 14, *GH16* group H code 16, *GI18* group I code 18, *GJ20* group J code 20, *GK22* group K code 22, *GL24* group L code 24, *GM26* group M code 26, *GN28* group N code 28, *GO30* group N code 30

4.1 Fungi Prevalence in Tiger Nut Samples

The result from the preliminary investigation showed that all the samples of tiger nut and tiger nut drink were contaminated with fungal species. The high incidence of fungal contamination may be due to poor storage conditions (time and moisture content levels) of tiger nut seeds which favored the proliferation of the fungi. In addition, most of the vendors lack adequate knowledge on basic hygiene practices as well as the ubiquitous nature of fungal spores particularly in open environments (Ukpabi and Ukenye 2015). Fungal species were also detected in the tiger nut drink, but the occurrence of the fungal count in the tiger nut drink was low, possibly due to the processing seeds during the tiger nut drink preparation.

Previous studies reported on the fungal population of the processed tiger nut drink in Kaduna to be within the range of 1.8×10^2 cfu/g-7.2×10^2 cfu/g. From this study, it is obvious that during processing the fungal load is reduced (Musa and Hamza 2013). *Aspergillus* spp. was predominantly identified in the samples. When compared to the results of Shamsuddeen and Aminu (2016), the fungal count of Musa and Hamza (2013) is similar. In this study, the average fungal count was 6.9×10^5. This may be explained by the difference in geographical zones, storage, and handling of the tiger nuts. The mean fungal count detailed by Friday and Joeguluba (2018) was 7.09×10^4, a higher figure than previously discussed, while the most similar fungal count was recorded by Ntukidem et al. (2020) to be 1.5×10^5. The results are also consistent with that of Tope (2020). It is possible that the transportation of tiger nuts from the north of Nigeria where they are grown to other regions such as the south-west and south-south creates conditions for fungal replication, accounting for the significant differences in fungal count between research done in the north and other regions. The growth of fungi may lead to the production

Table 3 Morphological features of the fungal isolates

S/N	Fungal isolates	Microscopical features	Macroscopical features	Probable organisms
1.	GA02	Dark green, umbonate, powdery white from the bottom glistening	Presence of long and smooth conidiospore, uniseriate with septate phialides	*Aspergillus* spp.
2.	GB04	Grayish and yellow region observed in outskirt part, middle part is raised, entire, velvety, and white from the bottom	Long chain and rough conidiospores, biseriate with septate metulae	*Aspergillus* spp.
3.	GC06	Gray, the middle part is cream and raised, glistening, wavy (undulate) wrinkled from the bottom.	Rough-walled conidiospores, septate flask-shaped phialides	*Penicillium* spp.
4.	GD08	Greyish black type cottony colony, raised, bottom part is whitish in color	Long crossed conidiospores and biseriate with septate metulae	*Aspergillus* spp.
5.	GE10	White and light brown color with touch of yellow region observed, pigment of gold around the colony, wavy, little wrinkled surface, reverse of the colony is brown	Long-chained conidiospores, uniseriate with septate phialides	*Aspergillus* spp.
6.	GF12	Light brown with touch of gold, the middle part is gold and raised, glistening with gold color, little wrinkled on the surface, white velvety from the bottom, reverse is brownish gold	Presence of rough conidiospores, biseriate with septate metulae	*Aspergillus* spp.
7.	GG14	Green with touch of light green, whitish in the middle, raised with bead on the surface	Long cross walled conidiospores, uniseriate with septate phialides	*Aspergillus* spp.
8.	GH16	Brown colony with white edges, the middle part is raised with parallel grooves on colony and flat colony and flat	Presence of long and rough conidiospores, uniseriate with septate phialides	*Aspergillus* spp.
9.	GI18	Lemon green, light green in the middle, raised, velvety white from the bottom	Rough conidiospores, uniseriate with non-septate phialides	*Aspergillus* spp.
10.	GJ20	Light lemon with white edges, the middle part is raised, wrinkled surface, and glistening	Presence of rough conidiospores, uniseriate with non-septate phialides	*Aspergillus* spp.
11.	GK22	Lemon green, middle part is white and lemon in color, raised, less wrinkled, and glistening	Long and rough conidiospores, uniseriate with septate phialides	*Aspergillus* spp.
12.	GL24	Brown colony with white edges, the middle part is raised, wrinkled on colony, cottony, and flat	Smooth-walled conidiospores, phialides borne directly on the vesicles (uniseriate), non-septate hyphae	*Aspergillus* spp.

(continued)

Table 3 (continued)

S/N	Fungal isolates	Microscopical features	Macroscopical features	Probable organisms
13.	GM26	Greenish black, the middle part is slightly raised, less wrinkled, granular colony from the bottom	Non-septate hyphae, unbranched spores that have a brownish pigment	*Mucor* spp.
14.	GN28	Greenish black, light yellow, the middle part is raised, creamy velvety from the bottom part, and glistening	Smooth-walled conidiospores, metulae borne directly on the vesicle (Biseriate), non-septate hyphae	*Aspergillus* spp.
15.	GO30	Green, the middle part is raised, glistening, velvety white from the bottom, white edges	Smooth-walled conidiospores, phialides borne directly on the vesicle (uniseriate), non-septate hyphae	*Aspergillus* spp.

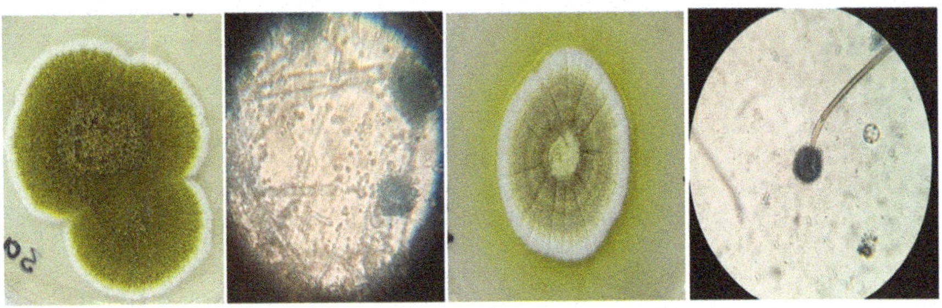

Plate 1 Macroscopic and microscopic characterization of fungal isolates obtained from tiger nut and tiger nut drink (**a**) fungal culture (**b**) spore stain

of mycotoxins and disease condition known as mycotoxicosis when humans consume these contaminated seeds or milk products (Umaru et al. 2014).

4.2 *Aflatoxin Detection with Ammonia Vapor*

From the results when yeast extract sucrose agar (YES) medium is exposed to ammonia vapor 3 out of 15 isolates were positive, while the other 12 were negative which indicates no change in color (see Table 4). Aflatoxin detection using ammonia vapor was also used by Achar et al. (2020) on peanuts and by Fani et al. (2014) on pistachios. It was also utilized on wheat flour by Fathy et al. (2016). It is a rapid aflatoxin detection assay that has been extensively employed on nuts. This implies that some of the fungal isolates could produce toxin and the samples could contain toxins which could constitute health risk to human especially children. The production of pink or red pigmentation following this assay has been reported by Moghadam et al. (2020) which is in agreement with our findings. Negedu et al. (2011) reported

Table 4 Determination of aflatoxin production by isolates using ammonia vapor

S/N	Fungal isolates	Ammonia vapor reaction
1.	GA02	–
2.	GB04	+
3.	GC06	–
4.	GD08	–
5.	GE10	+
6.	GF12	–
7.	GG14	–
8.	GH16	+
9.	GI18	–
10.	GJ20	–
11.	GK22	–
12.	GL24	–
13.	GM26	–
14.	GN28	–
15.	GO30	–

+ change in color, – no change in color

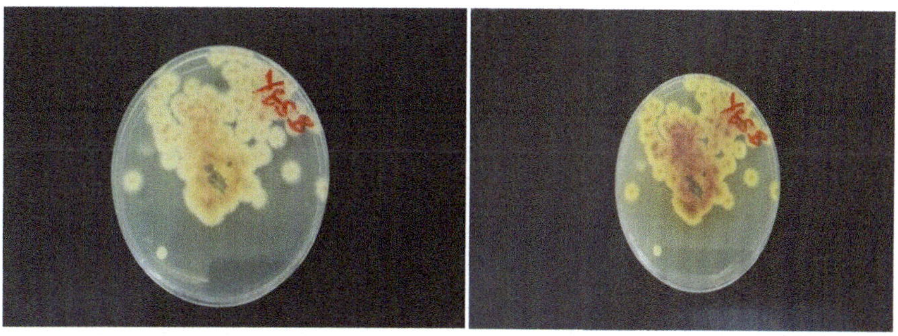

Plate 2 When exposed to ammonium hydroxide vapors, characteristic pink color change by aflatoxigenic *Aspergillus* spp. (right side) changes to characteristic pink color on reverse side of yeast extract sucrose (YES) medium

on agricultural products that can be contaminated with other forms of mycotoxins such as ochratoxins, aflatoxins, sterigmatocystin, fumonisins, zearalenone, deoxynivalenol, patulin, and others. Some of the crops include groundnuts, rice, tiger nut, maize, cassava, cotton seeds, soya beans, sorghum, and fruits. Beyond aflatoxins Sebastià et al. (2010) and Rubert et al. (2012) detected deoxynivalenol, ochratoxin A, and beauvericin in tiger nuts. Tiger nuts and other street snacks assessed by Rubert et al. (2013) contained zearalenones and aflatoxins. It is evident that consuming contaminated tiger nuts or milk products made from the contaminated seeds may pose risk of human exposure to mycotoxin.

5 Conclusion

This study showed that fungal isolates could occur in dry tiger nut seeds and some of the fungal flora may produce aflatoxins. Tiger nuts could become contaminated during the harvesting, storage, transportation, and processing of the seeds to the tiger nut drinks. There should be adequate sensitization on the proper handling and storage of tiger nuts or its products in order to assure consumers on the quality of the seeds and beverages, thereby reducing the risks of food infection or intoxications.

Acknowledgments The authors appreciate Covenant University Research, Innovation, and Discovery (CUCRID) for covering the cost of publication.

References

Achar PN, Quyen P, Adukwu EC, Sharma A et al (2020) 'Investigation of the Antifungal and Anti-Aflatoxigenic Potential of Plant-Based Essential Oils against *Aspergillus flavus* in Peanuts'. J Fungi 6(4): 383.

Adeyeye SAO (2016) Fungal mycotoxin in foods: A review. Cog Food Agri 2: 1-11.

Akomolafe OM, Awe TV (2017) Microbial contamination and polyethylene packaging of some fruits and vegetables retailed at Akure and Ado Ekiti, South Western Nigeria. J Stor Prod Posthar Res 8(6): 65-72.

Asante FA, Ellis WO, Oduro I, Saalia F K (2014) Effect of soaking and cooking methods on extraction of solids and acceptability of tigernut (*Cyperus esculentus* L.) milk. J Agri Stud 2(92): 76-86.

Belewu MA, Abodunrin OA (2006) Preparation of Kunnu from unexploited rich food source: Tiger Nut (*Cyperus esculentus*). Wor J Dairy Food Sci 1: 19- 21.

Chukwuma ER, Obiama N, Christopher OI (2010) The phytochemical composition and some biochemical effect of Nigerian tiger-nut (*Cyperus esculentus*) tuber. Pakis J Nutr 9(7): 709-715.

Daramola JA, Kester CT, Adeleye AO (2018) Microflora associated with smoked shrimps (*Farfantepenaeus notialis*) in some markets in Ota Metropolis, Ogun State, Nigeria. Nig J Fisheries 15: 1490-1496.

D'Mello JPF, Macdonald, AMC (1997) Mycotoxins. Ani Feed Sci Tech 69: 155-166.

Fani SR, Moradi M, Probst C, Zamanizadeh HR et al (2014) 'A critical evaluation of cultural methods for the identification of atoxigenic Aspergillus flavus isolates for aflatoxin mitigation in pistachio orchards of Iran', Eur J Plant Pathol 140(4): 631–642.

Fathy NA, Abdel- Hadi A, Abdul-Raouf U et al (2016) 'Qualitative detection of aflatoxins and aflatoxigenic fungi in wheat flour from different regions of Egypt. IOSR J Envi Sci 10(7): 20–26.

Friday OA, Joeguluba O (2018) 'Microbial Quality Evaluation of Tiger Nut Beverage (Kunun Aya) Processed Sold in University of Maiduguri'. *EC Nutrition 13: 138–142.*

Gambo A, Da'u A (2014) Tiger nut (Cyperus esculentus): composition, products, uses and health benefits – a review. Bayero J Pure Appl Sci 7(1): 56 - 61.

IARC (2015) Mycotoxin control in low- and middle-income countries. IARC Working Group report no. 9, eds. CP Wild, JD Miller, and JD Groopman. Lyon, France: International Agency for Research on Cancer.

Joint Food and Agriculture Organization; World Health Organization Expert Committee on Food Additives (JECFA) (2017) Co-Exposure of Fumonisins with Aflatoxins; Food Safety Digest; World Health Organization: Geneva, Switzerland, pp. 1-4.

Kebede H, Abbas HK, Fisher DK, Bellaloui N (2012) Relationship between aflatoxin contamination and physiological responses of corn plants under drought and heat stress. Toxins 4: 1385–1403.

Klich MA (2002) Identification of common Aspergillus species. (1st ed.). Centraalbureau coor Schimmel-culture, Utrecht: The Netherlands Publishers.

Kumar S, Shekhar M, Ali KA, Sharma P (2007) A rapid technique for detection of toxigenic and non-toxigenic strain of *Aspergillus flavus* from maize grain. Ind phytopatho 1: 31-34.

Maduka N, Ire FS (2018) Tigernut Plant and Useful Application of Tigernut Tubers (*Cyperus esculentus*) - A Review. Curr J Appl Sci Tech 29(3): 1-23.

Margherita F, Salvatore S, Gea OC (2012) Carcinogen Role of Food by Mycotoxins and Knowledge Gap, Carcinogen, Dr. Margarita Pesheva (Ed.), Available from:http://www. Intechopen.com/ books/carcinogen/carcinogen–role–offood–by–mycotoxins–andknowledge gap.

Marin S, Ramos AJ, Cano-Sancho G, Sanchis V (2013) Mycotoxins: Occurrence, toxicology, and exposure assessment. Food Chemi Toxic 60: 218-237.

Moghadam MM, Rezaee S, Mohammadi AH, Zamanizadeh HR et al (2020) The Potential of Aflatoxin Production in the Aspergillus Section Flavi Isolates of Pistachio in Iran. J Nutr Fasting Health 8(4): 254-263

Musa AA, Hamza A (2013) Comparative analysis of locally prepared "kununaya" (Tiger nut milk) consumed by students of Kaduna state university, Kaduna, Nigeria. Sci Worl J 8, 13-18.

Negedu A, Atawodi SE, Ameh JB, Umoh VJ, Tanko HY (2011) Economic and health perspectives of mycotoxins: a review. Conti J Biomed Sci 5(1): 5 -26.

Ntukidem V, Edima-Nyah A, Ndah L, Abraham N (2020) Assessment of Microbiological Safety and Organoleptic Properties of Tiger nut (*Cyperus esculentus*) Beverage Processed Locally and Sold in Uyo Metropolis of Akwa Ibom state, Nigeria. Inter J Food Nutr Saf 11(1): 37-50.

Nwinyi OC, Umane PO (2019) Review on probiotics potential, nutritional composition of bambara Nut (Vigna subterranea (L.)- An underutilized legume. IOP Conference Series: Earth and Environmental Science 331(1): 012057

Ogodo AC, Agwaranze DI, Nwaneri CB, Yakubu MN, Hussaini ZJ (2018) Comparative Study on the Bacteriological Quality of Kunun-Aya Sold in Wukari, Nigeria. Inter J Res Stud Microb Biotech 4(1): 23-29.

Oladele AK, Aina JO (2007) Chemical composition and functional properties of flour produced from two varieties of tigernut (*Cyperus esculentus*). Afr J Biotech 6(21): 2473-2476.

Olopade BK, Oranusi SU, Nwinyi OC, Njobeh PB, Lawal IA (2020) Modification of montmorillonite clay with Cymbopogon citratus for the decontamination of zearalenone in millet. AIMS Agri Food 4(3): 643-657

Peterson SW, Ito Y, Horn BW, Goto T (2001) *Aspergillus bombycris*, a new toxigenic species and genetic variation in its sibling species, *A. nomius*. Mycologia 93: 689-903.

Pitt JI, Hocking AD (2009) Fungi and Food Spoilage. (3rd ed.). New York: Springer Dordrecht, (Chapter1-4).

Rubert J, Soler C, Mañes J (2012) 'Occurrence of fourteen mycotoxins in tiger-nuts'. Food Cont 25(1): 374–379.

Rubert J, Fopohunda SO, Soler C, Ezekiel C (2013) 'A survey of mycotoxins in random street-vended snacks from Lagos, Nigeria, using QuEChERS-HPLC-MS/MS'. Food Cont 32(2): 673–677.

Sa'id AM, Abubakar H, Bello B (2017) Sensory and microbiological analysis of tiger nut (*Cyperus esculentus*) beverage. Pakis J Nutri 16(10): 731-737.

Saito M, Machida S (1999) A Rapid Identification Method for Aflatoxin –Producing Strains of *Aspergillus flavus* and *A. parasiticus by* Ammonia Vapour. Mycoscience 40: 205-208.

Sebastià N, Soler C, Soriano JM, Manes J (2010) 'Occurrence of Aflatoxins in Tigernuts and Their Beverages Commercialized in Spain'. J Agri Food Chem 58(4): 2609–2612.

Shamsuddeen U, Aminu H (2016) Occurrence of Aflatoxin in *Cyperus esculentus* (Tiger Nut) Sold and Consumed Raw in Kaduna. Inter J Sci Res Edu 4: 5189-5195.

Tope VA (2020) 'Physicochemical and microbial evaluation of tiger-nut milk sold in selected eateries in Awka, Anambra State theurapeutic effect of probiotics on pathogenic organisms View project Physicochemical and Microbial Evaluation of Tiger-nut milk View project Physicochemical and microbial evaluation of tiger-nut milk sold in selected eateries in Awka, Anambra State', Offl Pub Dir Res J Agri Food Sci 8(4): 111–115.

Ukpabi J, Ukenye EA (2015) An assessment of wholesome of imported Tigernut *Cyperus esculentus* used as snack food in Umuahia, Nigeria. Mal J Biosci 2(2): 132–138.

Umaru GA, Tukur IS, Akensire UA, Adamu Z et al (2014) Microflora of Kunun-zaki and Zobo Drinks in Relation to Public Health in Jalingo Metropolis, North-Eastern Nigeria. Inter J Food Res 1: 16-21.

Trends in Downstream Processing Approaches, Laccase Mediator Systems and Biotechnological Applications of Laccases

O. D. Akinyemi, E. F. Ahuekwe, O. Oziegbe, and O. C. Nwinyi

1 Introduction

Laccases (EC 1.10.3.2) are important industrial enzymes with great biotechnological potential. They are part of a large community of enzymes called polyphenol oxidases also called multicopper oxidases (MCOs) – enzymes that oxidize their substrates by transferring electrons to a trinuclear copper center from a mononuclear copper core (Bento et al. 2005). Ceruloplasmine (EC 1.16.3.1), nitrite reductase (EC 1.7.2.1), and ascorbate oxidase (EC 1.10.3.3) are some recognized members of the MCO family. Laccase catalyzes the one-electron oxidation of anilines, phenols, and aromatic thiols into their radicals, which results in the reduction to water of oxygen, oxidizing a broad spectrum of compounds, mostly phenolic, and even aromatic and aliphatic amines (Karaki et al. 2016); such free radicals undergo further oxidation or non-enzymatic reactions, including disproportionation, polymerization, or hydration (Shraddha et al. 2011). Laccase was first discovered in *Rhus vernicifera*, the Japanese lacquer tree, sap. In 1985, Bertrand discovered its feature as a metal which contains oxidase (Giardina et al. 2010). Over time, laccases were detected in various basidiomycetic and ascomycetic fungi as well as in different bacteria and plant species. As a result of their ability to catalyze the oxidation by one-electron of aromatic and phenol-containing compounds, laccases are used in numerous applications such as bleaching, delignification, and degradation processes. Hence, laccases are very useful in the actualization of the SDG 11 in the UN's agenda for safe, reliable, and sustainable cities and communities. Recently,

O. D. Akinyemi · E. F. Ahuekwe (✉)· O. Oziegbe · O. C. Nwinyi
Department of Biological Sciences, College of Science and Technology Covenant University, Ota, Nigeria
e-mail: oluwatobi.akinyemipgs@stu.cu.edu.ng; eze.ahuekwe@covenantuniversity.edu.ng; obinna.nwinyi@covenantuniversity.edu.ng

© The Author(s), under exclusive license to Springer Nature Switzerland AG 2022

A. O. Ayeni et al. (eds.), *Bioenergy and Biochemical Processing Technologies*, Green Energy and Technology, https://doi.org/10.1007/978-3-030-96721-5_15

laccases have also found use in biosensor design and obtainment of biofuel cells (Huber et al. 2018).

Laccase catalysis includes the reduction to water of one molecule of oxygen, which allows for wide spectrum oxidation of molecules (including compounds which are aromatic). Polyphenols, aromatic amines, and methoxy-substituted monophenols are among these substances. However, for laccases to oxidize non-phenolic compounds, there is need for the presence of mediators – low molecular weight organic compounds which undergo laccase oxidation. Without these mediators, laccases cannot oxidize non-phenolic substances, such as phenoxyl radicals and other non-phenolic antioxidants that may be polluting the environment (Arregui et al. 2019).

Inadequacies in the purification and recovery of laccase enzyme remain an issue in the economical production and application of laccases. High cost of equipment and long periods of existing purification processes necessitate research and development of more cost-friendly yet effective procedures. This will bring about the effective optimization of laccase production at an economical rate. This review focuses on identifying the currently developed methods to bring about this optimal and cost-friendly yield of laccase. The review also focuses on the existent types of mediators available to broaden laccase substrate range, as well as the immobilization techniques that are useful for laccase recovery. This review hopes to shed light on the current advances in these areas of laccase purification and applications, as well as give insight to the current challenges and future perspectives for research. This will aid more informed research strategies for the development of more effective laccases for diverse industrial applications.

2 Recovery and Purification of Laccases

Development of the application potential of laccase brings about an increase in efficient purification and extraction on a large scale. The downstream process (DSP) is a very important aspect of industrial production of biomolecules, especially the recovery, purification, and concentration of fermentation products. The DSP only comes after the desired product has been biosynthesized and is considered to be expensive and consumes a lot of time; DSP determines up to 80% of the cost of biomolecule production (Madadlou et al. 2017). DSP often requires various concentration and separation process stages and procedures; however, the feasibility of these procedures depends on the correct choice of process conditions and purification techniques (Antecka et al. 2019). Industrial laccase production also follows the same principle; although, in their natural state, the enzymes are relatively active, they do not meet expected industrial application specifications in terms of fermentation concentration and activity; hence the need for effective purification methods (Antecka et al. 2019).

Laccase purification from supernatants of culture of diverse microbes has brought about several publications exist. Chromatography, involving various ion exchange, affinity, and hydrophobic interaction mechanisms, is the technique that is commonly applied (Madadlou et al. 2017; Agrawal et al. 2019). Microfiltration, ultrafiltration, and acetone precipitation are some other techniques that have been applied in DSP of laccase. The purification process applied is often dependent on the intended application as well as commercial production (Junior et al. 2020). These processes help to highly concentrate products with a relative high purity. The operation time, however, for these processes is long, and the operating cost is relatively high for the yielded capacity; the manufacturing process as a whole, as a result of the high cost of the above processes, is not economically effective. Laccase industrial production therefore requires more effective and cheaper methods of DSP. An example of such techniques recently explored has proven to be economical for recovery of biomolecule is ATPE (aqueous two-phase extraction) (Antecka 2019; Junior et al. 2020).

ATPE refers to an extraction using aqueous two-phase systems (ATPS); it involves use of two compounds that are soluble in water. Both compounds are thoroughly mixed, these result in two phases that are immiscible – with water concentration between 70% and 90% (Grilo et al. 2016). The mixture of two polymers that are mutually immiscible, alcohol and salt, a salt and a polymer, two surfactant solutions, or ionic liquid and salt leads to their formation. The specific reactions between solution and phase-forming components result in this parathion process (Glyk et al. 2014; Antecka et al. 2019). ATPSs have advantages in protein purification including significant decrease in denaturation and high rate of mass transfer as a result of low surface tension between phases. In addition, the relatively low cost ensures excellent purity and enzymatic operation (Prinz et al. 2014; Rajagopalu et al. 2016). Diverse studies were conducted on purification of enzymes using ATPS based on polymers and salts; high concentration was observed of *Cerrena unicolor* laccase, when PEG 6000 and PEG 400 ATPE were used, with exceeding yields of 0.9 and 0.85, respectively. Further purification proved that PEG 6000 was more preferable, due to its less complicated salt phase. New series of ionic liquid-based constituents are under investigation. Junior et al. (2020) observed in PEG-rich phase up to 99.9% recovery of laccase in single step ATPS *Pleurotus sajor-caju* laccase purification. It is however very important, for industrial separation processes, to consider the chemical cost. For commercial applications, polymer ATPSs are recommended (Lee et al. 2017; Agrawal et al. 2019).

Another method of downstream processing that has shown great potential is foam fractionation which involves the continuous feeding of steam from dispersed gas into a liquid phase thereby forming a foam phase. Chemical compounds are attracted by the foam, as they show affinity for gas–liquid interface and leave the liquid bulk phase. Foam is collapsed upon extraction, and a liquid phase is newly formed which contains product in concentrated form (Antecka et al. 2019). Ostwald and Mischke (1940) carried out foam fractionation for purification of protein for the first time; it was used to separate yeast fermentation products from fermentation broth. In foam fractionation, although a recently rediscovered DSP technique, there has not been wide investigation of its application towards laccase purification. The source of the enzyme has been demonstrated to be essential for partitioning ability;

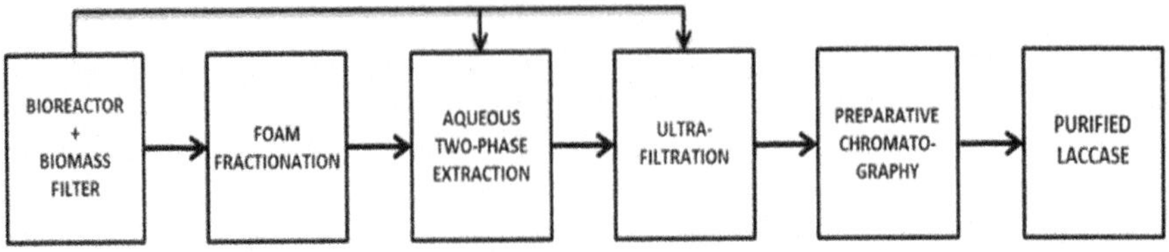

Fig. 1 Separation and concentration sequence of laccase as proposed by Antecka et al. (2019)

experimental results have been specifically obtained for *C. unicolor* and *Pleurotus sapidus* laccases (Blatkiewicz et al. 2017). In the foamate, recovery of activity up to 94% and activity losses of at most 2% were observed by Link et al. (2007) in the research to examine foam fractionation of *Trametes* sp. laccase and influence of pH. A comparison of different laccase purification methods – ultrafiltration, foam fractionation, and ATPE revealed ultrafiltration as the most effective method, having closely similar laccase yield as in ATPE (about 97.5%) but lesser activity loss; the least effective was observed in foam fractionation having about 24.9% yield and even more activity loss (Antecka et al. 2019). The following sequence of concentration of laccase was proposed for the application of all three methods (Fig. 1):

3 Laccase Immobilization

Different materials are useful as supports in immobilization of enzyme, including a variety of biopolymers and polymers of different sizes (nano- and micro-sizes) and unique physical and chemical characteristics, for example, functional group diversity and availability, inertness to the immobilized biomolecule, insolubility in the reaction environment, high ratio of surface-to-volume, and high porosity (Temoçin et al. 2018; Jankowska et al. 2020). Enzymes, upon immobilization, become more resilient and impervious to environmental change, enabling for simple reuse and recycling for a variety of uses (Shraddha et al. 2011; Bayramoglu et al. 2018). The above pros and their need in the improvement of enzyme action have mandated the need for efficient methods of immobilization; hence, various techniques of immobilization and substrates have been studied. Methods of immobilization are self-immobilization, covalent binding, adsorption, entrapment, and synergy of the above (Yang et al. 2017) (Fig. 2).

Recovery of activity is diverse according to method of choice for immobilization, the enzyme, and the parameters of preparation. The material and enzyme's capacity to form strong and long-lasting bonds can be influenced by the polymer used as a support for biomolecule immobilization (Balogh-Weiser et al. 2018). Immobilization of laccases has been accomplished with a variety of materials such as polymeric microspheres (Vera et al. 2020), electrospun materials (Ge et al. 2012; Balogh-Weiser et al. 2018; Jankowska et al. 2020), and mesoporous Al_2O_3 (Kołodziejczak-Radzimska et al. 2020). Due to their effective degradation, at a low

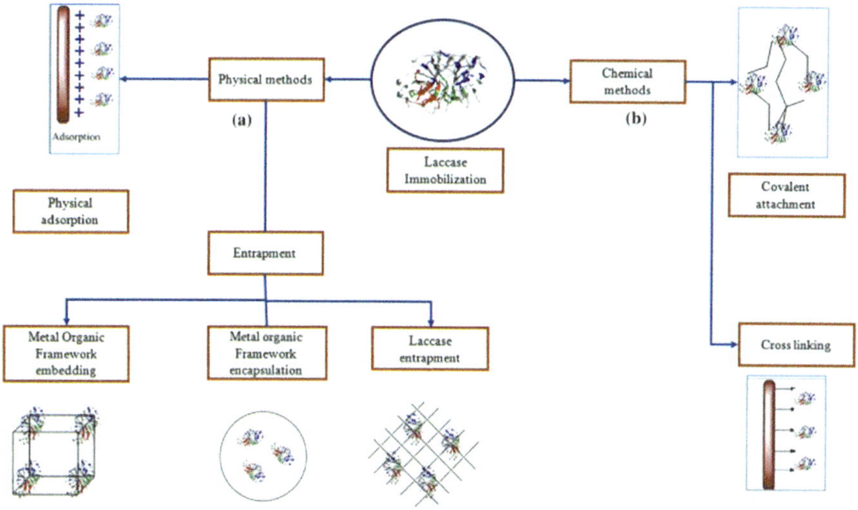

Fig. 2 Laccase immobilization techniques (Datta et al. 2021)

cost, of refractory chemicals, such as dyes, phenolic pollutants, antibiotics, and insecticides (Pezzella et al. 2014; Ammann et al. 2014; Yang et al. 2017; Vera et al. 2018), immobilized laccases have gotten a lot of attention. Laccase immobilization has also become a major role player in the design and application of biosensors for compound detection (Su et al. 2018; Datta et al. 2021).

4 Laccase Mediator Systems

Mediators serve as carriers of electrons to the compounds targeted from laccase; this extends laccase substrate range. Laccase may thus oxidize non-phenolic materials when a sufficient redox mediator is present, overcoming its restriction to phenolic components alone (Yang et al. 2012). A broad spectrum of pollution-causing substances, including those from personal care items, herbicides, and dyes, has been degraded using both natural and manufactured mediator systems. Mediators function by providing a step referred to as indirect oxidation which yields a radical in its oxidized form capable of oxidizing a broad spectrum of non-phenolic substrates and big molecules. The first artificial mediator used as a laccase-mediator system was 2, 20–azino–bis 3-ethylbenzothiazoline-6-sulfonic acid (ABTS) (Bourbonnais and Paice 1990); other synthetic materials include violuric acid (VLA), the oxime, and 1-hydroxybenzotriazole (HBT) – these have also shown great capacity in the decomposition of recalcitrant aromatic compounds (Blanquez et al. 2019).

Laccase mediators that are synthetic like HBT and VLA, however, have a degree of toxicity and high cost that accompanies their use. Hence, natural laccase mediators are regarded as more suitable, and researches have and are being conducted to

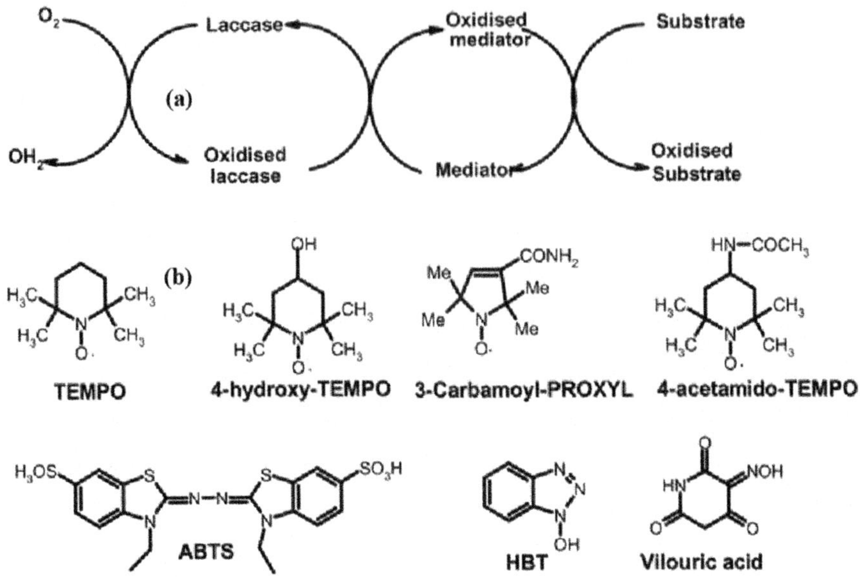

Fig. 3 (**a**) Mechanism of substrate oxidation by laccase mediator systems; (**b**) some synthetic laccase mediators (Wells et al. 2006)

find more natural mediators (vanillin, syringaldehyde, and acetosyringone); lignin-related phenolic compound has been found to show their capacity to behave as natural laccase mediators as well as the effectiveness with which they do so (Camarero et al. 2005). A point of concern in the research and application of natural mediators is their limitation to commercial application of laccase; extra research is therefore necessary to assess further the potential commercial applications of natural mediators to laccases. Oxidation by laccase of the substrate with a mediator may not always go in a different way. For example, Malachite green degradation has been shown to yield similar products in ABTS presence and/or absence for some enzymes and different for others (Fig. 3).

There are different catalytic mechanisms displayed by different types of mediators. Substrate oxidation mediated by ABTS works using transfer of electron. First is oxidation ABTS to $ABTS^{+}$ (radical cation) and to $ABTS^{2+}$ (dication) with 472 mV and 885 mV redox potentials, respectively. Unlike ABTS, when laccase is oxidized and then deprotonated, HBT and violuric acid (N-OH type mediators) produce the N-oxy radical, which then abstracts the benzylic hydrogen atom from the substrate. Phenolic mediators, in a similar manner follow abstraction mechanism by radical hydrogen, but in its case have a radical phenoxy as intermediate (Wong 2009; Hu et al. 2009) (Table 1).

Table 1 Comparison of kinetic parameters of laccase mediators (Li et al. 1999)

	$^ak_{cat}$ (min^{-1})			^aK_m (µM)				
Species	Violuric acid	1-HBT	Syringaldazine	ABTS	Violuric acid	1-HBT	Syringaldazine	ABTS
Botrytis cinerea	40 ± 2	10 ± 1	52 ± 3	23 ± 2	11,000 ± 1000	12,000 ± 4000	0.8 ± 0.2	28 ± 8
Pycnoporus cinnabarinus	370 ± 20	22 ± 2	180 ± 10	920 ± 50	9000 ± 1000	29,000 ± 7000	4 ± 1	18 ± 5
Myceliophthora thermophila	27 ± 1	0.12 ± 0.02	1100 ± 250	440 ± 300	18,000 ± 2000	10,000 ± 6000	2.9 ± 0.4	96 ± 5
Trametes villosa I	260 ± 20	84 ± 6	3000 ± 100	2700 ± 100	5000 ± 1000	15,000 ± 3000	3.9 ± 0.5	58 ± 8

[a] ***The Km/kcat values are the result of fitting the Michaelis-Menten equation with nonlinear regression on 6 to 10 experimental data that spanned the whole rate-[substrate] profile (from the initial, linear phase to the saturated phase), where K_m is Michaelis constant and k_{cat} is catalyst rate constant***

5 Applications of Laccase

Many biotechnological processes involve laccase due to their being able to oxidize a wide spectrum of compounds (non-phenolic and phenolic). Laccases are especially useful in major procedures including clean-up of industrial effect, water purification, etc. (Imran et al. 2012). Laccase application in different industries include the following:

5.1 *Laccase Application in Food Industry*

Laccase enzyme is used in the food industry in the manufacturing of juices and wine and for baking (Minussi et al. 2002). High concentration of phenolic compounds in fruit juices, wine, and beer and the natural polymerization and co-oxidation reactions that occur with them over long periods of time give rise to undesirable colors and aroma – this is referred to as enzymatic darkening. Laccase treatment aids the efficient reduction of this phenolic content. Compared to other conventional treatment methods for this purpose (e.g., adding ascorbic acid), laccases were observed to be more efficient, in fruit juices, for flavor stability and color (Ribeiro et al. 2010) (Fig. 4).

In baking, it has been found to improve gluten strength, dough machinability (improved strength and stability as well as a reduction in stickiness), freshness, softness, and dough crumb structure and products of baking (Minussi et al. 2002). This improvement has been observed even when low quality dough is used. Laccase plays an important function in the production of baking goods that are free of gluten using cereal flours such as oats and starches (rice, potato, and maize). However, laccase is not approved as a food additive and therefore has to be filtered out of product

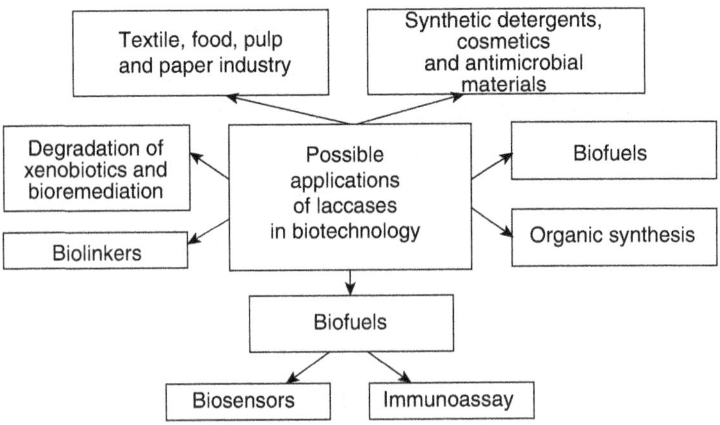

Fig. 4 Applications of laccase (Imran et al. 2012)

by using membrane filtration, for example (Brijwani et al. 2010), or better still the use of immobilized laccase in order to overcome legal barriers (Imran et al. 2012).

5.2 *Laccase Application in Textile Industry*

In the textile industry, vast quantities of water and inorganic and organic chemicals are used for wet processing; some of these chemicals include synthetic dyes that are commonly used in diverse sectors, such as the textile industry (Imran et al. 2012). The textile and paper industries manufacture vast quantities of synthetic dyes that maintain their durability upon light exposure and microbial degradation and in the presence of diverse chemicals. There are currently more than 10,000 synthetic dyes used worldwide (Blanquez et al. 2019). The high resistance of these dyes causes them to remain in sewage plants or on soil and eventually washed up into rivers causing pollution. Specific treatment procedures are existent, which can degrade recalcitrant dyes – oxic and anoxic processes are mixed in these processes. However, anoxic processes bring about generation of amines from azo dyes which are carcinogenic posing health risks; hence, in textile industry, laccase application is highly important (Asgher et al. 2013; Blanquez et al. 2019). Fungal and bacterial laccases have proven to successfully decolorize and degrade dyes and inks that are flexographic (Fillat et al. 2012).

5.3 *Application of Laccase in Pharmaceutical Industry*

Laccases are also used to make pharmaceuticals like actinocin, which is derived from 4-methyl-3-hydroxyanthranilic acid. As an anticancer agent, actinocin works by preventing tumor cells from transcribing deoxyribonucleic acid (DNA). Another example is vinblastine – also an anticancer medication, which is effective in leukemia therapy. *Catharanthus roseus* plant produces vinblastine naturally in small quantity. Precursors of vinblastine – katarantine and vindoline – are produced in larger amounts. Laccase has been observed to have 40% conversion of these into vinblastine (Imran et al. 2012). Laccases are also important in biodegradation of PPCP (personal care product and pharmaceutical) organic contaminants with rising ubiquity and antibiotics which may not have been metabolized and cause environmental pollution (Yang et al. 2017). They have been used as ingredients in the production of PPCPs which possess antioxidant, anticancer, detoxifying, antimicrobial, or other things (Upadhyay et al. 2016; Senthivelan et al. 2016).

5.4 *Application of Laccase in Biodegradation of PAH and Other Contaminants*

Rapid industrialization in the world today, including agricultural processes with the extensive use of pesticides and herbicides, has brought about increase in contamination of the environment – soil, water, and air pollution. Some of these carcinogenic and mutagenic pollutants include benzene, polycyclic aromatic hydrocarbons (PAH), polychlorinated biphenyls (PCB), toluene, xylene (BTEX), ethyl benzene, trinitrotoluene (TNT), ethane (DDT), 1,1,1-trichloro2,2-bis (4-chlorophenyl), and pentachlorophenol (PCP). The ability of laccase to play essential roles in the biodegradation of these recalcitrants and increasingly discovered information about that through research have created rising interest in the use of laccase in bioremediation.

Diazinon, classified as a moderately dangerous chemical (class 2) by the World Health Organization (WHO), is an organophosphorus pesticide that was observed to be degraded by *A. bisporus* and *T. versicolor* laccases co-immobilized on poly-(glycidylmethacrylate) microsphere (Vera et al. 2020). Zeng et al. (2017) also researched the *T. versicolor* laccase breakdown of the isoproturon, an herbicide, and also that of its laccase-mediator systems. Laccase from *Aspergillus niger* was discovered to affect significantly oxidation indices of oxygenated PAH (OPAH) and polycyclic aromatic hydrocarbon (PAH) material, especially anthracene-9,10-dione, 9-fluorenone, and phenanthrene, in wasted oil for cooking after frying youtiao, nuggets, and pie made from pumpkin (Teng et al. 2019).

5.5 *Application of Laccase in Forest Products Industry*

Laccase has shown potential to help extract from materials made from wood and water phenolic residues, pitch, and coloring; the technology of laccase is applicable in almost all stages of the paper product supply chain, from pulping to secondary fiber recovery and effluent treatment. The majority of laccase and laccase-mediator systems applications in forestry has to do with the paper and pulp industry, where special focus is majorly on the study of laccase and/or laccase-mediated systems for treatment mill water and biobleaching (Widsten and Kandelbauer 2008). Laccase-mediated systems are used with oxygen-based chemical oxidants and chlorine for lignin degradation and separation needed for paper preparation on industrial scale. This helps solve issues like toxicity, cost, and recycling as it results in replacement of ClO_2 partially in pulp mills (Shraddha et al. 2011).

5.6 Application of Laccase in Petroleum Industry

Laccase in the petroleum industry plays major roles in bioremediation, especially in cases of oil spills. After 100 days of remediation, research of bacteria, and laccase immobilized in the intertidal zones, bioremediation of major oil spills found about 66.5 percent effectiveness of bacteria-laccase consortium immobilized for heavy oil. (Dai et al. 2020). It was also observed in that research that n-alkanes having long-chain C26-C35 and PAHs having above 3 ring were efficiently broken down. Laccase was also found to be expressed most abundantly – alongside peroxidases and catalase – in strains of fungi which inhabit the rhizosphere of areas where there have been oil spills which showed tolerance to a concentration of crude oil of about 20% (Asemoloye et al. 2018). The strains were identified by Asemoloye et al. (2018) as *Trichoderma harzianum* asemoJ, *Talaromyces purpurogenus* asemoF, *A. flavus* asemoM, and *A. niger* asemoA. In the research, while 87 U/ml and 145 U/ml of peroxidases and catalase were produced respectively, the volume of laccase produced was up to 180 U/ml.

5.7 Applications of Laccase in Biosensors and Enzymatic Biofuels

Biosensors, embedded medical devices, and other gadgets employ enzymatic biofuel cells (EBFCs) as a portable source of environmentally benign energy. Due to its comparatively high redox potential in contrast to other commercial fuel cells, laccase has been frequently employed as an EBFC cathode (Ghosh et al. 2019). On functionalized screen-printed carbon electrodes (SPCEs), laccase from *Coriolus hirsuta* was immobilized by Othman and Wollenberger (2020) using different methods of enzyme modification and immobilization. Carboxyl functionalized multiwalled carbon nanotubes (COOH–MWCNT) immobilized laccase had the highest electrochemical response. The biosensors produced were seen to be relatively stable showing no activity loss even beyond 20 days of storage. Christwardana (2017) combined physico-entrapment and crosslinking for the immobilization on carboxylated carbon nanotube of low activity laccase in order to increase its biofuel performance using polyethylenimine and glutaraldehyde. Glutaraldehyde was observed to have higher catalytic activity.

6 Current Challenges and Future Perspectives

Edible mushroom production is a major area in laccase production and application in the food industry. The cultivation of these edible fungal species provides a significant opportunity for laccase production, as evidenced by various studies (Chanakya et al. 2015). However, there are few studies on the increase in laccase concentration in these fungi during industrial production without affecting their nutritional value. This is a major opportunity in terms of optimization of producing fungus while also optimizing laccase and other ligninolytic enzyme production. As a result, it is required to refine the enzyme recovery process from residual compost, as well as to investigate various main recovery approaches and purification steps that allow for high purities. Because the crude extract is complicated and poorly defined, this is a significant difficulty. In this regard, including ultrasonication into the main laccase recovery utilizing ATPS may be a viable alternative for increasing yield. Postemsky et al. (2017) have been observed that, in addition to enhancing activity, it can enhance yield, particularly of phenolic chemicals; therefore, treating the residual compost crude extract and evaluating its effect on activity of laccase could be an excellent strategy (Postemsky et al. 2017).

Laccase application appears to be limited in comparison to its potential. Reduced manufacturing costs and the development of technology to effectively regulate the reactions on specific substrates, including polyphenols, to be handled by laccase should be the research focus. Laccase manufacturing costs and their vast substrate specificity are two key obstacles to their industrial application (Zerva et al. 2019). While laccases' broad substrate range might be beneficial for biodegradation, difficulty may arise as a result, for their application commercially in biocatalysis due to the production of by-products as a result of free radicals. Furthermore, laccases' wide spectrum of reaction with substrates is a disadvantage in the realm of biosensor production. Biosensors, particularly those applied in biomedical procedures, necessitate a significant level of specificity with regard to targeted substance, which most commercially available laccases cannot currently achieve (Mayolo-Deloisa et al. 2015; Zerva et al. 2019). Given the high cost of manufacturing, the necessity for the discovery or creation of new enzymes with desired properties appears to be critical. Because the information on laccase manufacturing costs is limited, additional study into the development of manufacturing techniques is required.

Laccase production under submerged cultivation in the lab has been frequently documented in current years. There is a limited representation, however, of laccases commercially, with low purity degrees and not well-defined extracts, complicating the repeatability of many of the methods used in production and purification (Mayolo-Deloisa et al. 2020). As a result, increased interest in the manufacture and concentration of laccases commercially is required for generation of novel products.

Laccase's potential for pretreatment of lignocellulosic residues in order to incorporate them into biorefinery processes and produce biofuels has been widely documented in the literature (Agrawal et al. 2019). However, in order to get the maximum degradation yields, it is important to consider further optimization of circumstances, as the fungus' efficacy is often larger than that of the enzyme. This might be indicative of additional enzymes capable of modifying lignin being present; therefore, the potential of employing enzyme combinations, increasing laccase affinity by chemical modifications, or altering the conditions of the processes being performed should be investigated.

PEGylation – attaching PEG to laccase (a single molecule) – could also be used to increase stability. It is a well-known method for modifying protein medicines, and its effectiveness has been well established. Despite the fact that there are few data on laccase PEGylation, they show that it improves its catalyzing ability and activity (Mayolo-Deloisa et al. 2015; Su et al. 2018). If the procedure warrants it, specifically if enzyme is employed to identify specific chemicals using biosensors, this can be a viable option. One of laccase's uses in this regard is the identification of different chemicals in specific meals. Laccase is also employed in the production of chemicals, as well as detection of these compounds, and it was recently discovered that it can remove morphine from environments that are aqueous in nature (Huber et al. 2018). This demonstrates the enzyme's enormous potential industrially and in recovery of polluted regions, highlighting the significance and importance of the data given here.

7 Conclusion

Laccase purification and application remains very essential to the industrial advancement in the world today; hence laccase research is continuously on the increase to develop economical and effective purification procedures, efficient laccase mediator systems, and durable laccase immobilization techniques. With improved downstream processing and continued increase in research, there is an increased possibility of enhanced yield, stability, and recovery of laccase. This will further widen the application reach of laccases in various industries as it continues to play a major part in the provision of reliable and sustainable communities worldwide.

Acknowledgments The authors appreciate Covenant University Research, Innovation, and Discovery (CUCRID) for conference and publication support.

References

Agrawal K, Bhardwaj N, Kumar B, et al (2019) Process optimization, purification and characterization of alkaline stable white laccase from *Myrothecium verrucaria* ITCC-8447 and its application in delignification of agro-residues. Int. J. Biol. Macromol 125:1042–1055.

Ammann E, Gasser C, Hommes G, Corvini P (2014) Immobilization of defined laccase combinations for enhanced oxidation of phenolic contaminants. Appl. Microbiol. Biotechnol 98: 1397–1406.

Antecka A, Blatkiewicz M, Boruta T et al (2019) Comparison of downstream processing methods in purification of highly active laccase. Bioprocess Biosyst Eng 42: 1635–1645.

Arregui L, Ayala M, Gómez-Gil X, et al (2019) Laccases: structure, function, and potential application in water bioremediation. Microb. Cell Factories 18:200.

Asgher M, Ahmad Z, Iqbal H (2013) Alkali and enzymatic delignification of sugarcane bagasse to expose cellulose polymers for saccharification and bio-ethanol production. Ind Crops Prod 44: 488-495.

Asemoloye M, Ahmad R, Jonathan S (2018) Transcriptomic responses of catalase, peroxidase and laccase encoding genes and enzymatic activities of oil spill inhabiting rhizospheric fungal strains. Environ. Pollut. 235: 55–64. https://doi.org/10.1016/j.envpol.2017.12.042.

Balogh-Weiser D, Nemeth C, Ender F et al (2018) Electrospun Nanofibers for Entrapment of Biomolecules. ChemCatChem 10(16): 3490–3499.

Bayramoglu G, Karagoz B, Arica Y (2018) Cyclic-carbonate functionalized polymer brushes on polymeric microspheres: Immobilized laccase for degradation of endocrine disturbing compounds. J Ind Eng Chem 60: 407–417.

Bento I, Martins L, Lopes G, Carrondo M, Lindley P (2005) Dioxygen reduction by multi-copper oxidases; a structural perspective. Dalton Trans 21: 3507–3513.

Blánquez A, Rodríguez J, Brissos V et al (2019) Decolourization and detoxification of textile dyes using a versatile *Streptomyces* laccase-natural mediator system. Saudi J. Biol. Sci. 26: 913–920.

Blatkiewicz M, Antecka A, Górak A et al (2017) Laccase concentration by foam fractionation of *Cerrena unicolor* and *Pleurotus sapidus* culture supernatants. Chem Eng Process 38(3): 455–464.

Bourbonnais R, Paice M (1990) Oxidation of Non-phenolic Substrates – an expanded role for laccase in lignin biodegradation. FEBS Lett. 267: 99–102.

Brijwani K, Huss R, Vadlani P (2010) Fungal Laccases: Production, Function, and Applications in Food Processing. Enzyme Res. 2010(5):149748.

Camarero S, Ibarra D, Martínez M et al (2005) Lignin-derived compounds as efficient laccase mediators for decolourization of different types of recalcitrant dyes. Appl. Environ. Microbiol. 71: 1775–1784.

Chanakya H, Malayil S, Vijayalakshmi C (2015) Cultivation of *Pleurotus spp.* on a combination of anaerobically digested plant material and various agro-residues. Energy Sustain Dev 27: 84–92

Christwardana M (2017) Combination of physico-chemical entrapment and crosslinking of low activity laccase-based biocathode on carboxylated carbon nanotube for increasing biofuel cell performance. Enzyme Microb. Technol. 106: 1-10.

Dai X, Liv J, Yana G et al (2020) Bioremediation of intertidal zones polluted by heavy oil spilling using immobilized laccase-bacteria consortium. Bioresour. Technol. 309: 123305.

Datta S, Veena R, Samuel M et al (2021) Immobilization of laccases and applications for the detection and remediation of pollutants: a review. Environ. Chem. Lett. 19: 521–538.

Fillat A, Gallarda O, Vidal T et al (2012) Enzymatic grafting of natural phenols to flax fibres: Development of antimicrobial properties. Carbohydr. Polym. 87(1):146–152.

Ge L, Zhao Y, Mo T et al (2012) Immobilization of glucose oxidase in electrospun nanofibrous membranes for food preservation. Food Control 26: 188–193.

Ghosh B, Saha R, Bhattacharya D et al (2019) Laccase and its source of sustainability in an enzymatic biofuel cell. Bioresource Technology Reports 6: 268-278.

Giardina P, Faraco V, Pezzella C et al (2010) Laccases: a never-ending story. Cell. Mol. Life Sci. 67: 369–385.

Glyk A, Scheper T, Beutel S (2014) Influence of Different Phase-Forming Parameters on the Phase Diagram of Several PEG-Salt Aqueous Two-Phase Systems. J. Chem. Eng. Data 59: 850–859.

Grilo A, Aires-Barros M, Azevedo A et al (2016) Partitioning in aqueous two-phase systems: fundamentals, applications and trends. Sep. Purif. Rev. https://doi/abs/10.1080/15422119.2014.983128.

Huber D, Bleymaier K, Pellis A et al (2018) Laccase catalyzed elimination of morphine from aqueous systems. N Biotechnol 42: 19–25.

Hu R, Chao P, Zhang Q et al (2009) Laccase-mediator system in the decolorization of different types of recalcitrant dyes. J. Ind. Microbiol. Biotechnol. 36: 45–51.

Imran M, Asad M, Hadri S et al (2012) Production and industrial applications of laccase enzyme. J. Mol. Cell Biol. 10(1): 1-11.

Jankowska K, Zdartaa J, Grzywaczyka A et al (2020) Electrospun poly (methyl methacrylate)/polyaniline fibres as a support for laccase immobilization and use in dye decolourisation. Environ. Res. 184: 109332.

Junior A, Vieira A, Cruz I et al (2020) Sequential degradation of raw vinasse by a laccase enzyme producing fungus *Pleurotus sajor-caju* and its ATPS purification. Biotechnol. Rep. 25; e00411.
Karaki N, Aljawish A, Humeau C et al (2016) Enzymatic modification of polysaccharides: mechanisms, properties, and potential applications: a review. Enzyme Microb. Technol. 90: 1–18.
Kołodziejczak-Radzimska A, Budna A, Ciesielczyk F, Moszynski D, Jesionowski T (2020) Laccase from Trametes versicolor supported onto mesoporous Al2O3: stability tests and evaluations of catalytic activity. Process Biochem 95:71–80. https://doi.org/10.1016/j.procbio.2020.05.008
Lee S, Khoiroh I, Ooi C et al (2017) Recent Advances in Protein Extraction Using Ionic Liquid-based Aqueous Two-phase Systems. Sep. Purif. Rev. 46: 291–304.
Li K, Xu F, Eriksson K (1999) Comparison of fungal laccases and redox mediators in oxidation of a nonphenolic lignin model compound. Appl. Environ. Microbiol 65(6): 2654–2660.
Link D, Zorn H, Gerken B et al (2007) Laccase isolation by foam fractionation – new prospects of an old process. Enzyme Microb. Technol. 40: 273–277.
Mayolo-Deloisa K, González-González M, Simental-Martínez J et al (2015) Aldehyde PEGylation of laccase from *Trametes versicolor* in route to increase its stability: effect on enzymatic activity. J. Mol. Recognit. 28: 173–179.
Mayolo-Deloisa, K, González-González M, Rito-Palomares M (2020) Laccases in Food Industry: Bioprocessing, Potential Industrial and Biotechnological Applications. Front. Bioeng. Biotechnol. 8:222.
Madadlou A, O'Sullivan S, Sheehan D (2017) Fast protein liquid chromatography. Methods Mol Biol 1485: 365–373.
Minussi R, Pastore G, Duran N (2002) Potential applications of laccase in the food industry, Trends Food Sci. Technol., 13: 205–216.
Ostwald W, Mischke W (1940) Untersuchungen über Zerschäumung mit besonderer Rücksicht auf Fragen der. angewandten Chemie I. Kolloid Zeitschrift 90:17–25.
Othman A, Wollenberger U (2020) Amperometric biosensor based on coupling aminated laccase to functionalized carbon nanotubes for phenolics detection. Int. J. Biol. Macromol. 153: 855-864.
Pezzella C, Russo M, Marzocchella A et al (2014) Immobilization of a *Pleurotus ostreatus* laccase mixture on perlite and its application to dye decolourisation. Biomed Res. Int. 1–11.
Postemsky P, Bidegain M, González-Matute R et al (2017) Pilot-scale bioconversion of rice and sunflower agro-residues into medicinal mushrooms and laccase enzymes through solid-state fermentation with *Ganoderma lucidum*. Bioresour. Technol. 231: 85–93.
Prinz A, Hönig J, Schüttmann I et al (2014) Separation and purification of laccases from two different fungi using aqueous two-phase extraction. Process Biochem 49: 335–346.
Rajagopalu D, Show P, Tan Y et al (2016) Recovery of laccase from processed *Hericium erinaceus* (Bull. : Fr) Pers. Fruiting bodies in aqueous two-phase system. J. Biosci. Bioeng. 127: 600-609.
Ribeiro D, Henrique S, Oliveira L et al (2010) Enzymes in juice processing: a review. International J. Food Sci. Technol. 45: 635–641.
Senthivelan T, Kanagaraj J, Panda R (2016) Recent trends in fungal laccase for various industrial applications: an eco-friendly approach- a review. Biotechnol. Bioprocess Eng. 21: 19–38.
Shraddha, Ravi S, Simran S et al (2011) Laccase: Microbial Sources, Production, Purification and Potential Biotechnological Applications. Enzyme Res. 217861.
Su J, Noro J, Fu J et al (2018) Exploring PEGylated and immobilized laccases for catechol polymerization. AMB Express 8:134. doi: https://doi.org/10.1186/s13568-018-0665-5
Temoçin Z, İnal M, Gökgöz M et al (2018) Immobilization of horseradish peroxidase on electrospun poly (vinyl alcohol)–polyacrylamide blend nano-fiber membrane and its use in the conversion of phenol. Polym. Bull. 75: 1843–1865.
Teng C, Wu S, Gong G (2019) Bio-removal of phenanthrene, 9-fluorenone and anthracene-9,10-dione by laccase from *Aspergillus niger* in waste cooking oils. Food Control 105: 219-225.
Upadhyay P, Shrivastava R, Agrawal P (2016) Bioprospecting and biotechnological applications of fungal laccase. 3 Biotech 6: 15.
Vera M, Nyanhongo G, Guebitz G et al (2018) Immobilization of *Myceliophthora thermophila* laccase on poly (glycidyl methacrylate) microspheres enhances the degradation of azinphos-methyl. J. Appl. Polym. Sci. 136: 1–10.

Vera, M, Nyanhongo, G, Guebitz, G et al (2020) Polymeric microspheres as support to co-immobilized *Agaricus bisporus* and *Trametes versicolor* laccases and their application in diazinon degradation. Arab. J. Chem. 13: 4218–4227.

Wells A, Terria M, Eve T et al (2006) Green oxidations with laccase mediator systems. Biochem. Soc. Trans. 34(2): 304–8.

Widsten P, Kandelbauer A (2008) Laccase Applications in the Forest Products Industry A Review. Enzyme Microb. Technol. 42: 293–307

Wong D (2009) Structure and action mechanism of ligninolytic enzymes. Appl. Biochem. Biotechnol. 157: 174–209.

Yang Y, Fan F, Zhuo R, Ma F, Gong Y, Wan X (2012) Expression of the laccase gene from a white rot fungus in Pichia pastoris can enhance the resistance of this yeast to H_2O_2-mediated oxidative stress by stimulating the glutathione-based anti-oxidative system. App. Environ. Micro. 78: 5845–5854.

Yang J, Lin Y, Yang X et al (2017) Degradation of tetracycline by immobilized laccase and the proposed transformation pathway. J. Hazard. Mater. 322: 525–531.

Zeng S, Qin X, Xia L et al (2017) Degradation of the herbicide isoproturon by laccase-mediator systems. Biochem. Eng. J. 119: 92–100.

Zerva A, Simic S, Topakas E et al (2019) Applications of Microbial Laccases: Patent Review of the Past Decade (2009–2019). Catalysts 9: 1023.

Atherosclerosis and Scientific Interventions: A Review

E. E. Alagbe, T. E. Amoo, I. O. Oboh, A. O. Ayeni, A. A. Ayoola, and O. Agboola

1 Introduction

A special report by Bonow et al. (2002) indicated that as of 2002, the World Health Organization (WHO) had recognized cardiovascular diseases (CVDs) as an epidemic and concluded that efforts to combat these diseases have had its fair share of setbacks. One of the reasons for this failure was the lack of adequate knowledge of the disease and the lack of implementation of scientific solutions. As of 2010, statistics show that the probability of dying from a CVD is about 30%, indicating that fatalities from other diseases such as cancers are relatively insignificant (Nichols et al. 2013). Unfortunately, the younger population is not an exception to the risk of this disease as studies by Wang et al. (2017) have indicated that between 2010 and 2014, CVDs amongst individuals aged <45 years increased significantly. In light of the current SARS-CoV-2 pandemic, CVDs are known to be the second-highest comorbidity that has led to deaths.

CVDs are a group of disorders associated with regions of the heart and blood vessels. Currently, CVDs are the leading cause of death worldwide, with a death toll estimate of 17.9 million as of 2016, of which over three-quarters of the deaths are recorded among middle- and low-income countries (Thumala et al. 2017). Studies have also indicated that the presence of underlying diseases such as diabetes, hypertension, hyperlipidaemia or any established diseases increases the risks of CVDs, and behavioural risk factors such as smoking, obesity, poor nutrition, inactivity and excessive use of alcohol also increase CVD risk (Thumala et al. 2017). CVDs are

E. E. Alagbe (✉) · T. E. Amoo · A. O. Ayeni · A. A. Ayoola · O. Agboola
Department of Chemical Engineering, Covenant University, Ota, Nigeria
e-mail: edith.alagbe@covenantuniversity.edu.ng

I. O. Oboh
Department of Chemical Engineering, University of Uyo, Uyo, Nigeria

© The Author(s), under exclusive license to Springer Nature Switzerland AG 2022

A. O. Ayeni et al. (eds.), *Bioenergy and Biochemical Processing Technologies*, Green Energy and Technology, https://doi.org/10.1007/978-3-030-96721-5_16

mainly age-related diseases, and statistics show that males over 80 years old have a five-time risk of CVDs compared to their younger counterparts, specifically those within the age bracket of 40–59 years (Mozaffarian et al. 2016), due to the increased content of reactive oxygen species in the blood, linked to persistent inflammation, hence disease progression (North and Sinclair 2012; Curtis et al. 2018; Chinedu et al. 2018).

Atherosclerosis, a type of *arteriosclerosis* that is caused by the clogging or hardening of the arteries, induced by plaques, is a major cause of coronary and cerebrovascular diseases, which constitute major emergencies worldwide (Rosamond 2010). Over time, as research on the disease broadened, understanding of atherosclerosis improved. Previously, the disease was known to be caused by the accumulation of lipids in the innermost layer of the arterial wall (Libby et al. 2013; Motayo et al. 2021). Now, it is believed to be due to persistent inflammatory conditions caused by lipid buildup, which is controlled by immune response (Joris and Majno 1978; Weber and Noels 2011). When this occurs, there are two possible outcomes: first, haemorrhage in the plaque and, second, blood clot formation (thrombus) on the plaque's surface. This can occur in large and medium-sized arteries and could lead to either a heart attack or stroke (Pralhad and Schultz 2004).

Rocha and Libby (2009), in their study, project a substantial increase in the number of deaths caused by atherosclerosis in the years to come. There have been concerns that the number of aged people from ages >60 years would increase from 11% in 2000 to 22% of the world population by 2050 (WHO 2015), hence a need for future concern. There have also been reports on the increase in co-morbidities that drive CVDs in people (Grundy et al. 1999), which are set to experience an increment of 55% between 2013 and 2035 (IDF 2015).

Current trends show a progressive gradual fall in morbidity rates as a result of CVDs, which is mainly due to scientific interventions, like plant-based interventions (Elizabeth et al. 2018), and a more critical study of the development of the disease. Recent studies focus on the disease progression such as simulations, in-vitro and in-vivo studies have served as a beacon in these dark times.

2 Pathophysiology of Atherosclerosis

The pathophysiology of atherosclerosis was studied extensively by Lusis (2000), Weber and Noels (2011) and Douglas and Channon (2014), and briefly, it is explained that as damage is done to the arterial wall triggered by high wall shear stress (WSS) REF, an immune response is triggered in which immune cells are sent to the site where the damage has occurred. This process results in the formation of debris in the tunica intima (atheroma). The accumulation of this atheroma leads to swelling and reduces blood flow and pressure downstream of the site, resulting in stenosis.

At the time the lesion is initiated at the endothelial site, there are no obstructions to blood flow; hence, the victim is usually asymptomatic. Usually, the lesions begin at childhood and advances in size and numbers over time (Turan et al. 2017). This plaque's growth continues throughout the life span of the patient as a dynamic process with the plaque sizes continually increasing, cumulating into stenosis of the artery. This leads to a considerable rise in blood velocity and movement. Ironically, a reduction in total blood flow is observed (Turan et al. 2017).

2.1 *Newtonian and Non-Newtonian Blood Rheology*

For simplicity, blood rheology may be assumed Newtonian such that the viscosity of the blood remains constant irrespective of shear being applied at any specified temperature. In reality, blood does change with increased shear. Therefore, it is more realistic to model blood as a non-Newtonian fluid where the effect of shear on its viscosity is accounted for. Shear thinning is considered the most common phenomenon associated with non-Newtonian fluid behaviour (Chhabra 2020), as observed in blood, characterized by an apparent viscosity that decreases gradually as the shear rate increases. Some models commonly used in characterizing blood are the power law model, Cross model and Carreau-Yasuda model.

2.2 *Blood Vessel Characteristics*

A major characteristic for the fluid/porous media interaction in microcirculation is the repeated bends and twists in the microcirculation known as tortuosity, which indicates the impedance to diffusion associated with flow by local boundaries or local viscosity (Khaled and Vafai 2003). This property of the system is especially important in medical applications in cases where CVD is a problem.

Blood vessels can be modelled as either rigid or flexible pipes, depending on the focus of the simulation data available. In most cases, tissues are considered porous media due to their composition of dispersed cells separated by connected voids, allowing for minerals, nutrients and other blood components to reach all the cells within a tissue (Khaled and Vafai 2003). The porous media nature of the tissues have found application in the transport of drugs, tissue regeneration, transport of residual solvents which are used in fabrication of biodegradable scaffolds (Khaled and Vafai 2003). Unfortunately, this would not happen when a stenosed vessel is present.

2.3 *Mass Transport and Growth in Tissues*

In order to demonstrate cell growth and nutrient diffusion within biological tissues, Galban and Locke (1997, 1999) presented a theoretical framework that considered mass diffusion as the only means of mass transport of nutrients in a vessel.

Mathematical models have been developed for the generation of chondrocytes and the consumption of nutrients in order to get a view of the characteristics of cell growth in a biodegradable polymer matrix. They presented three cell growth kinetics representing different growth mechanisms: the modified Contois, Moser and nth order heterogeneous reaction. The research compared the theoretical cell growth results from the different models with experimental data and illustrated, to a large extent, the general trends in cell growth behaviour – including plaque growth.

2.4 *Blood Characteristics and Modelling in Micro-vessels*

The flow of blood in human arteries is typically classified as a multi-phase non-Newtonian pulsatile flow in a tapered, elastic duct. The pulsatile character of flow is usually responsible for its time-dependent viscoelastic and thixotropic nature (Yilmaz and Gundogdu 2008). The rheology of blood is usually a function of composition and structure. Pal (2003) demonstrated that the plasma in the blood exhibited a Newtonian characteristic viscosity of 1.2 mPa.s at a temperature of 37 °C (De Gruttola et al. 2005).

Blood tends to behave like a Newtonian fluid when the shear rate exceeds a limiting value. Crowley and Pizziconi (2005) indicated that the range of the limiting value for shear stress should be from 100 to 300 s^{-1}. However, some other studies have also shown that blood behaves as a non-Newtonian fluid, and where viscosity varies with the flow rate, the whole blood viscosity depends on the shear rate in a non-linear manner (Baskurt and Meiselman 2003). At the systolic phase, flow and pressure increase, causing the red blood cells to dissociate and deform for efficient flow. Hence, the faster the flow rate, the more deformation efficiency will lead to lower viscosity. On the other hand, in the diastolic phase, associated blood flow is usually slower; hence, blood is usually more viscous at this stage (Cowan et al. 2012).

2.5 *Past Works on the Modelling of Plaque Growth in Micro-circulation*

Young (1968) studied the effect of time-dependent plaque growth on blood flow through a 3D tube-like vessel. The Navier-Stokes equation was used to model flow behaviour. Plaque growth was assumed to be axially symmetric with laminar and steady flow in a constant diameter artery. The fluid was assumed to be Newtonian and incompressible. It was reported that as plaque height increased, flow was impeded and that the more area covered by this plaque, the more are the chances of impedance to flow. This in turn increases the wall shear stress (WSS) and the degree of stenosis, which eventually leads to a blocked artery and triggers off a CVD, which can be fatal.

Johnston et al. (2006) studied the pulsatile flow effect on WSS and massless particle flow pattern, comparing Newtonian blood with the generalized power law model, in a real coronary model artery (without any sign of plaque development). They observed that based on *importance factor* estimate, there are situations, depending on the geometry of the artery in the flow cycle, where the importance of the non-Newtonian model cannot be overlooked. More so, the phase in the cycle that has greater emphasis was in the range of 0.05–0.25 s of the cardiac cycle. It was also observed that upon a comparison of pulsatile flow and steady-state simulations, WSS distributions are similar in both Newtonian and non-Newtonian cases, except in regions where low velocity is either slow or reverse. Studies for massless particle distribution indicate that the non-Newtonian model could cause less particulate concentration within the arterial model as a couple of the particles are reversed out of the domain during the cardiac cycle. The reason for this is that the Newtonian model allows for maximum flow velocity at the centre of flow geometry in relation to the non-Newtonian blood model. This work gives a good insight into the variations in flow properties on WSS and particle transport. However, more insights could have been drawn if the effect of fluid-solid interaction associated with pulsatile flow was incorporated into the model as well as the effect that the development of plaque within the arterial model would have on its significance on the Newtonian and non-Newtonian blood models.

Chan et al. (2007) studied non-Newtonian blood flow through a stenosed artery, incorporating fluid-structure interactions. The aim of the study was to numerically model the interactions in terms of fluid and structural responses between the non-Newtonian nature of blood in a vessel and small elastic deformation. Using the Carreau and power law models to predict blood behaviour, blood was assumed to be non-Newtonian in nature. The velocity profile, WSS and WSS distribution were quantified and studied. The degree of stenosis was 45% with unsteady flow characteristics. The equations were solved using FLOTRAN and ANSYS codes. The velocity profile study indicated that close to the centreline, the Carreau model velocity dropped further. This behaviour was attributed to the higher viscosity of the fluid at a low shear rate characteristic of the centreline region. The power law model velocity experienced a distinct decrease in centreline velocity in comparison with the Newtonian model. Re-circulation zones were also less prominent for the power law model. Relative to the power and Newtonian models, the Carreau model exhibited the highest WSS distributions, which was attributed to the high viscosity near the wall; however, the power model has less WSS distribution. High WSS values are a pointer to the likelihood of a CVD.

Liu and Tang (2010) studied the three-dimensional mathematical model to investigate the geometrical orientation of the atherosclerotic plaques in coronary arteries and the influence of inlet flow rate, wall shear stress and blood viscosity on the growth of atherosclerotic plaques. The blood was assumed to be incompressible, laminar and Newtonian in nature. In the modelling, a rigid wall model was utilized, and the Navier-Stokes equation was used as the governing equation of the model. A no-slip condition was applied to the velocities at the wall boundary. To determine the progression of the plaque over time, a linear model was used to relate the increase in arterial wall thickness to wall shear stress. The Navier-Stokes equation was solved using the finite element method with piece-wise quadratic functions for velocity. A piece-wise linear function was used for pressure over a tetrahedral mesh, applying the generalized minimal residual) method (GMRES) to solve the system iteratively, and COMSOL Multiphysics for numerical computations was performed. This model was used every 3 months for the progression of the plaque. On the study of the effect of viscosity, viscosities of 0.0245, 0.0295 and 0.0345 dyne.s/cm^2 were studied with a fixed inlet flow rate of 1.5 cm^3/s. It was observed that higher viscosities represented superior WSS (wall shear stress), so the effect of inlet flow rates also revealed that higher inlet flow rates represented superior WSS (1.3, 1.5 and 1.8). Atherosclerosis progression was studied over time intervals of 18 months for a period of 10.5 years, using the model. The effect of inlet flow rate (1.3, 1.5 and 1.8) on plaque progression was studied, and it was observed that the overall stenosis size was 30.7%, 27.9% and 24.2%, respectively, and it was observed that rapid flow did not favour the development of the plaque over time. The effect of viscosity studies (0.0245, 0.0295 and 0.0345 dyne.s/cm^2) indicated that higher viscosities did not favour the formation of plaques over time. Moreover, it was observed that the rate of plaque progression reduced over time as the plaque grew. Although the authors justified the assumption for the selection of Newtonian blood model as well as a rigid wall domain for the studied model, it can, however, be said that the study of non-Newtonian and elastic wall models on WSS distribution and plaque progression would give better insight into the study as the geometry and flow conditions used in this study differ from other studies.

Filipovic et al. (2013) studied the computer simulation of 3D plaque formation and progression in the carotid artery. The fluid behaviour was guided by the Navier-Stokes equation as well as the continuity equation, and mass transfer within the lumen of the artery and through the arterial wall was guided by the convection-diffusion equation. The low-density lipoprotein (LDL) transport in the lumen and through the vessel tissue was coupled with the Kedem-Katchalsky equations. Fluid-structure interaction was used to estimate effective wall stress analysis. It was assumed that pressure was a function of WSS. The governing equations were changed into a finite element system of incremental-iterative equations and were solved over time steps. The inflammatory process is modelled using three reaction-diffusion partial differential equations. The governing partial differential equations for plaque formation were based on mass balance and Darcy's law. It was observed that almost all cross-section areas are increasing during the follow-up time, and it was concluded that there was a significant correlation between a large increase in cross-section areas and low WSS. The study did not consider the effect that pulsatile flow would have on fluid-solid interaction in terms of the elasticity of the blood vessel as this would give a more realistic specific result for the arterial model used.

Saveljic et al. (2018) studied the haemodynamic parameters of mass flow and wall shear stress (WSS) using patient-specific data on vessel geometry as well as blood analysis data (low-density lipoprotein (LDL), high-density lipoprotein (HDL), cholesterol and triglycerides) in the coronary artery for a period of 8 months. Flow was considered both in the lumen and within the arterial wall in the case of the plaque growth model. In this study, blood was modelled as a Newtonian fluid. The region of low WSS was identified to be ranging from 0.21 to 0.59 Pa. Also, plaque concentration was determined at the region of highest solute (LDL) concentration of 2.5×10^{-3} mg/mL. The largest diameter reduction of 65.7% was observed. Their work was able to show that sites of low WSS had a correlation with sites of plaque accumulation. Hence, knowing bio-molecular parameters such as LDL and cholesterol, it would be possible to predict sites of plaque development in the arterial domain using computer simulations. The overall simulation involved the coupling of the Navier-Stokes equation with the Kedem-Katchalsky, Darcy and inflammatory equations. Although the authors were able to show important information about how accurate the models coupled with computer simulations could be in predicting future vesicular occlusion, more comprehensive studies are needed, such as the effect that non-Newtonian blood properties would have on the accuracy of the predictions. So the fluid-structure interaction was also neglected as this could also play a significant role in the accuracy of the computation.

3 Inferences from Past Works

The statistics on mortalities associated with CVDs, as indicated by Nichols et al. (2013), show the importance of CVDs, most especially atherosclerosis, in the quality of public health, most especially during periods of disease outbreaks. This work has also classified blood rheology models used in the simulation of blood-flow-associated rheology. The most popular blood rheology is the Newtonian, power law, Cross, Casson and Carreau-Yasuda models. Each of these rheology has its strengths and weaknesses in characterizing blood flow in terms of the accuracy of WSS, mass transport and pressure distribution in the arterial vessel domain in comparison with experimental data.

The approximation of blood rheology as Newtonian can dampen the accuracy of simulation results, especially in pulsatile situations where shear stress ranges may exit the region of Newtonian accuracy. The power law is simple to use - with only two model constants, the Cross models was an improvement such that it described the shear-thinning behaviour of blood where the power law index constants where less than unity. The Casson model improved the Cross and power law models by considering situations describing the blood yield stress required to initiate the flow of blood initially at rest. The Carreau-Yasuda model was observed to be the most common model used to characterize the non-Newtonian property of blood probably because it covers the entire rheological changes of blood associated with the most ranges of shear rates where blood flow can be observed. Although there has been a relationship between temperature and blood viscosity, as defined in the work of De Gruttola et al. (2005), this has not been widely considered in haemodynamic simulations associated with atherosclerosis and how body temperature influences blood rheology vis-à-vis atherosclerosis initiation and progression. The assumption of Newtonian blood characteristics in stenosed regions of vessels is limiting, as has been brought to the fore by Cho and Kensey (1991), who pointed out the large instability range of shear rates. In other research, the Casson model has been used to correct the possible associated errors.

The assumption of rigid vessels most especially in arteries has been refuted by Yilmaz and Gundogdu (2008), who stipulated that arteries are elastic in nature and experience phases of expansion and retraction during the cardiac cycle. The assumption of a steady flow is only reasonable in smaller blood vessels such as veins because of the damping of the pulsatile blood flow in these vessels, as elaborated by Yilmaz and Gundogdu (2008).

The assumption of only asymmetric stenosis in most works seem impractical as studies of Long et al. (2001) have indicated that there are significant changes in post-stenotic flow phenomena, especially in terms of the development of a downstream separation zone and turbulence.

Liu and Tang (2010), Filipovic et al. (2013) and Saveljic et al. (2018) attempted to study atherosclerotic inflammatory progression over time in plaque growth and inflammatory models. However, they did not predict the time the plaque would grow to the level (under different haemodynamic conditions) of occlusion in the blood vessels that could lead to fatality risks. Predicting this could help in determining the periods in a patient's treatment that would require special attention and surgical intervention.

4 Conclusion

In getting a holistic insight into atherosclerosis simulation, the importance of the assumptions and the incorporation and integration of the major models cannot be overemphasized. Although some of the assumptions that seem questionable are usually a result of testing various models and their effects on haemodynamics, this is not enough because variation in rheological, flow and mass characteristics could cause significant deviations in results. With the development of inflammatory and plaque growth models, more patient-specific studies (data) are required for the purpose of characterizing and getting accurate results that could help in making predictions during the progression of the disease so that expert procedures and interventions could be carried out. In most studies, the non-Newtonian characteristic of blood has been significantly downplayed, hence making non-Newtonian haemodynamic research require more attention. The models that have been used in model arterial stenoses (plaque geometry, that is, shape or model equation) by various authors need to be compared to give insight into the deviations in results caused by differences in plaque geometry.

Acknowledgement The authors are appreciative of the financial support of Covenant University, Ota, Nigeria, in the publication of this manuscript.

Declaration of Conflict of Interest The authors declare no conflict of interest in the contents of this manuscript.

References

Baskurt OK, Meiselman HJ. (2003). Blood Rheology and Hemodynamics. Seminars in Thrombosis and Hemostasis. Vol. 29. No. 05.

Bonow RO, Smaha LA, Smith SC, Mensah GA, Lenfant C. (2002). World Heart Day 2002: The Intl. burden of cardiovascular disease: Responding to the emerging global epidemic. Circulation, *106*(13), 1602-1605

Chan WY, Ding Y, Tu JY. (2007). Modeling of non-Newtonian blood flow through a stenosed artery incorporating fluid-structure interaction. Anziam J., *47*, C507-C523.

Chhabra RP. (2020). Introduction: Bubbles, Drops, and Particles in Non-Newtonian Fluids. 2nd Edition. CRC Press

Chinedu SN, Emiloju OC, Iheagwam FN, Rotimi SO, Popoola JO. (2018). Phylogenetic relationship and genetic variation among thaumatococcus daniellii and megaphrynium macrostachyum ecotypes in Southwest Nigeria. Asian J. of Plant Sciences. 17(1), 27-36

Cho YI, Kensey KR. (1991). Effects of the non-Newtonian viscosity of blood on flows in a diseased arterial vessel. Part 1: Steady flows. Biorheology, *28*(3-4), 241-262.

Cowan AQ, Cho DJ, Rosenson RS. (2012). Importance of blood rheology in the pathophysiology of atherothrombosis. Cardiovascular Drugs and Therapy, *26*(4), 339-348.

Crowley TA, Pizziconi V. (2005). Isolation of plasma from whole blood using planar microfilters for lab-on-a-chip applications. Lab on a Chip, *5*(9), 922-929.

Curtis AB, Karki R, Hattoum A, Sharma UC. (2018). Arrhythmias in Patients ≥80 Years of Age: Pathophysiology, Management, and Outcomes. J. of the American College of Cardiology, *71*(18), 2041-2057.

De Gruttola S, Boomsma K, Poulikakos D. (2005). Computational simulation of a non-Newtonian model of the blood separation process. Artificial Organs, *29*(12), 949-959.

Douglas G, Channon KM. (2014). The pathogenesis of atherosclerosis. Medicine, *Medicine*, *42*(9), 480–484

Elizabeth OO, Nonso IF, Adebola NI, John OJ. (2018). Comparative study on chemical composition and antioxidant activity of Annona Muricata plant parts cultivated in covenant university, Ota, Ogun state, Nigeria. Current Research in Nutrition and Food Science, 6(3), 807-815.

Filipovic N, Teng Z, Radovic M, Saveljic I, Fotiadis D, Parodi O. (2013). Computer simulation of three-dimensional plaque formation and progression in the carotid artery. Medical and Biological Engineering and Computing, *51*(6), 607-616.

Galban CJ, Locke BR. (1997). Analysis of cell growth in a polymer scaffold using a moving boundary approach. Biotechnology and Bioengineering, *56*(4), 422-432.

Galban CJ, Locke BR. (1999). Analysis of cell growth kinetics and substrate diffusion in a polymer scaffold. Biotechnology and Bioengineering, *65*(2), 121-132.

Grundy SM, Benjamin IJ, Burke GL, Chait A, Eckel RH, Howard BV, Mitch W, Smith SC, Sowers JR. (1999). Diabetes and cardiovascular disease: A statement for healthcare professionals from the american heart association. Circulation, *100*(10), 1134-1146.

International Diabetes Federation (2015). IDF Diabetes Atlas, 6th edn. http://www.idf.org/diabetesatlas

Johnston BM, Johnston PR, Corney S, Kilpatrick D. (2006). Non-Newtonian blood flow in human right coronary arteries: Transient simulations. J. of Biomechanics, 39(6), 1116-11128.

Joris I, Majno G. (1978). Atherosclerosis and Inflammation. The thrombotic process in atherogenesis, 104, 227–233.

Khaled ARA, Vafai K. (2003). The role of porous media in modeling flow and heat transfer in biological tissues. Intl J. of Heat and Mass Transfer, *46*(26), 4989-5003.

Libby P, Lichtman AH, Hansson GK. (2013). Immune Effector Mechanisms Implicated in Atherosclerosis: From Mice to Humans. Immunity, *38*(6), 1092-1104

Liu B, Tang D. (2010). Computer simulations of atherosclerotic plaque growth in coronary arteries. MCB, 7(4), 193–202.

Long Q, Xu XY, Ramnarine KV, Hoskins P. (2001). Numerical investigation of physiologically realistic pulsatile flow through arterial stenosis. J. of Biomechanics, *34*(10), 1229-1242.

Lusis AJ. (2000). Atherosclerosis. Nature, 407, 233-241.

Motayo BO, Oluwasemowo OO, Olusola BA, Aworunse OS, Oranusi SU. (2021). Evolution and genetic diversity of SARS-CoV-2 in Africa using whole genome sequences. Intl. J. of Infectious Diseeases, 103, 282-287

Mozaffarian D, Benjamin EJ, Go AS, Arnett DK., Blaha MJ, Cushman M, Das SR, de Ferranti S, Després JP, Fullerton HJ, Howard VJ, Huffman MD, Isasi CR., Jiménez MC, Judd SE, Kissela BM, Lichtman JH, Lisabeth LD, Liu S, … Turner MB. (2016). Heart Disease and Stroke Statistics-2016 Update: A Report From the American Heart Association. Circulation, *133*(4), e38-e60.

Nichols M, Townsend N, Scarborough P, Rayner M. (2013). Cardiovascular disease in Europe: Epidemiological update. European Heart J., *34*(39), 3028-3034.

North BJ, Sinclair DA. (2012). The intersection between aging and cardiovascular disease. In Circulation Research, 110(8), 1097-1108.

Pal R. (2003). Rheology of concentrated suspensions of deformable elastic particles such as human erythrocytes. J. of Biomechanics, *36*(7), 981-989.

Pralhad RN, Schultz DH. (2004). Modeling of arterial stenosis and its applications to blood diseases. Mathematical Biosciences, 190(2), 203–220.

Rocha VZ, Libby P. (2009). Obesity, inflammation, and atherosclerosis. Nature Reviews Cardiology, 6(6), 399–409.

Rosamond. (2010). Erratum: Heart disease and stroke statistics-2008 update: A report from the American Heart Association statistics committee and stroke statistics subcommittee Circulation, 122(1), e25-e1460.

Saveljic I, Nikolic D, Milosevic Z, Isailovic V, Nikolic M, Parodi O, Filipovic N. (2018). 3D Modeling of Plaque Progression in the Human Coronary Artery. Multidisciplinary Digital Publishing Institute Proceedings, 2(8), 388.

Thumala, D., Kennedy, B. K., Calvo, E., Gonzalez-Billault, C., Zitko, P., Lillo, P., ... & Slachevsky, A. (2017). Aging and health policies in Chile: new agendas for research. Health Systems & Reform, 3(4), 253-260.

Turan Y, Kozan A, Başkaya MK. (2017). Management of Lipid Metabolism. Primer on Cerebrovascular Diseases, 870–873.

Wang X, Gao M, Zhou S, Wang J, Liu F., Tian F, Jin J, Ma Q, Xue X, Liu J, Liu Y, Chen Y. (2017). Trend in young coronary artery disease in China from 2010 to 2014: A retrospective study of young patients≤45. BMC Cardiovascular Disorders, 17(1), 1-8.

Weber C, Noels H. (2011). Atherosclerosis: Current pathogenesis and therapeutic options. In Nature Medicine, *17*(11), 1410-1422.

World Health Organization (2015). Facts about aging. http://www.who.int/ageing/about/facts/en/

Yilmaz F, Gundogdu MY. (2008). A critical review on blood flow in large arteries; relevance to blood rheology, viscosity models, and physiologic conditions. Korea-Australia Rheology J., *20*(4), 197-211.

Young DF. (1968). Effect of a time-dependent stenosis on flow through a tube. J. of Manufacturing Science and Engineering.

Comparison of Pectinase Activity in the Flavedo and Albedo of Citrus and *Thaumatococcus daniellii* Fruits

G. D. Ametefe, F. N. Iheagwam, F. Fashola, O. I. Ibidapo, E. E. J. Iweala, and S. N. Chinedu

1 Introduction

The catalysis of the polymers of pectin in plant cell walls is facilitated by pectinases. Pectin is the substrate for the production of the pectin enzyme or pectinases (Cherekar and Pathak 2020). The growing demand for pectinases in industries is responsible for the increasing application of this enzyme in many areas, thus making microbial pectinases account for approximately 25% of the world sales in food enzymes (Ajayi et al. 2018; Almulaiky et al. 2020; Ruiz et al. 2012). As of 2020, the market for food enzymes was estimated to approach US$ 41.4 billion (El Enshasy et al. 2018). It should be noted that pectinases are applied in the processing of food such as juices (Ametefe et al. 2017; Mahmoodi et al. 2017). Hence, pectinase is also a food enzyme. Despite the importance of these food enzymes, the high cost of

G. D. Ametefe (✉)
Department of Biochemistry, College of Science and Technology, Covenant University, Ota, Ogun State, Nigeria
e-mail: george.ametefe@stu.cu.edu.ng

F. N. Iheagwam · E. E. J. Iweala · S. N. Chinedu
Department of Biochemistry, College of Science and Technology, Covenant University, Ota, Ogun State, Nigeria

Covenant University Public Health and Wellbeing Research Cluster, Covenant University, Ota, Ogun State, Nigeria

F. Fashola
Department of Biochemistry, College of Science and Technology, Covenant University, Ota, Ogun State, Nigeria

Department of Biotechnology, Federal Institute of Industrial Research, Lagos State, Nigeria

O. I. Ibidapo
Department of Biotechnology, Federal Institute of Industrial Research, Lagos State, Nigeria

© The Author(s), under exclusive license to Springer Nature Switzerland AG 2022

A. O. Ayeni et al. (eds.), *Bioenergy and Biochemical Processing Technologies*, Green Energy and Technology, https://doi.org/10.1007/978-3-030-96721-5_17

production involved in the use of these enzymes is a concern. This could be linked to the importation of these food enzymes, which is believed to negatively affect their utilization by small-scale industries (Aboul-Fotouh et al. 2016). The main methods for the production of pectinases are solid-state and submerged fermentation (Ametefe et al. 2017; Rangarajan et al. 2010). The main difference between both methods of production of pectinases is in the moisture content needed for fermentation, with submerged fermentation requiring more moisture than solid-state fermentation (Manan and Webb 2017a, b; Oumer and Abate 2018; Soccol et al. 2017; Subramaniyam and Vimala 2012).

Encouragingly, the production of pectinases and most food enzymes using agricultural wastes and cheaply available microbes is a great relief for local production in underdeveloped and developing nations of the world (Chinedu et al. 2010; Aboul-Fotouh et al. 2016; Azzaz et al. 2013; Khattab et al. 2019; Murad and Azzaz 2011). For instance, Pretel et al. (2008) reported that of the total citrus fruit production worldwide, fresh consumption constituted about 65%, with 35% being consumed after industrial processing. Though the study did not mention the accompanying agricultural wastes (the peels) generated from the consumption of the fruits, both fresh consumption and industrial processes perhaps justify the obvious agrowaste generation from citrus fruits in our environment. Microorganisms such as *Aspergillus* species, *Penicillium chrysogenum*, and *Trichoderma* spp. have been utilized for the production of pectinases (Okafor et al. 2010). Yeasts such as *Saccharomyces cerevisiae* and *Wickerhamomyces anomalus* have also been used for the production of pectin enzymes (Martos et al. 2013). However, *Aspergillus niger* (a fungus) has emerged as the preferred microbe for the production of pectin enzymes (pectinases) because the microorganism has been reported to show high pectinase production ability (Akhter et al. 2011; Sudeep et al. 2020). In addition, the long-held tradition of placing this microorganism in the class of "generally regarded as safe" (GRAS) microorganisms by the United States Food and Drug Administration further informs its usage in the production of pectinase (El Enshasy et al. 2018; Iwashita 2002). Furthermore, a study by Chinedu et al. (2017) compared the production of polygalacturonase (a subclass of pectinase) from *Solanum macrocarpon* L. fruit with that obtained from fungi for the commercial production of the enzyme, hence making fungi the basis of comparison for commercial pectinase production. Agricultural wastes such as corn cobs and banana and orange peels are known to produce this enzyme (Okafor et al. 2010; Widowati et al. 2017). These fruit peels comprise the flavedo (outer peel) and albedo (inner peel) (Multari et al. 2021). Despite the information in the literature on the production of pectinases using readily available cheap raw materials, there is a dearth of information on the comparison of pectinase production using the albedo of some locally sourced fruits in Nigeria, like *Citrus aurantium* and *Thaumatococcus daniellii*. However, studies have utilized the peels of citrus fruits in producing pectinases (Ajayi et al. 2018; Chinedu et al. 2017). Hence, this study is aimed at comparing the following: pectin content in the albedo and flavedo of *Citrus sinensis*, pectinase activity from *Aspergillus niger* in the albedo and flavedo of *Citrus sinensis*, and pectinase activities in the albedo of *Citrus sinensis*, *Citrus aurantium*, and *Thaumatococcus daniellii* as substrates.

2 Materials and Methods

2.1 Chemicals and Reagents

High-purity salts and analytical grade reagents from Qualikems, Delhi, India, were used for the study. They include potassium nitrate (KNO_3), hydrochloric acid (HCl), sodium chloride (NaCl), sodium hydroxide (NaOH), dinitrosalicylic acid (DNSA), and potassium chloride (KCl).

2.2 Substrate Collection and Preparation

Oranges (*Citrus sinensis* and *Citrus aurantium*) were sourced from Nigerian Army Shopping Arena, Lagos State, Nigeria, while *Thaumatococcus daniellii* fruits were obtained from the Owode-Yewa 'forest' plant area (Atan-Ota) in Ogun State, Nigeria. The flavedo and albedo were manually removed, washed, oven-dried (58 ± 3 °C for 72 hours), and milled into 0.06-inch particle size.

2.3 *Pectin Extraction from* Citrus sinensis *(Sweet Orange) Albedo and Flavedo*

The method of Kanmani et al. (2014) was employed. The sweet oranges were cut into parts with a knife, and the peels were removed. The flavedo and albedo were then separated and oven-dried 72 h at 58 ± 3 °C and ground 0.097 inches and stored at room temperature for subsequent use. Ninety mL of distilled water was added to 5 grams of each of the substrates and subsequently followed by the addition of 10 mL of citric acid at pH 3.5. The mixture was heated for 1 hour. The filtration of the resultant extract was undertaken with a muslin cloth. The filtrate obtained was then coagulated using 95% ethanol and left for 2 hours for the pectin to float. The gelatinous pectin flocculants were then skimmed off. Ethyl alcohol was used to filter and wash the extracted pectin to reduce impurities to the barest minimum (Khule et al. 2012). The precipitate from the above processes was dried at 40 ± 3 °C in an oven, followed by the calculation of yield (%).

$$\%\text{Pectin} = P / I_{fp} \times 100$$

where % pectin = yield of pectin (in percent), P = the weight of extracted pectin, and I_{fp} = the initial weight of the peel powder.

2.4 Isolation and Screening of Pectinase Producing Aspergillus *spp.*

Aspergillus niger was isolated from deteriorated fruit wastes and the accompanying soil samples in the dump sites of the Federal Institute of Industrial Research, Lagos State, Nigeria, following the procedure of Okonji et al. (2019) and the use of streptomycin as the antibiotic for the culture of fungi. Isolated *A. niger* was thereafter identified (excluded from yeasts strains) and confirmed via microscopy and molecular technique, respectively (Udenwobele et al. 2014; Khattab et al. 2016; Wartu et al. 2017). The screening of *A. niger* isolate for pectinase production was undertaken by adopting the method of Chinedu et al. (2010), with the substitution of cellulose for pectin in the screening process.

2.5 Assessment of Pectinase Production, Extraction, and Activity

The solid-state fermentation method following the procedure of Ametefe et al. (2017) was used to produce and extract pectinase with slight modification. Two grams of each of the substrates were moistened with nutrient medium (2 mL; pH 4.0) and autoclaved at 121 °C for 15 minutes. Upon cooling at room temperature, 1% citrus pectin was aseptically added and mixed before inoculation with *A. niger*. The incubation of the mixture was undertaken for 6 days at 30 °C, and the enzyme extraction was facilitated using NaCl solution (0.1 M). The activity of pectinase was assayed using the dinitrosalicylic acid (DNSA) method by Miller (1959).

2.6 Analyses of Data

The data obtained were presented as the mean ± standard deviation of triplicate values. The one-way analysis of variance (ANOVA) was employed for the data analyses. Statistical significance was determined using Bonferroni correction at a 95% confidence interval. All statistical analyses were done on Microsoft Excel v. 2016 (Washington, United States).

3 Results and Discussion

The pectinase extracted using *Citrus sinensis* flavedo as substrate had a significantly lower activity ($p > 0.05$) relative to its albedo (Fig. 2). The decline in pectinase activity (Fig. 2) could have been due to the significant decrease ($p < 0.05$) in pectin

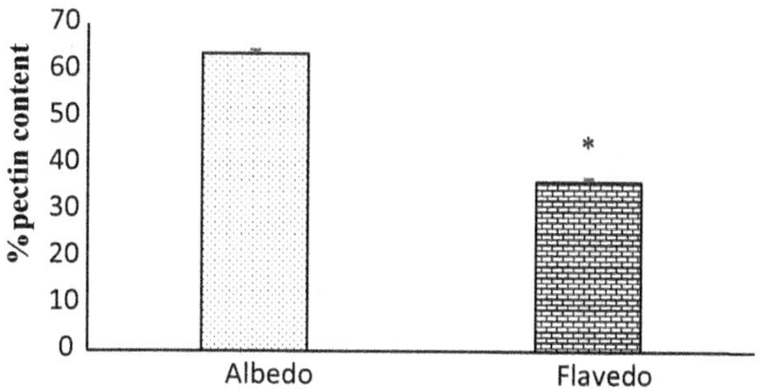

Fig. 1 Differences in the pectin content of *Citrus sinensis* albedo and flavedo. Bars = mean ± standard deviation (triplicate). *$p < 0.05$ = significant decrease in the flavedo pectin content relative to albedo

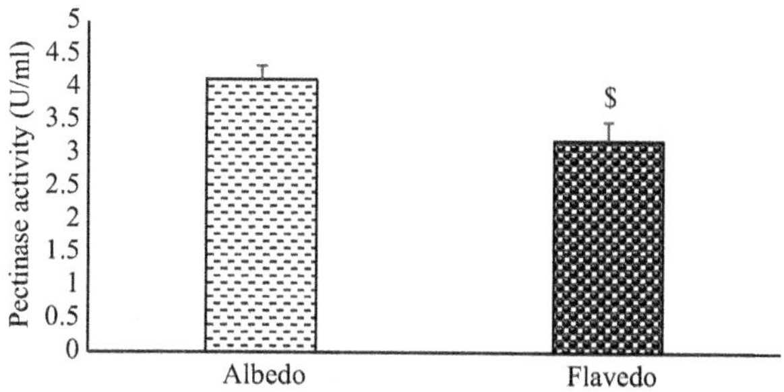

Fig. 2 Effect of *Citrus sinensis* albedo and flavedo as substrates on pectinase activity. Bars = mean ± standard deviation (triplicate). $^{\$}p < 0.05$ = significant decrease in pectinase activity relative to albedo

content in flavedo compared to its albedo (Fig. 1) (Liu et al. 2006), hence, experimentally, making albedo the pectin-rich portion of *Citrus sinensis* peels.

Similar findings on the use of cellulases to release fruit albedo showed pectinase application for the completion of the polysaccharide hydrolysis in the albedo (Ben-Shalom et al. 1986; Rouhana and Mannheim 1994; Soffer and Mannheim 1994). Another study demonstrated that a vital enzyme, polygalacturonase (a pectin enzyme), led to the degradation of albedo (or mesocarp) (Pretel et al. 2005, 2008). Therefore, these findings have possibly further justified the higher pectinase activity in the albedo compared to the flavedo (Fig. 2). The findings show higher interaction of pectin enzymes with the albedo of fruits, resulting in the degradation of the polysaccharides, hence justifying the preference for albedo as a better substrate for pectinase production.

In another study, the pectin quantity in the albedo was higher than that obtained in the flavedo, corroborating the findings of this study (Zanella and Taranto 2015). In addition, the findings from Liu et al. (2006) disclosed that the pectin content in the flavedo of orange peels was 27% of the total extract, further justifying the albedo as the portion with a higher concentration of pectin (the substrate for pectinase production).

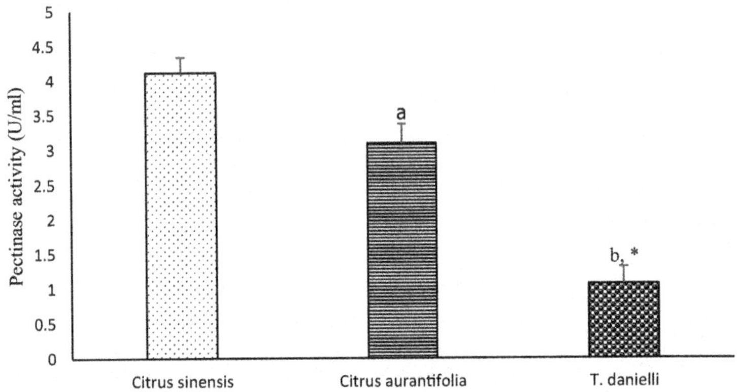

Fig. 3 Effect of the albedos of *Citrus sinensis*, *C. aurantifolia*, and *T. daniellii* on pectinase activity. Bars = mean ± standard deviation (triplicate). [a][b]$p < 0.05$ = significant decrease relative to *C. sinensis*, and *$p < 0.05$ = significant decrease in pectinase activity in *T. daniellii* relative to *C. aurantifolia*

The utilization of the albedos of *C. sinensis*, *C. aurantifolia*, and *T. daniellii* in the determination of pectinase activity revealed *C. sinensis* as the best, followed by *C. aurantifolia* and *T. daniellii* (Fig. 3). A decline ($p < 0.05$) in the activity of the enzyme in the albedo of *T. danielli* relative to the albedo of *C. aurantifolia*, therefore showing the superiority of pectin content in the albedos of the citrus species investigated relative to the albedo of *T. daniellii*. However, the albedo of *C. sinensis* still showed statistical significance ($p < 0.05$) in the enzyme activity over the albedo of *C. aurantifolia*. The significance of pectinase activities in both the citrus species has been further justified in the study by Mahmoodi et al. (2017) of the pectin compositions in different fruits, and the percentage of pectin in citrus fruit was found to be higher than the pectin from apple, lemon, melon peel, carrot, banana, and cranberry fruits (Aina et al. 2012; Canteri-Schemin et al. 2005; Raji et al. 2017; Sharma et al. 2013). It is, however, necessary to point out that the difference in the mechanical peeling of the fruits (for obtaining the flavedo and albedo) and the possible variations in the particle sizes of the powdered samples could have influenced the differences between the percentage of pectin extracted from the studied fruits and that in other works of literature in the field.

The observed dissimilarities in the pectinase activities of the substrates could be attributed to the concentration of pectin, the species of plant from which the substrates were sourced, and the area of growth and harvest of the substrates (Liu et al. 2006; Pretel et al. 2005; Pretel et al. 2008). In another study, the cutting of the flavedo of grapefruit enhanced enzymatic reaction with the albedo for the proper peeling of the fruit (Bruemmer et al. 1978). The possible varying compositions of the albedos in the investigated citrus fruit species influenced the application of enzymes in the peeling of the fruits, further signifying that the variations in the pectin content would have had a corresponding effect, resulting in the different pectinase activities (Fig. 3) obtained (Pretel et al. 2008; Pretel et al. 1997). However, the albedos of

C. aurantifolia and *T. daniellii* could be used to augment the local production of pectinases. There is a shortage of information in the comparison of the pectinase activities in the albedos of the citrus species investigated. However, the superior pectin content in the albedo compared to the flavedo (Fig. 1) could suggest a higher content of the albedo in *C. sinensis* than that in C. *aurantifolia* and *T. daniellii* for pectinase production. The production of pectinase using the albedo of *T. daniellii* has also been shown to produce pectinase for the first time. Additionally, the implication of this study is that pectin, the substrate for pectinase production, has been further established in the peels of both lime (*C. aurantifolia*) and sweet orange (*C. sinensis*), as further supported in the findings of Dominiak et al. (2014) and Georgiev et al. (2012).

4 Conclusion

The study showed the superiority of pectinase activity in the albedo over that in the flavedo of *C. sinensis*, which was also mirrored in the significant pectin content in the albedo as compared to that in the flavedo of the same plant species. The albedo of *C. sinensis* showed the highest pectinase activity relative to both the albedos of *C. aurantifolia* and *T. daniellii*. Literature search showed that, this study addressed the dearth of information on the production of pectinase using the albedos of *T. daniellii* and *C. aurantifolia* and comparisons as exemplified in the investigation. The findings from this study will further supplement the local production of pectinases. Future work will be geared toward the extraction of pectin from the *C. aurantifolia* and *T. daniellii* substrates investigated; and, subsequent production of the corresponding enzymes (pectinases) from the pectin extracts (as substrates).

Acknowledgement The authors thank the Covenant University Centre for Research, Innovation, and Discovery (CUCRID), Ota, Nigeria, for being responsible for financing this manuscript. The authors also thank the Department of Biotechnology, Federal Institute of Industrial Research, Oshodi, Nigeria, for allowing the use of their laboratory.

Funding There was no aid from any funding agency for this research.

References

Aboul-Fotouh GE, El-Garhy GM, Azzaz HH et al (2016) Fungal cellulase production optimization and its utilization in goat's rations degradation. Asian J Anim Vet Adv 11: 824-831.

Aina V, Barau MM, Mamman O et al (2012) Extraction and characterization of pectin from peels of lemon (citrus limon), grape fruit (*Citrus paradisi*), and sweet orange (*Citrus sinensis*) British J Pharmacol Toxicol 3(6):259–262.

Ajayi AA, Salubi AE, Lawal B et al (2018) Optimization of pectinase production by *Aspergillus niger* using central composite design. African J Clin Exp Microbiol 19(4): 314–319.

Akhter M, Morshed MA, Uddin A et al (2011) Production of pectinase by *Aspergillus niger* cultured in solid-state media. Int J Biosci (IJB) 1(1): 33-42.

Almulaiky YQ, Albishri AA, Khalil NM et al (2020) Polygalactutonase by *Aspergillus niger* using seaweed waste under submerged fermentation: production, purification, and characterization. Biomed J Sci Tech Res 25(4): 19416-19422.

Ametefe GD, Dzogbefia VP, Apprey C et al (2017) Optimal conditions for pectinase production by *Saccharomyces cerevisiae* (ATCC 52712) in solid-state fermentation and its efficacy in orange juice extraction. IOSR-JBB 3(6): 78-86.

Azzaz HH, Murad HA, Kholif AM et al (2013) Pectinase production optimization and its application in banana fiber degradation. Egypt J Nutr Feeds 16(2): 117-125.

Ben-Shalom M, Levi A, Pinto R (1986) Pectolytic enzyme studies for peeling of grapefruit segment membrane. J Food Sci 51: 421-423.

Bruemmer JH, Griffin AW, Onayami O (1978) Sectionizing grapefruit by enzyme digestion. Proceeding of the Florida State Horticultural Society 91: 112-114

Canteri-Schemin MH, Fertonani HCR, Waszczynskyj N, Wosiacki G (2005) Extraction of pectin from apple pomace. Braz Arch Biol Technol 48:259–266.

Chinedu SN, Eni AO, Adeniyi AO et al (2010) Assessment of growth and cellulase production of wild-type microfungi isolated from Ota, Nigeria. Asian J Plant Sci 9: 118-125.

Chinedu SN, Dayo-Odukoya OP, Iheagwam FN (2017) Partial purification and kinetic properties of polygalacturonase from *Solanum macrocarpum* L. fruit. Biotechnol 16(1): 27-33.

Cherekar MN, Pathak AP (2020) Production and characterization of a holoalkaline pectinase from *Halomonas pantellerinsis* strain SSL8 isolated from Sambhar Lake, Rajasthan. Curr Trends Biotechnol Pharm 14(3): 319-326.

Dominiak M, Søndergaard KM, Wichmann J et al (2014) Application of enzymes for efficient extraction, modification, and development of functional properties of lime pectin. Food Hydrocoll 40: 273-282.

El Enshasy HA, Elsayed EA, Suhaimi N et al (2018) Bioprocess optimization for pectinase production using *Aspergillus niger* in a submerged cultivation system. BMC Biotechnol 18: 17.

Georgiev Y, Ognyanov M, Yanakieva I et al (2012) Isolation, characterization and modification of citrus pectins. J Biosci Biotechnol 1(3): 223-233.

Iwashita K (2002) Recent studies of protein secretion by filamentous fungi. J Biosci Bioeng 94(6): 530–535.

Kanmani P, Dhivya E, Aravind J et al (2014) Extraction and analysis of pectin from citrus peels: augmenting the yield from *Citrus limon* using statistical experimental design. IJEE 5(3): 303-312.

Khattab SMR, Abdel-Hadi AM, Abo-Dahab NF et al (2016) Isolation, characterization, and identification of yeasts associated with foods from Assiut city, Egypt. Br Microbiol Res J 13(1): 1-10.

Khattab MSA, Azzaz HH, Abd-El-Tawab AM et al (2019) Production optimization of fungal cellulase and its impact on ruminal degradability and fermentation of diet. Int J Dairy Sci 14: 61-68.

Khule RN, Nitin BM, Dipak SS et al (2012) Extraction of pectin from citrus fruit peel and use as natural binder in paracetamol tablet. Der Pharm Lett 4(2): 558-564.

Liu Y, Shi J, Langrish TAG (2006) Water-based extraction of pectin from flavedo and albedo of orange peels. Chem Eng J 120(3): 203-209.

Mahmoodi M, Najafpour GD, Mohammadi M (2017) Production of pectinases for quality apple juice through fermentation of orange pomace. J Food Sci Technol 54(12): 4123–4128.

Manan MA, Webb C (2017a) Modern microbial solid-state fermentation technology for future biorefineries for the production of added-value products. Biofuel Res J 4(4): 730–740.

Manan MA, Webb C (2017b) Design aspects of solid-state fermentation as applied to microbial bioprocessing. JABB 4(1): 91.

Martos MA, Zubreski ER, Garro OA et al (2013) Production of pectinolytic enzymes by the yeast *Wickerhanomyces anomalus* isolated from citrus fruits peels. Biotechnol Res Int Article ID 435154. https://doi.org/10.1155/2013/435154

Miller GL (1959) Modified DNS method for reducing sugars. Anal Chem 31(3): 426-428.

Multari S, Licciardello C, Caruso M et al (2021) Flavedo and albedo of five citrus fruits from Southern Italy: physicochemical characteristics and enzyme-assisted extraction of phenolic compounds. Food Measure 15: 1754–1762.

Murad HA, Azzaz HH (2011) Microbial pectinases and ruminant nutrition. Res J Microbiol 6(3): 246-269.

Okafor UA, Okochi VI, Chinedu SN et al (2010) Pectinolytic activity of wild-type filamentous fungi fermented on agro-wastes. Afr J Microbiol Res 4(24): 2729-2734.

Okonji RE, Itakorode BO, Ovumedia JO et al (2019) Purification and biochemical characterization of pectinase produced by *Aspergillus fumigatus* isolated from soil of decomposing plant materials. J Appl Biol Biotechnol 7(3): 1-8.

Oumer OJ, Abate D (2018) Comparative studies of pectinase production by *Bacillus subtilis* strain Btk 27 in submerged and solid-state fermentation. Biomed Res Int Article ID 1514795, doi:https://doi.org/10.1155/2018/1514795.

Pretel MT, Lozano P, Riquelme F et al (1997) Pectic enzymes in fresh fruit processing: optimization of enzymic peeling of oranges. Process Biochem 32(1): 43-49

Pretel MT, Amorós A, Botella MA et al (2005) Study of albedo and carpelar membrane degradation for further application in enzymatic peeling of citrus fruits. J Sci Food Agric 84(1): 86-90.

Pretel MT, Sanchez-Bel P, Egea I et al (2008) Enzymatic peeling of citrus fruits: factors affecting degradation of the albedo. TFSB 2(1): 52-59.

Raji Z, Khodaiyan F, Rezaei K et al (2017) Extraction optimization and physicochemical properties of pectin from melon peel. Int J Biol Macromol 98:709–716.

Rangarajan V, Rajasekeran M, Ravichandran R et al (2010) Pectinase production from orange peel extract and dried orange peel solid as substrates using *Aspergillus niger.* Int J Biotechnol Biochem 6(3): 445-453.

Rouhana A, Mannheim CH (1994) Optimization of enzymatic peeling of grapefruit. LWT – Food Sci Technol 27(2): 103-107.

Ruiz HA, Rodriguez-Jasso RM, Rodriguez R et al (2012) Pectinase production from lemon peel pomace as support and carbon source in solid-state fermentation column-tray bioreactor. Biochem Eng J 65: 90-95.

Sharma A, Shrivastava A, Sharma S et al (2013) Biotechnology for environmental management and resource recovery. Berlin: Springer pp. 107–124.

Soffer T, Mannheim CH (1994) Optimization of enzymatic peeling of oranges and pomelo. LWT – Food Sci Technol 27(3): 245-248.

Soccol CR, da-Costa ESF, Letti LAJ et al (2017) Recent developments and innovations in solid-state fermentation. Biotechnology Res Innov 1(1): 52-71.

Subramaniyam R, Vimala R (2012) Solid-state and submerged fermentation for the production of bioactive substances: A comparative study. Review article. Int J Sci Nat (IJSN) 3(3): 480-486.

Sudeep K C, Upadhyaya J, Joshi DR et al (2020) Production, characterization and industrial application of pectinase enzyme isolated from fungal strains. Ferment 6(2): 59.

Udenwobele DI, Nsude CA, Ezugwu AL et al (2014) Extraction, partial purification, and characterization of pectinases isolated from *Aspergillus* species cultured on mango (*Mangifera indica*) peels. Afr J Biotechnol 13(24): 2445–2454.

Wartu JR, Whong CMZ, Abdullahi IO et al (2017) Phylogenetics of aflatoxigenic moulds and prevalence of aflatoxin from in-process wheat and flour from selected major stores within northern Nigeria. Sci World J 12(4): 83-87.

Widowati E, Utami R, Mahadjoeno E et al (2017) Effect of temperature and pH on polygalacturonase production by pectinolytic bacteria *Bacillus licheniformis* strain GD2a in submerged medium from Raja Nangka (*Musa paradisiaca* var. *formatypica*) banana peel waste. IOP Conf Ser: Mater Sci Eng 193: 012018.

Zanella K, Taranto OP (2015) Influence of the drying operating conditions on the chemical characteristics of the citric acid extracted pectins from 'pera' sweet orange (*Citrus sinensis* L. Osbeck) albedo and flavedo. J Food Eng 166: 111–118.

Production of Rhamnolipid Biosurfactant from Waste Cooking Oil Using *Pseudomonas putida* in a Batch Reactor

O. O. Sadare, T. Mokhutsane, and M. O. Daramola

1 Introduction

Surfactants are surface-active chemicals with a world production of more than 15 million tons per annum (Henkel et al. 2012). They are used in many industrial processes such as food processing, improved oil recovery, and pharmaceuticals (Henkel et al. 2012). The application of surfactants is dependent on their ability to reduce surface tension and improve wetting ability, solubility, and ability to foam (Mulligan 2004). The majority of surfactants on the market are produced from petrochemical sources, and thus they are termed synthetic surfactants (Henkel et al. 2012). With the increased environmental consideration from consumers and producers, increased interest is moving toward biologically derived surfactants (biosurfactants) (Henkel et al. 2012). Biosurfactants are surface-active chemicals with varying structures, produced by microorganisms (Janek et al. 2013). They are amphipathic molecules with a hydrophobic domain (made of long-chain fatty acids and their additives) and a hydrophilic domain (which can be a carbohydrate, amino acid, phosphate, or

O. O. Sadare
Department of Chemical Engineering, Faculty of Engineering, Built Environment and Information Technology, University of Pretoria, Hatfield, Pretoria, South Africa

T. Mokhutsane
School of Chemical and Metallurgical Engineering, Faculty of Engineering and the Built Environment, University of the Witwatersrand, Johannesburg, South Africa

M. O. Daramola (✉)
Department of Chemical Engineering, Faculty of Engineering, Built Environment and Information Technology, University of Pretoria, Hatfield, Pretoria, South Africa

School of Chemical and Metallurgical Engineering, Faculty of Engineering and the Built Environment, University of the Witwatersrand, Johannesburg, South Africa
e-mail: michael.daramola@up.ac.za

© The Author(s), under exclusive license to Springer Nature Switzerland AG 2022

A. O. Ayeni et al. (eds.), *Bioenergy and Biochemical Processing Technologies*, Green Energy and Technology, https://doi.org/10.1007/978-3-030-96721-5_18

cyclic peptide). The specific chemical distinction of biosurfactants makes them advantageous over petrochemically derived surfactants in that they are nontoxic to the environment, biodegradable, highly selective, and active in extreme temperatures and pH ranges (Mulligan 2004; Cameotra and Pruthi 2003). Bacteria, fungi, and yeasts are capable of producing biosurfactants as extracellular materials or compounds incorporated in the cell membrane. Rhamnolipids, trehalose lipids, and surfactin are the most effective of the known biosurfactants. They can effectively reduce the surface tension of water and interfacial tension of water/oil mixtures considerably, even at low concentrations (Kanna et al. 2014). Glycolipids are the most studied and most promising class of biosurfactants. They are low-molecular-mass biosurfactants containing carbohydrates and long-chain aliphatic or hydroxyl aliphatic acids. Rhamnolipids are the most extensively studied of the glycolipid class and have been predominantly recognized among the *Pseudomonas aeruginosa* organisms (Reis et al. 2011). The cost of cleaning up contaminated soil and groundwater exceeds several millions of rands, a situation that has drawn researchers' attention to bioremediation options (Maier and Soberon-Chavez 2000).

Bioremediation involves enhancing material accessibility, environmental conditions, and microorganism activity to speed up the degradation of contamination elements in a system (Fernandes 2009). Rhamnolipids have been discovered to have the potential in serving numerous bioremediation applications. They are capable of biodegrading both aliphatic and aromatic hydrocarbons and can degrade hydrocarbons in the presence of toxic metals. They have further shown high affinities for various metals, such as copper, cadmium, zinc, and lead. Furthermore, rhamnolipids are biodegradable and can be a carbon source for other microorganisms as they do not accumulate in a system (Maier and Soberon-Chavez 2000). Rhamnolipids have also been found to possess biological control abilities. They can control zoosporic plant pathogens in that they can cause the termination of their motility and the disintegration of their population (Maier and Soberon-Chavez 2000). In a liquid growth medium, *P. aeruginosa* mainly produces two kinds of rhamnolipids: rhamnosyl-rhamnosyl-β-hydroxydecanoyl-β-hydroxydecanoate (dirhamnolipid) and rhamnosyl-β-hydroxydecanoyl-β-hydroxydecanoate (monorhamnolipid) (Reis et al. 2011). Rhamnolipids have the ability to lower surface tension to 29 mN/m, emulsify hydrocarbons, and act as a stimulus for the growth of *P. aeruginosa* on n-hexadecane (Fernandes 2009). Besides *Pseudomonas aeruginosa*, other microorganism species, such as *Pseudomonas putida*, *Pseudomonas fluorescens*, *Escherichia coli*, and *Pseudomonas oleovorans*, have the ability to produce rhamnolipids. However, *Pseudomonas putida* has been reported to produce the highest concentration among this group at a reported value of 60 mg/L (Henkel et al. 2012). *Pseudomonas aeruginosa* is recognized as the main producer of rhamnolipids (Tuleva et al. 2002). However, owing to its pathogenic nature, other species, such as *Pseudomonas putida*, are considered alternatives due to their nonpathogenic nature. Therefore, this study has chosen to further investigate the rhamnolipid production of *Pseudomonas putida*.

Different studies have been conducted on the production of biosurfactants using different nitrogen sources and carbon sources (Cameotra and Pruthi 2003; Kanna et al. 2014; Wittgens et al. 2011; Tan and Li 2018; Tiso et al. 2020). According to Cameotra and Pruthi (Cameotra and Pruthi 2003), nitrate was utilized as the principal nitrogen source for the production of biosurfactants in *Pseudomonas* strain 44T1 and *Rhodococcus* strain ST-5 grown on olive oil and paraffin. Kanna (Kanna et al. 2014) also investigated the influence of different nitrogen sources, for example, ammonium nitrate, ammonium sulfate, and urea, on biosurfactant production using *P. putida* and found that ammonium sulfate produced the highest biosurfactant concentration of 2.5 g/L after 140 h. Wittgens et al. (Wittgens et al. 2011), Tan and Li (Tan and Li 2018), and Tiso et al. (Tiso et al. 2020) utilized glucose, sugar, and ethanol, respectively, as carbon sources to independently grow *Pseudomonas putida* for the production of rhamnolipids. The current limitation hindering the extensive application of biosurfactants is the high cost of their production.

The use of expensive substrates, comparatively low product yields, and costly downstream processes make biosurfactant production uncompetitive to synthetic surfactants. Waste streams are becoming viable raw materials for biosurfactant production. This is mainly because they are inexpensive, they do not compete with human food, and their reuse is advantageous to the environment (Henkel et al. 2012). Waste cooking oil has been identified as a low-cost product that can be utilized as a substrate for the production of rhamnolipids. Hence, its potential as a carbon source was explored in this study.

Against this background, this study attempts to come up with an eco-friendly, readily available, and cheap way to produce biosurfactants by utilizing waste cooking oil as a substrate and to improve product yield by analyzing the effects of varying nitrogen sources.

2 Materials and Method

Pseudomonas putida ATCC™ 49128™† was purchased from Microbiologics in San Diego, California, USA. The waste cooking oil used as a carbon source for the media was obtained from Ukweza Catering in the Johannesburg area of South Africa. The nitrogen sources, urea (≥99.0% purity) and NH_4NO_3 (≥99.0% purity), were obtained from Sigma Aldrich (Pty) (Merck), South Africa. The COD kit was purchased from Merck SA (Pty). The standard rhamnolipid (90%) purity was purchased from Sigma Aldrich. All chemicals were used without further purification.

2.1 Microorganism and Growth Conditions

In order to increase the number of *Pseudomonas putida* cells, a subculturing experiment was conducted in a simple growth media containing a Luria-Bertani (LB) medium (10 g/L tryptone, 5 g/L yeast extract, 10 g/L NaCl) complemented with 10 g/L of glucose, as described by Wittgens et al. (Wittgens et al. 2011). The grown cells of *Pseudomonas putida* were transferred into the growth media. Microorganisms with the ability to produce biosurfactants are broadly known to use both water-immiscible hydrocarbons and carbohydrate-containing mineral salt media to grow (Kanna et al. 2014). LB medium is the most commonly used culture medium to grow *P. aeruginosa* strains (Wei et al. 2005), but it was not conducive for the growth of *P. putida* in this experiment. Therefore, cell growth was achieved in simple media containing distilled water and nitrogen and carbon sources. *P. putida* was grown at a temperature of 30 °C and rotation speed of 215 rpm, and growth was measured over a 7-day period. Two sets of growth media containing two different nitrogen sources, namely, urea and NH_4NO_3, were prepared. The experiments were conducted in batch reactors. Three Erlenmeyer flasks (bioreactors) had 100 mL growth media containing 1.5 g/L urea. Three other Erlenmeyer flasks had 100 mL growth media containing 2 g/L NH_4NO_3. The concentration of the carbon source (waste cooking oil) was varied in each set of three Erlenmeyer flasks was varied, each containing 2%, 6%, and 10% (v/v) of waste cooking oil. The cell samples were collected every 24 h and weighed using a wet-weight method to determine cell growth. The consumption of waste cooking oil by the microbe was quantified with the use of a carbon oxygen demand (COD) kit. Three milliliters of the growth medium sample were slowly pipetted down the side of the tilted reactor cell. The reactor cell was then sealed and the content vigorously mixed. The reactor cell was heated at 148 °C in a thermoreactor for 2 hours. The cells were then allowed to cool to room temperature before measuring absorbance in a photometer. The activity of the biosurfactant was verified by determining surface tension using the "capillary tube" method. The height to which the liquid was raised in the capillary tube was measured. The density was determined by measuring the mass of a specific volume of liquid. Using eq. (1), the height and density measured were used to calculate the surface tension:

$$T = \frac{\phi ghr}{2} \tag{1}$$

where T is the surface tension (N/m), ρ is density (kg/m^3), g is the gravitational constant (9.81 m/s^2), h is the liquid height in the capillary tube (m), and r is the capillary tube radius (m).

2.2 Rhamnolipid Quantification

One-milliliter (1 mL) cell-free solution was added to 4.5 mL of 6:1 dilute sulfuric acid (volume/volume). The mixture was then heated at 95 °C in a water bath for 10 minutes. Once heated, the mixture was allowed to cool to room temperature. One-tenth milliliter (0.1 mL) of a 3% (v/v) solution of thioglycolic acid was added to the cooled mixture. The new mixture was then incubated for 3 hours in the absence of light. The absorbance of the incubated mixture was determined using a spectrophotometer at a wavelength of 430 nm. The rhamnolipid concentration was obtained from a calibration curve obtained from the pre-calibration of the atomic absorption spectrophotometer (AAS) using standard rhamnolipid.

3 Results and Discussion

3.1 Cell Growth

Figure 1(a, b) depicts the effects of varying the initial concentrations of waste cooking oil in the NH_4NO_3 medium and urea medium, respectively, on cell growth. It could be observed that *P. putida* grows well when using waste cooking oil as a carbon source and that the initial substrate concentration has an influence on cell growth. Furthermore, cell growth increased from 1.2 g to 2.02 g when the initial concentration of waste cooking oil in the NH_4NO_3 growth medium was increased from 2% (v/v) to 6% (v/v). However, there was a reduction in cell growth from 2.02 g to 1.79 g when the initial concentration of waste cooking oil was increased to 10%. According to Caro (Caro et al. 2008), it is expected that the higher the initial substrate concentration is, the higher is the cell growth. However, from the results in Fig. 1, this is not the case since the 6% (v/v) waste cooking oil media in the NH_4NO_3 growth medium had the highest cell growth, and the lowest cell growth is observed in the urea growth medium. The cell growth behavior in the ammonium nitrate growth medium suggests that the combination of 6% waste cooking oil and 2 g/L NH_4NO_3 reached a maximum cell growth of 1.79 g. However, in the urea medium, the results show that there is growth inhibition in the 6% (v/v) waste cooking oil medium, which resulted in the cell mass decreasing from 0.25 g to a value of 0.057 g after 7 days. The maximum cell mass obtained in the NH_4NO_3 medium after 3 days using 6% waste cooking oil was 2.0 g. In the urea medium, a maximum cell

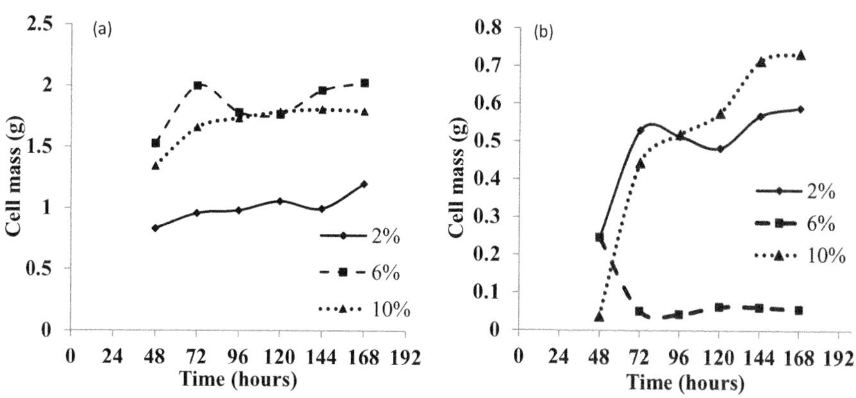

Fig. 1 Effect of waste cooking oil concentrations on cell growth in the (**a**) NH_4NO_3 medium and (**b**) urea medium. Experimental conditions: temperature (30 °C), stirring speed (215 rpm)

mass of 0.73 g was obtained after 7 days using 10% waste cooking oil media. The nitrogen source used for *P. putida* influenced rhamnolipid production, which affects cell growth as the rhamnolipid assists in the uptake of the waste cooking oil to feed the cells. From Fig. 1, it is clear that NH_4NO_3 promotes cell growth compared to urea. The cell mass in NH_4NO_3 media is at least 39% higher at 144 h than cell mass in the urea medium. Therefore, NH_4NO_3 is the most effective nitrogen source for cell growth.

3.2 Rhamnolipid Production

The production of biosurfactants is repeatedly observed when microorganisms are grown on hydrophobic/immiscible substrates, and the decrease in the surface tension of the growth medium has been used as an indication of effective biosurfactant production (Cameotra and Singh 2009). The ineffective production of rhamnolipids when a hydrophobic substrate is used is due to the solubility of the substrate and thus the lack of the necessity of the cells to secrete biosurfactants to enhance solubility (Wei et al. 2005). The presence of a biosurfactant was verified by analyzing its activity in terms of surface tension using the capillary tube method. Figure 2 shows the changes in the surface tension in the (a) NH_4NO_3 medium and (b) urea medium. It is clear that the surface tension between the hydrophobic waste cooking oil and water was reduced in all growth media, indicating the presence of biosurfactants. The reduction in surface tension is due to two phenomena: the cell uptake of waste cooking oil and the activity of the biosurfactant produced by the cells. The greatest reduction in surface tension in the NH_4NO_3 medium was obtained with an initial waste cooking oil concentration of 2%. The surface tension decreased from 0.33 N/m to 0.038 N/m after 7 days. In the urea medium, a maximum reduction in

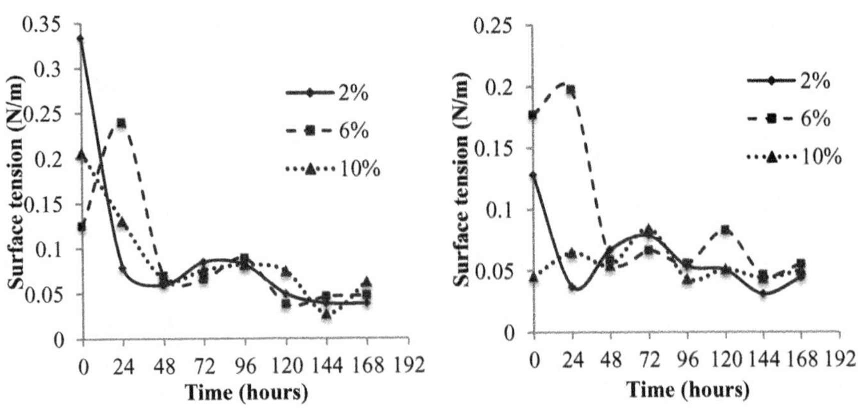

Fig. 2 The surface tension of the (**a**) NH_4NO_3 medium and (**b**) urea medium over time. Experimental conditions: temperature (30 °C), stirring speed (215 rpm)

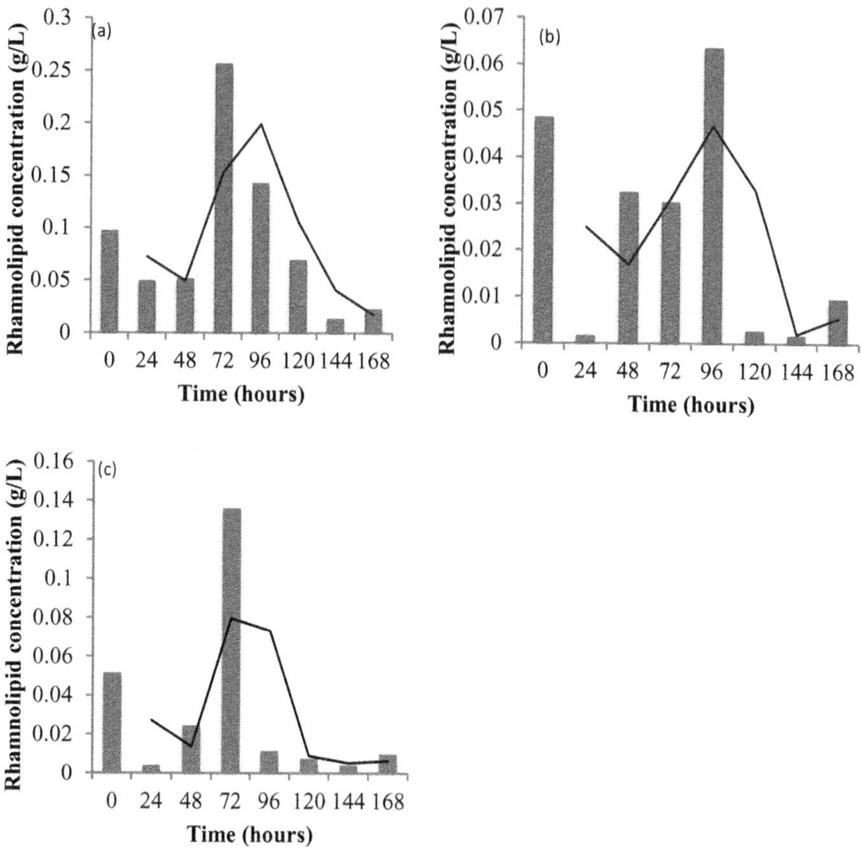

Fig. 3 Rhamnolipid concentration in the NH_4NO_3 medium with initial waste cooking oil concentration of (**a**) 2% v/v, (**b**) 6% v/v, and (**c**) 10% v/v. Experimental conditions: temperature (30 °C), stirring speed (215 rpm)

surface tension occurred with an initial waste cooking oil concentration of 6%. The surface tension was reduced from a maximum of 0.2 N/m to a minimum of 0.05 N/m after 7 days. The behavior of the surface tension is inconsistent with most published results, such as Kuiper (Kuiper et al. 2004), Janek (Janek et al. 2013), and Kanna (Kanna et al. 2014), which reported a rapid decrease in surface tension at the beginning, followed by constant surface tension. Fig. 2 shows fluctuations in the surface tension, which can be explained by considering the production of rhamnolipids, as shown in Figs. 3 and 4.

Figure 3 and Figure 4 show the rhamnolipid production in NH_4NO_3 and urea media, respectively. The rhamnolipid biosurfactant was produced using simple salt media containing urea and NH_4NO_3 as nitrogen sources and waste cooking oil as a carbon source. Rhamnolipid production relies greatly on nitrogen exhaustion by *P. aeruginosa*. Different nitrogen sources behave as either inhibitors or activators of

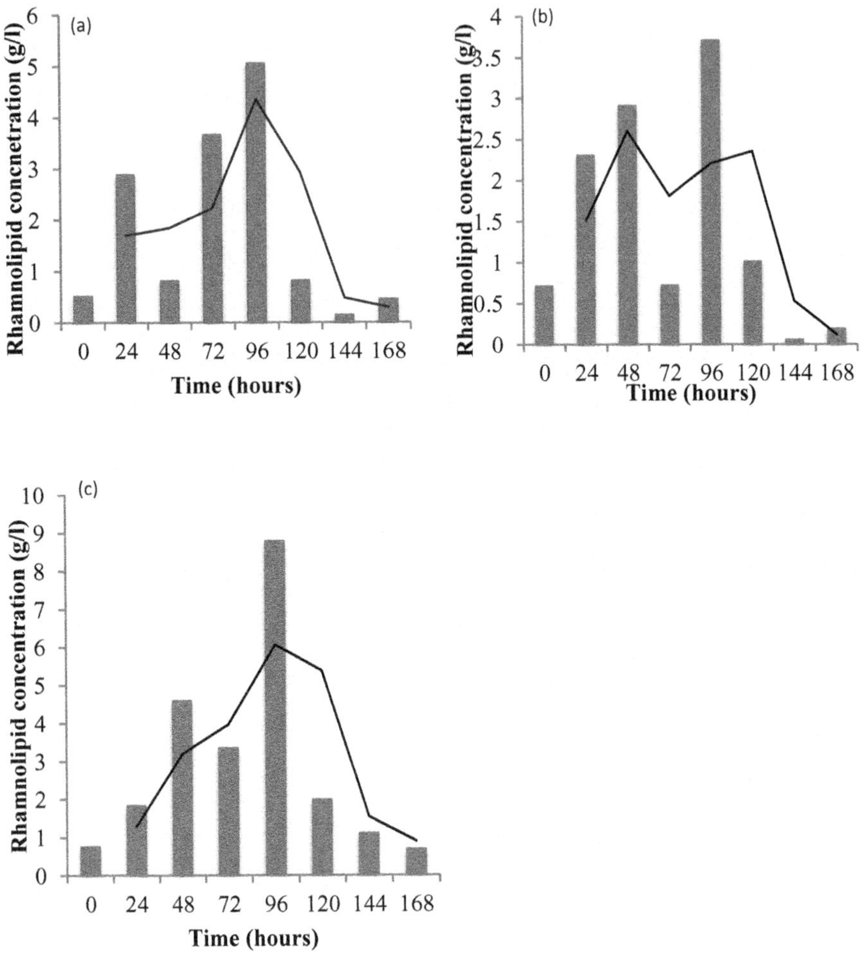

Fig. 4 Rhamnolipid concentration in urea media with initial waste cooking oil concentration of (**a**) 2% v/v, (**b**) 6% v/v, and (**c**) 10% v/v. Experimental conditions: temperature (30 °C), stirring speed (215 rpm)

rhamnolipid production (Reis et al. 2011). In the NH_4NO_3 medium, a maximum rhamnolipid concentration of 0.25 g/L was achieved after 72 hours using 2% waste cooking oil. In the urea medium, a maximum rhamnolipid concentration of 8.8 g/L was achieved after 96 h using 10% waste cooking oil. Rhamnolipid is predominantly produced by *Pseudomonas aeruginosa*. Wei (Wei et al. 2005) studied the production of rhamnolipid by *P. aeruginosa* using seven different carbon sources. Olive oil produced the highest rhamnolipid concentration of 3.6 g/L. Nitschke (Nitschke et al. 2005) was able to reach a rhamnolipid concentration of 11.7 g/L after 144 h from 2% (w/v) soybean waste using *P. aeruginosa.* The rhamnolipid concentration of 8.8 g/L achieved in this experiment shows better biosurfactant production than that in Wei (Wei et al. 2005) but a reduced production compared to what was reported by Nitschke (Nitschke et al. 2005). This suggests that nonpathogenic *P. putida* grown using urea as a nitrogen source can compete with the pathogenic *P. aeruginosa* to produce rhamnolipid.

According to Cameotra and Pruthi (Cameotra and Pruthi 2003), nitrate is the principal nitrogen source for the production of biosurfactants in *Pseudomonas* strain 44T1 and *Rhodococcus* strain ST-5 grown on olive oil and paraffin, respectively (Cameotra and Pruthi 2003). Comparing Figs. 3 and 4, it is clear that urea promotes rhamnolipid production more effectively than NH_4NO_3. This means that these results are not consistent with the study conducted by Cameotra and Pruthi (Cameotra and Pruthi 2003), which may be due to *P. putida* assimilating the nitrogen source differently compared to the *Pseudomonas* strain 44T1 used by Cameotra and Pruthi (Cameotra and Pruthi 2003). Kanna (Kanna et al. 2014) studied the influence of different nitrogen sources on the production of biosurfactants using *P. putida* and found that from ammonium sulfate, ammonium nitrate, and urea, ammonium sulfate produced the highest biosurfactant concentration of 2.5 g/L after 140 h. Ammonium nitrate and urea achieved a rhamnolipid concentration of 1.8 g/L and 1.2 g/L, respectively. Under the experimental conditions in this study, rhamnolipid production using NH_4NO_3 was lesser compared to that achieved in the study conducted by Kanna (Kanna et al. 2014). However, the rhamnolipid concentration obtained in the urea medium was higher compared to what was reported by Kanna (Cameotra and Pruthi 2003).

Figure 3 and Figure 4 show fluctuations in the rhamnolipid concentration, which means rhamnolipids were being produced and consumed by the *P. putida* cells in the medium. Furthermore, the rhamnolipid concentration increased up to a peak and then decreased thereafter. This trend is inconsistent with a study conducted by Tuleva (Tuleva et al. 2002) and Wei (Wei et al. 2005), which showed a steady increase in rhamnolipids before plateauing. However, a study conducted by Henkel (Henkel et al. 2012) shows a similar trend to the one presented in Figs. 3 and 4. The decreases in rhamnolipid concentration could be due to the cells absorbing the biosurfactant on the cell surface to assist in taking up the waste cooking oil and also due to cells absorbing the rhamnolipid to be used intracellularly. The decrease in rhamnolipids, owing to the aforementioned reasons, could explain the fluctuation in the surface tension shown in Fig. 2. As the rhamnolipids are taken up by the cells, the surface tension increased slightly because the waste cooking oil is also being taken up by the cells, contributing to lower surface tension. This trend in biosurfactant concentration could work effectively for bioremediation because the production of biosurfactants could assist in making contaminants available for biodegradation, and the consumption of the biosurfactants may also assist in preventing the accumulation of rhamnolipids in the environment.

4 Conclusion

The current study investigated the production of rhamnolipid biosurfactants from waste cooking oil in a batch reactor using *Pseudomonas putida*. The following conclusions were drawn:

I. *P. putida* was able to use waste cooking oil as the only carbon source under batch conditions to grow, as well as to produce, rhamnolipid biosurfactants.
II. Cell growth was highest in the ammonium nitrate medium, containing an initial waste cooking oil concentration of 6%. The highest rhamnolipid concentration of 8.8 g/L was achieved in the urea medium after 96 h using 10% (v/v) waste cooking oil.
III. Under the experimental conditions of this study, ammonium nitrate improved the growth of *Pseudomonas putida*, while urea effectively enhanced the production of rhamnolipids. The highest (8.8 g/L) rhamnolipid concentration obtained at 10% (v/v) waste cooking oil is sufficient to make nonpathogenic *P. putida* competitive with pathogenic *P. aeruginosa* in terms of biosurfactant production. However, the fluctuations in rhamnolipid concentration make the growth conditions in this study unsuitable for industrial application due to the consumption of desired products. However, the rhamnolipid production trend observed in this study could make these conditions suitable for bioremediation purposes. This is due to the occurrence of biodegradation, accompanied by rhamnolipid production, without the accumulation of biosurfactants in the environment since the rhamnolipid is regularly consumed.

Acknowledgement The authors acknowledge the School of Chemistry and the School of Physics at the University of Witwatersrand for assisting in preparing the custom-made glassware used in this study.

References

Henkel M, Muller MM, Kugler JH, Lovaglio RB, Contiero J, Syldatk C, Hausmann R 2012 Rhamnolipids as biosurfactants from renewable resources: Concepts for next-generation rhamnolipid production. Process Biochemistry **47** 1207-1219.

Mulligan CN 2004 Environmental application for biosurfactants. Environmental Pollution 183–198.

Janek T, Lukaszewicz M, Krasowska A 2013 Identification of biosurfactant produced by the arctic bacterium *Pseudomonas putida* BD2. Colloids and Surfaces B: Biointerfaces, **110** 379-386.

Cameotra SS, Pruthi V 2003 Effects of nutriens on the optimal production of biosurfactant by *Pseudomonas putida* – A Gujarat oil field isolate. Journal of Surfactants and Detergents, **6**(1), 65-68.

Kanna R, Gummadi, SN, Kumar GS 2014 Production and characterisation of biosurfactant by *Pseudomonas putida* MTCC 2467. J. Biol. Sci. **14**(6), 436–445.

Reis RS, Pereira AG, Neves BC, Freire DMG 2011 Gene regulation of rhamnolipid production in *Pseudomonas aeruginosa* – A review. Bioresource Technology **102** 6377–6384.

Maier RM and Soberon-Chavez G 2000 *Pseudomonas aeruginosa* rhamnolipids: biosynthesis and potential applications. Applied Microbiol Biotechnology 625–633.

Fernandes NMM 2009 *Characterisation of the avtivity of biosurfactants produced by Pseudomonas species isolated from food.* Johannesburg: MSc dissertation, submitted to the University of the Witwatersrand, South Africa.

Tuleva BK, Ivanov GR and Christova NE 2002 Biosurfactant production by a new *Pseudomonas putida* strain. Naturforsch. **57c** 356–360.

Wittgens A, Tiso T, Arndt TT, Wenk P, Hemmerich J, Müller C, Wichmann R, Küpper B, Zwick M, Wilhelm S, Hausmann R, Syldatk C, Rosenau F, Blank LM 2011 Growth independent rhamnolipid production from glucose using the non-pathogenic Pseudomonas putida KT2440. Microbial Cell Factories **10** (80) 1-17 http://www.microbialcellfactories.com/content/10/1/80

Tan YN and Li Q 2018 Microbial production of rhamnolipids using sugars as carbon sources. Microb Cell Fact **17** 89-101. https://doi.org/10.1186/s12934-018-0938-3

Tiso T, Ihling N, Kubick S, Biselli A, Schonhoff A, Bator I, Thies S, Karmainski T, Kruth S, Willenbrink A-L, Loeschcke A, Zapp P, Jupke A, Jaeger K-E, Büchs J and Blank LM 2020 Integration of Genetic and Process Engineering for Optimized Rhamnolipid Production Using *Pseudomonas putida.* Front. Bioeng. Biotechnol. **8** (976) 1–23. https://doi.org/10.3389/fbioe.2020.00976.

Wei YH, Chou CL and Chang JS 2005 Rhamnolipid production by indigenous *Pseudomonas aeruginosa* J4 originating from petrochemical wastewater. Biochemical Engineering Journal **27** 146-154.

Caro A, Boltes K, Leton P, Garcia-Calvo E 2008 Biodesulfurization of dibenzothiophene by growing cells of Pseudomonas putida CECT 5279 in biphasic media. Chemosphere **73** 663-669.

Cameotra SS and Singh P 2009 Synthesis of rhamnolipid biosurfactant and mode of hexadecane uptake by *Pseudomonas* species. Microbial Cell Factories **8**(6) 1-7.

Kuiper I, Langendijk EL, Pickford R, Derrick JP, Lamers GEM, Thomas-Oates JE, Lugtenberg BJJ, Bloemberg GV 2004 Characterization of two *Pseudomonas putida* lipopeptide biosurfactants, putisolvin I and II, which inhibit biofilm formation and break down existing biofilms. Molecular Microbiology **51**(1) 97-113.

Nitschke M, Costa SG, Haddad R, Goncalves LA, Eberlin MN, Contiero J 2005 Oil wastes as unconventional substrates for rhamnolipid biosurfactant production by *Pseudomonas aeruginosa* LBI. Biotechnol. Prog. **21**(5) 1562–1566.

Estimation Model for Cow Dung-aided Water Hyacinth Digestion

Ochuko Mary Ojo, Josiah Oladele Babatola, and Adesoji Adeniran Adesina

1 Introduction

Increase in population has led to over-dependence on fossil fuels such as natural gas and petroleum. But these energy sources are non-renewable and are rapidly depleting. Furthermore, the exploitation, processing and combustion of these fossil fuels constitute a dangerous threat to the feeble environment (Owamah and Izinyon, 2015). Anaerobic digestion (AD) of waste organic substances has been distinguished to be among the limited biotechnological processes that can generate biofuel, reduce environmental pollution and improve agricultural productivity through the use of its digestate as compost for organic farming (Zhang et al., 2016). Water hyacinth (*Eichhornia crassipes*) is a freshwater aquatic plant and is a native of Brazil. It is known to be one of the most widespread aggressive aquatic plants in the world. It is also regarded as a disadvantageous weed in many countries as it grows recklessly and has the tendency to deplete nutrients and oxygen quickly from water bodies (Ojo, 2017). Although biogas production from substrates of plant origin such as WH has been extensively studied, the low biodegradability of these substrates and consequent low yield of biogas has often made the digestion process uneconomical in comparison to highly biodegradable animal wastes (El-Mashad and Zhang, 2010).

Co-digestion, which is known as the simultaneous anaerobic digestion of two or more feedstocks in a single digester, has been employed by several authors for the improvement of biogas yield (Haider, 2015; Ojo, 2017; Ojo et al., 2018; Ojo et al.,

O. M. Ojo (✉) · J. O. Babatola
Department of Civil Engineering, The Federal University of Technology, Akure, Nigeria
e-mail: omojo@futa.edu.ng

A. A. Adesina
ATODATECH LLC, Los Angeles, CA, USA

© The Author(s), under exclusive license to Springer Nature Switzerland AG 2022

A. O. Ayeni et al. (eds.), *Bioenergy and Biochemical Processing Technologies*, Green Energy and Technology, https://doi.org/10.1007/978-3-030-96721-5_19

2019; Ojo and Babatola, 2020). Although some researchers have established the viability of co-digesting WH with other more familiar animal feedstocks (such as cow dung) for improved biogas production, none of them have established an optimum mix ratio for efficient biogas production while evaluating some of the parameters that affect biogas production. Al-Imam et al. (2013) evaluated biogas production from cow dung (CD), poultry manure (PM) and WH, but a co-digestion of the different feedstocks was not carried out. It was observed from that study that biogas production from CD, PM and WH was 0.034 m3/kg, 0.058 m3/kg and 0.014 m3/kg respectively. Adegunloye et al. (2013) investigated the ratio variation of WH on the production of pig dung (PD) biogas. However, the blended WH and PD was only weighed in ratios 1:1 and 1:3. The aim of this study is to determine the best mix of WH to be co-digested with CD in order to maximize the quantity of biogas produced from a batch-fed anaerobic digester.

2 Methodology

2.1 Digester Design

Twenty-five-litre-capacity laboratory-sized batch-fed digesters were used in this study. Eq. 1 was used to calculate the total volume of the digester (*TVD*):

$$TVD = 25\text{L} \tag{1}$$

But the volume of the digestion chamber (*VDC*) was taken as $\frac{3}{4}$ of the TVD in order to allow for gas collection.

Eqs. 2 and 3 were used to calculate *VDC*:

$$VDC = \frac{3}{4}\text{x}TVD \tag{2}$$

$$VDC = \frac{3}{4}\text{x}25 = 18.75\text{L} \tag{3}$$

One fourth of the *TVD* was allowed for gas collection; hence, the volume of the gas chamber (VG) was calculated using Eqs. 4 and 5:

$$\text{VG} = \frac{1}{4}\text{x}TVD \tag{4}$$

$$\text{VG} = \frac{1}{4}\text{x}25 = 6.25\text{L} \tag{5}$$

2.2 Substrate Mix Ratios

The following 11 mix ratios of WH:CD were used in this study: 10:0, 9:1, 8:2, 7:3, 6:4, 5:5, 4:6, 3:7, 2:8, 1:9 and 0:10. WH was harvested from a remote pond in Akure, while CD was obtained from the Animal Production and Health farm of the Federal University of Technology, Akure.

The substrates were mixed with the appropriate amount of water specified by Fekadu (2014) and Ojo (2017) and depicted in Eq. 6 in order to achieve the recommended 8% of the total solid content in the fermentation slurry:

$$Y = \frac{mTS - 8\%X}{8\%} \tag{6}$$

where mTS is the mass of total solids,
X is the mass of fresh substrate,
Y is the mass of water added to get 8% total solids in the digester.

The mixing of substrates was done manually, while the experimental set-up is shown in Fig. 1.

A rotameter flowmeter of model LZM-4 T with a capacity of 0.1–1 L/Min equipped with a measuring tube was used for the gas flow measurements. Fig. 2 shows the rotameter used in the experiments. The measuring tube had an adjustable cross-section, bordering at the entrance of the tube and broadest at the exit of the tube. The float located inside the flow tube was engineered so that its diameter was nearly identical to the flow tube's inlet diameter. When the gas from the digester was introduced into the tube, the float was lifted from its initial position at the inlet,

Fig. 1 Experimental set-up

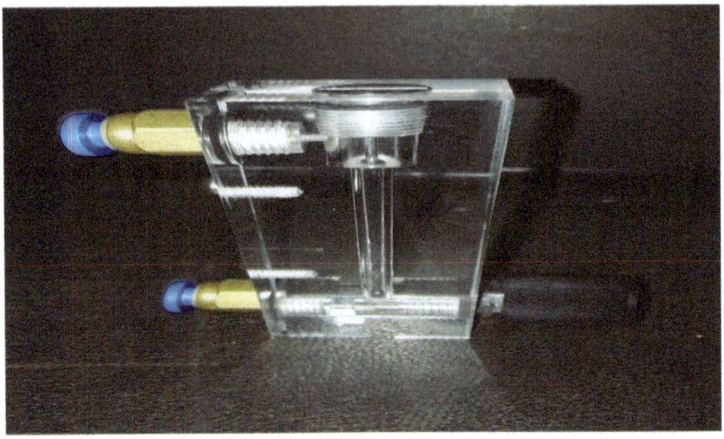

Fig. 2 Rotameter used in the experiments

allowing the gas to pass between it and the tube wall. As the float rose, more biogas passed by the float because the tapered tube's diameter was increasing. Ultimately, a point was reached where the flow area was large enough to allow the entire volume of the gas to flow past the float. The float was now stationary at that level within the tube as its weight was being supported by the fluid forces that caused it to rise. This position corresponded to a point on the tube's measurement scale and provided an indication of the biogas flow rate.

The equation for the biogas flow rate was calculated using the ideal gas equation in Eq. 7:

$$\frac{P_1V_1}{T_1} = \frac{P_2V_2}{T_2} \tag{7}$$

The gas flow rate was calculated using Eq. 8:

$$V_2 = \frac{P_1V_1T_2}{P_2T_1} \tag{8}$$

V_1 is the rotameter reading (normal litres per minute), and V_2 is the actual flow rate (litres per minute).

P_1 is the pressure at normal conditions, and P_2 is the actual pressure.

T_1 is the temperature at normal conditions, and T_2 is the actual temperature of air.

The normal conditions used were 0 °C (273.15 °K) and 1 atmosphere (1.01325 bar).

2.3 Data Analysis

The data obtained were analyzed using Sigma Plot 2010 software for non-linear regression with four parameters, as described in Ojo (2017). The sigmoid or S-curve obtained is described by Eq. 9:

$$y = a\left[1 - e^{(-bt)}\right]^{c} \tag{9}$$

where a is ultimate biogas production, b is pseudo-biogas production velocity (rate constant), c is shape factor and t is retention time.

Biogas production rate is given by Eq. 10:

$$G^{1} = \frac{dy}{dt} = a\left(1 - e^{-bt}\right)^{c-1}.be^{-bt} = abe^{-bt}\left(1 - e^{-bt}\right)^{c-1} \tag{10}$$

Optimum (maximum) biogas production rate will be reached using Eq. 11:

$$\frac{d^{2}y}{dt^{2}} = abe^{-bt}\left[(c-1)\left(1 - e^{-bt}\right]^{c-2}.be^{-bt}\right] + \left[\left(1 - e^{-bt}\right)^{c-1} ab\left(-be^{-bt}\right)\right] = 0 \tag{11}$$

which simplifies to Eq. 12:

$$acb^{2}e^{-bt}(c-1)\left(1 - e^{-bt}\right)^{c-2} e^{-bt} = acb^{2}e^{-bt}\left(1 - e^{-bt}\right)^{c-1} \tag{12}$$

The time taken for maximum biogas production, t_{max}, is given by Eq. 13 and is further simplified to Eq. 14:

$$(c-1)e^{-bt} = \left(1 - e^{-bt}\right)att = t_{max} \tag{13}$$

$$t_{max} = \frac{(lnC)}{b} \tag{14}$$

At optimum time, the maximum biogas production rate (G^{1}_{max}) is given by Eq. 15:

$$G^{1}_{max} = abe^{-b\left(\frac{lnC}{b}\right)}\left[1 - e^{-b\left(\frac{lnC}{b}\right)}\right]^{c-1} = -abc[1-c]^{c-1} \tag{15}$$

Table 1 Cumulative biogas production for CD-aided WH digestion

Feedstocks	Mix ratios									
WH	10	9	8	7	6	5	4	3	2	1	0
CD	0	1	2	3	4	5	6	7	8	9	10
Cumulative volume of biogas produced (litres)	32.17	53.17	67.04	81.04	89.3	99.29	116.38	128.11	141.72	132.77	126.95

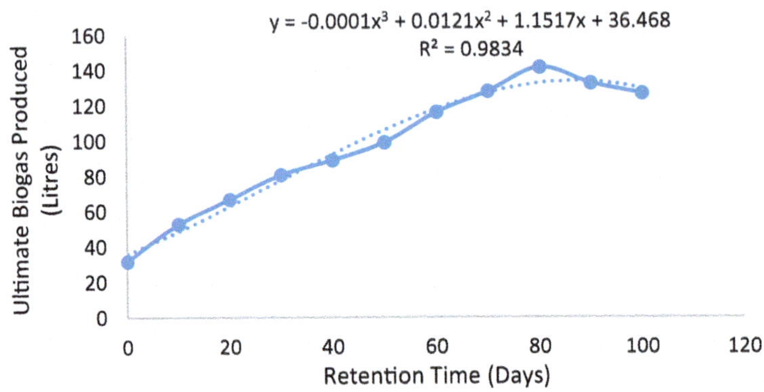

Fig. 3 Ultimate biogas production for CD-aided WH digestion

3 Results and Discussions

The ultimate biogas yield for each mix was obtained by calculating the average of the cumulative biogas volume from day 35 to 40. The cumulative volume of biogas produced for the different mixes is presented in Table 1.

Single-substrate digestion of CD produced an ultimate gas volume of 126.95 L, corresponding to 20.3 L/kg of CD. Fig. 3 portrays the ultimate biogas yield for CD-aided WH digestion. The results showed a range of 53.18–141.72 L, with 2 WH:8 CD recording the highest value corresponding to 22.68 L/kg.

The ultimate biogas production curve for CD-aided WH digestion is shown in Fig. 3. The third-order polynomial curve has an R^2 value of 0.9834, which is quite significant. The graph shows that the highest ultimate biogas of 141.72 L was produced by 2 WH:8 PD and impressively makes it the best PD-aided WH digestion mix.

Table 2 CD-aided WH digestion maximum biogas production rate

Composition (%CD)	a value	b value	c value	t_{max} (days)	Maximum biogas production rate (L/day)
0	34.16	0.1637	17.85	17.6054	2.116757
10	58.18	0.157	21.46	19.52988	3.440793
20	72.44	0.1617	22.47	19.24664	4.407624
30	86.14	0.1685	23.44	18.72074	5.456434
40	94.66	0.1669	21.21	18.30121	5.952935
50	104.4	0.1696	20.57	17.82921	6.676728
60	122	0.1765	22.34	17.59988	8.10361
70	134.3	0.1789	23.66	17.68466	9.030292
80	149.4	0.1761	22.9	17.78045	9.895529
90	139.8	0.1749	22.11	17.70172	9.203971
100	133.3	0.1789	23.33	17.60615	8.965811

a ultimate biogas production; b pseudo-biogas production velocity (rate constant); c shape factor; t_{max} time for maximum biogas production

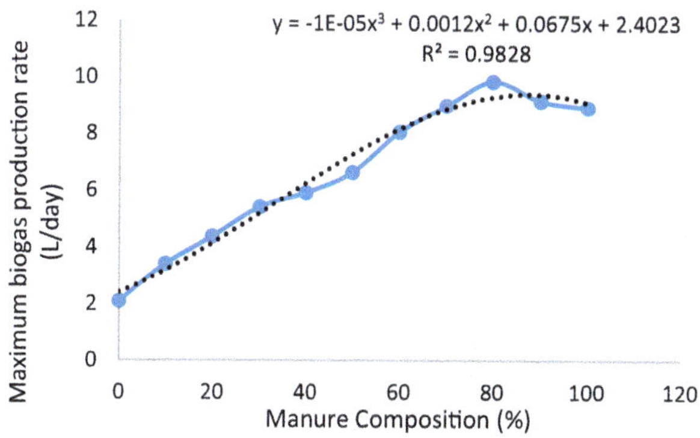

Fig. 4 Maximum biogas production rate for CD-aided WH digestion

3.1 *CD-Aided WH Digestion Maximum Biogas Production Rate*

Table 2 shows the results of the estimation model for the CD-aided WH digestion maximum biogas production rate. Mix ratios 10 WH:0 CD, 5 WH:5 CD, 4 WH:6 CD, 3 WH:7 CD, 2 WH:8 CD, 9 WH:1 CD and 0 WH:10 CD produced maximum gas between the 17th and 18th day, while mix ratios 7 WH:3 CD and 4 WH:6 CD produced maximum gas on the 18th day. Mix ratios 9 WH:1 CD and 2 WH:8 CD produced maximum gas on the 19th day. The third-order polynomial curve for the different mixes of the CD-aided biogas production rate is shown in Fig. 4. The curve has an impressive R^2 value of 0.9828, and the standard deviation of the rate constant is 0.007422. From the results obtained, the best mix for CD-aided WH digestion is 2 WH:8 CD. This mix produced maximum biogas of 9.895529 L/day, and this was produced on day 17 of the digestion process.

4 Conclusion

The effect of the mix ratios of co-digested WH and CD on biogas production was evaluated. Eleven mix ratios of WH to CD were assessed, and each experimental run was done over a retention period of 40 days. The data obtained were analyzed using a non-linear parameter estimation model. From the results obtained, the best mix for CD-aided WH digestion is 2 WH:8 CD. This mix ratio produced 22.68 L of biogas per kg of feedstock at a maximum biogas production rate of 9.895529 L/day. The co-digestion of WH and CD in the appropriate mix ratio should be encouraged in order to maximize biogas production from anaerobic digesters.

Acknowledgement The authors wish to acknowledge the staff of Water Resources and Environmental Engineering Laboratory of the Department of Civil Engineering, Federal University of Technology, Akure, for creating an enabling environment to carry out the experiments in this study.

References

Adegunloye DV, Olosunde SY, Omokanju AB (2013) Evaluation of Ratio Variation of Water Hyacinth (Eichhornia Crassipes) on the Production of Pig Dung Biogas. Int. Res J Biol Sci 2(3): 44-48.

Al-Imam MFI, Khan MZH, Sarkar MAR, Ali SM(2013) Development of biogas processing from cow dung, poultry waste, and water hyacinth IJONAS 2(1): 13-17.

El-Mashad HM, Zhang R (2010) Biogas Production from Co-digestion of Dairy Manure and Food Waste. Bioresour Techno 101:4021-4028.

Fekadu M (2014) Biogas production from Water Hyacinth (eichhornia crassipes) Co-digestion with cow-dung. MSc Dissertation, Department of Biology, Haramaya University

Haider MR, Zeshan YS, Malik RN, Visvanathan C (2015) Effect of mixing ratio of food waste and rice husk co-digestion and substrate to inoculum ratio on biogas production. Bioresour. Techno 190: 451–457.

Ojo OM. (2017) Biomethanation of Water Hyacinth and Selected Animal Dungs for Biogas Production. Unpublished Ph.D Thesis in the Department of Civil Engineering, The Federal University of Technology, Akure.

Ojo OM, Babatola JO, Adesina AA., Akinola, AO, Lafe O (2018) Synergistic Effect of co-digesting different mix ratios of Water Hyacinth and Cow-dung for Biogas production. FUTAJEET 12 (1): 54 – 59

Ojo OM, Babatola JO, Akinola AO, Lafe O, Adelodun AA (2019) Co-Digestion of Water Hyacinth and Poultry Manure for Improved Biogas yield. ABUAD J Eng Res Dev 2 (1): 42-48

Ojo OM, Babatola JO (2020) Association between Biogas Quality and Digester Temperature for Selected Animal Dung-Aided Water Hyacinth Digestion Mixes. J Appli Sci Environ Manage 24 (6): 966-959

Owamah HI, Izinyon OC (2015) Development of simple-to-apply biogas kinetic models for the Co-digestion of Food Waste and Maize Husk. Bioresour Techno 194: 83– 90.

Zhang Q, Hu J, Lee D (2016) Biogas from Anaerobic Digestion Processes: Research Updates. Renew Energy 98:108-119.

Phytochemicals and Anti-Microbial Properties of Neem (*Azadirachta indica*) Seed Oil Extract

M. E. Ojewumi, O. R. Obanla, G. P. Ekanem, and J. U. Nsionu

1 Introduction

Plant parts, such as seeds, fruits and leaves, that have medicinal properties have been widely used by humans to combat diseases and sicknesses. For over 2000 years, Azadirachta indica and other plants have been well known in many countries as medicinal plants having a wide range of application (Biswa et al., 2002; Ojewumi & Owolabi, 2012). Essential oils from plants have been used in the pharmaceutical and food industries for several years due to their anti-microbial properties. Since ancient times, some of these plants have been a very good source of aromatic extract. Aside from the hazards they cause to men and non-living things, synthetic pesticides and insecticides are also usually expensive, making them inaccessible to rural dwellers around the world (Ojewumi 2018; Akinmoladun et al., 2007). Active compounds derived biologically from selected plants species, such as lemongrass, Ocimum gratissiumum, Moringa oleifera, Mentha spicata, Acasia fistula, Acarcia Arabica, Azadirachta indica, hyptis sauveolen and Eleusive indica, have been very effective for the control of various types of insects, especially mosquitoes (Adeniyi & Ayepola, 2008; Bhowmik et al., 2010; Biswas et al., 2002; Chuakul, Saralamp, & Boonpleng, 2002; Claustra et al., 2005; Ojewumi et al., 2018a; Ojewumi et al., 2017a). Every part of the A. indica plant has been proven in herbal and Ayurveda medicines to be a remedy against numerous ailments (Khare, 2008; Koul, Isman, & Ketkar, 1990; Vaidya & Devasagayam, 2007). A. indica extracts have been used for various diseases in herbal medicine for a very long time (Champagne et al., 1992; Kumar, Mishra et al., 2010; Kumar, Sharma, & Devi, 2018). The seeds and leaf extract of the plant are effective against different types of malarial parasites

M. E. Ojewumi (✉) · O. R. Obanla · G. P. Ekanem · J. U. Nsionu
Chemical Engineering Department, Covenant University, Sango Ota, Ogun State, Nigeria
e-mail: modupe.ojewumi@covenantuniversity.edu.ng

© The Author(s), under exclusive license to Springer Nature Switzerland AG 2022

A. O. Ayeni et al. (eds.), *Bioenergy and Biochemical Processing Technologies*, Green Energy and Technology, https://doi.org/10.1007/978-3-030-96721-5_20

Fig. 1 Fresh unprocessed neem seeds

(Akin-Osanaiya et al., 2013; Daniel & Dishi, 2011; Maithani et al., 2011; Ucheya, Ucheya, & Amiegheme, 2011). The alcoholic extract components of both its leaves and seeds are effective against sensitive strains and chloroquine-resistant malarial parasites (Kumar et al., 2010; Oseni & Akwetey, 2012). Studies recently revealed, crude extract of neem seed and its purified fractions to show inhibitory ability on the growth of asexual and sexual stages of different drug resistant strains of the human malarial parasite P. falciparum (Maithani et al., 2011).

Various techniques have been used to extract oil from seed kernels of various sources. These include aqueous and Soxhlet extraction (Ojewumi et al., 2019a; Yusoff et al., 2016) supercritical fluid extraction (Martins, De Melo, & Silva, 2016; Nguyen et al., 2011), cold mechanical pressing and solvent extraction methods (Bhutada et al., 2016). Aqueous extraction is preferred to solvent extraction since it entails the use of water, which is economical, portable and safer compared to solvent (Ojewumi et al., 2019b; Ojewumi et al., 2019c) (Figs. 1 and 2).

2 Material and Methods

2.1 Raw Materials

Fresh neem seeds (*A. indica*) were sourced from Sokoto State, Northern Nigeria, and were processed in the Chemical Engineering Department Laboratory.

Fig. 2 Sorted processed neem seeds

2.2 Seed Processing

Fresh seeds were de-pulped from their fruits and de-hulled to remove the husk from the seeds. Hand-picked clean seeds were oven dried at 60 °C for 72 hours to remove moisture content. The seeds were pulverized (turned into power) using Corona milling machine in order to obtain a very large surface area for good extraction process.

2.3 Extraction

The extraction of neem oil was achieved using the Soxhlet extraction method. Two hundred grams of neem seed was weighted into a round-bottom flask containing 500 ml of hexane. The flask was then coupled to the extractor and the condenser attached at the rear end, and a fractionating column and a receiving glass tube were also attached. Cooling water was supplied continuously from the pipe to the condenser for cooling. A heating mantle was used to supply heat to the round-bottom flask for 2 hours. The procedure was repeated three times, and 600 g of leaves were used. The apparatus was allowed to cool down after each extraction process before proceeding to the next extraction process. The oil extracts were left to evaporate until dry and then stored in an airtight glass container until needed (Ojewumi et al., 2018b; Ojewumi et al., 2019d).

2.4 Preparation of Microbial Culture

Six microbes (four bacteria and two fungi) – *Pseudomonas aeruginosa, K. pneumoniae, Staphylococcus aureus, Escherichia coli, C. albicans* and *Rhizopus* – were obtained from the Microbiology Unit of the Department of Biological Sciences, Covenant University, Ota, Ogun State, Nigeria. Isolates were checked for purity and reconfirmed using basic biochemical tests.

2.5 Phytochemical Analysis

Phytochemical screening for the secondary metabolites was carried out using the ethods of Edeoga, Okwu and Mbaebie (2005); Elizabeth et al. (2019); Ojewumi et al. (2017a); Ojewumi et al. (2017b) and Ojewumi et al. (2017c)

Test for carbohydrates: 1 ml of Molisch's reagent was added to 2 ml of the oil extract, after which a few drops of concentrated sulphuric acid were added. A purple colouration depicts the presence of carbohydrates.

Test for glycosides: 3 ml of chloroform and 10% NH3 solution were added to 2 ml of the oil extract. A pink colouration depicts that glycosides are present.

Test for terpenoids (Salkowski test): 2 ml of chloroform and 3 ml of concentrated H_2SO_4 were added to 5 ml of the extract to form a layer of reddish-brown colouration.

Test for phenols: 2 ml of 10% ferric chloride was added to 2 ml of the leaf oil extract. A bluish-green color formed, showing the presence of phenols.

Test for acids: Sodium bicarbonate solution was added to 1 ml of the oil extract. The formation of effervescence depicts the presence of acids.

Test for steroids: To 1 ml sample extract, 1 ml chloroform and equal volume of concentrated H_2SO_4 acid was added through the walls of the test tube. Red colour in the upper layer and yellow colour with green fluorescence were observed, showing the presence of steroids.

Test for cardiac glycosides: Glacial acetic was added to 1 ml of the leaf extract. A few drops of ferric chloride with concentrated H_2SO_4 were added. A reddish brown at the junction of two layers was observed, which indicates the presence of cardiac glycosides.

2.6 Determination of Anti-Microbial Analysis

Test for the Anti-Microbial Activity of the Oil Extracts

Sterile nutrient agar plates were prepared and allowed to solidify. Aliquot 0.1-ml *S. aureus*, *E. coli*, *K. pneumoniae*, *P. aeruginosa*, *C. albicans* and *Rhizopus* was introduced into different plates, and a sterile cotton swab was used to spread the inocula evenly on the surface of the agar. The plates were left on the bench for 1 hour for the inocula to diffuse into the agar. A sterile cork borer of 5 mm was used to make five ditches each on the different plates. Then 100-mg/ml, 50-mg/ml, 25-mg/ml, 12.5-mg/ml and 6.25-mg/ml extracts were made; 0.5 ml of the extracts was dropped in each of the appropriately labelled plates; and control was set up for each plate by adding 0.5 ml of the appropriate solvent into the fifth ditch. The plates were left on the bench for a few minutes for the diffusion of the extract into the agar to take place. It was then incubated at 37 °C for 24 hours. After incubation, the zone of clearance was measured around each ditch using a metric ruler by taking measurements from the edge of the plate to the point where the growth of the organism started. The diameter of the zone of inhibition, which represents antibacterial activity, was recorded (Anibijuwon et al., 2012). The agar used was the Mueller-Hinton agar (MHA). The control used for the bacteria was gentamicin.

Determination of the Minimum Inhibitory Concentration (MIC) of the Oil

The broth dilution method was used to determine MIC. Various concentration of extract were used, which ranged from 5 mg/ml – 200 mg/ml each concentration 0.1 ml was added to each 9 ml of nutrient broth containing 0.1 ml of standardized test organism of bacterial cells. The tubes were incubated aerobically for 24 hours at 37 °C.

Determination of Minimum Bactericidal Concentration (MBC) of the Oil

Samples from the same tubes used in MIC (minimum inhibitory concentration) determination which did not show any visible growth after the period of incubation was streaked on nutrient agar plates. The least or lowest concentration of the extract indicating a bacterial effect after 2 hours of aerobic incubation at 37 °C was regarded as the minimum bactericidal concentration (MBC).

Table 1 Phytochemical properties of neem oil

S/no	Constituents	Hexane extract
1	Carbohydrates	Absent
2	Glycosides	Absent
3	Cardiac glycosides test	Absent
4	Terpenoids	Present
5	Triterpenoids	Present
6	Phenols	Present
7	Acids	Present
8	Steroids	Present

3 Results and Discussion

The phytochemical screening of the crude oil extract was carried out. Table 1 shows the results obtained.

Table 2 revealed the anti-microbial activities of neem seed crude oil extract with gentamicin and nystatin used as control for bacteria and fungi, respectively.

Table 3 shows the lowest concentration of the oil extract that will inhibit the visible growth of a microorganism after overnight incubation.

The minimum bactericidal concentration (MBC), which is the lowest concentration of an antibacterial agent that is required to kill a particular bacterium, was analyzed and reported in Table 4.

3.1 Composition of Neem Oil Extract Using Gas Chromatography-Mass Spectrometry (Fig. 3 and Table 5)

3.2 Phytochemical Analysis

From the results obtained, Terpenoids, Triterpenoids, Acids and Phenol were found out to be present in the oil extract. Other metabolites not present in this research have been identified by other researchers in different parts of Neem plants (seeds, leaves, stem and flowers)

Table 2 Anti-microbial analysis results on neem oil

		Zone of inhibition (mm)	
Organisms	Extract	Control gentamicin	Commercial extract
S. aureus	4	12	3
E. coli	4	5	5
K. pneumoniae	4	10	5
P. aeruginosa	4	11	4
		Control nystatin	
C. albicans	40	30	25
Rhizopus	20	33	17

Table 3 Minimum inhibitory concentration determination

	Concentration of neem seed extract (mg/ml)				
Organisms	100	50	25	12.5	6.25
E. coli	–	–	–	+	+
S. aureus	–	–	+	+	+
K. pneumoniae	–	–	–	+	+
P. aeruginosa	–	+	+	+	+
C. albicans	–	–	–	+	+
Rhizopus	–	–	–	+	+

Key: (+) growth was noticed
(–) growth was absent (inhibited)

Table 4 Minimum bactericidal concentration determination of *M. oleifera* seed extract

	Concentration of neem seed extract (mg/ml)				
Organisms	100	50	25	12.5	6.25
S. aureus	–	–	+	+	+
E. coli	–	–	+	+	+
K. pneumoniae	–	+	+	+	+
P. aeruginosa	–	+	+	+	+
C. albicans	–	+	+	+	+
Rhizopus	–	+	–	+	+

Key: (+) growth observed
(–) growth inhibited

3.3 Anti-Microbial Analysis

The results obtained in Table 2 showed that oil extract had little or no bacteria inhibition on *S. aureus*, *E. coli*, *K. pneumoniae*, *P. aeruginosa*. However, it showed high inhibition on *C. albicans* and *Rhizopus*. The antibacterial activity of oil extracts was low compared to the zone of inhibition of the fungal isolates, with 40 and 20 mm for C. albicans and Rhizopus respectively. Since the extracts of the oil had high

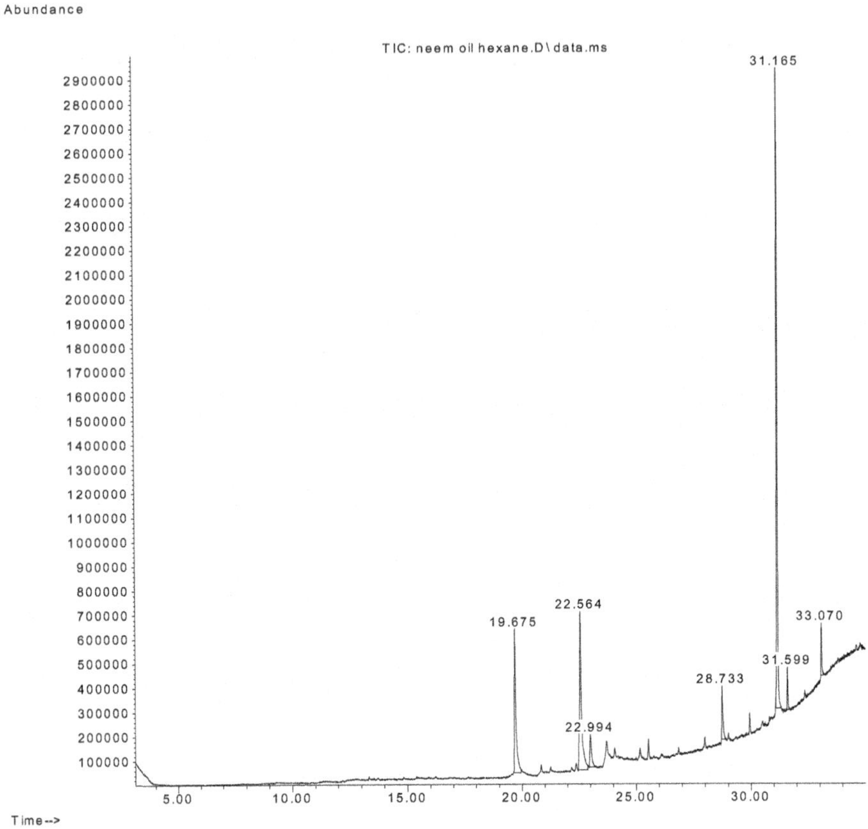

Fig. 3 Spectrum of neem oil extracted using hexane

inhibitory characteristics on these two fungi, effective drugs can be developed from the extracts. This is of great benefit for the medical world when it Comes to eradicating the diseases caused by fungi

3.4 MIC and MBC Analysis

Based on the MIC results, MBC was performed against the entire selected microorganisms using different concentrations of the neem crude oil extract. The extract was discovered to have consequential antibacterial activity against the microorganisms tested; the results of this research support the traditional claims regarding neem seed. The result of the MIC of the extract showed that *S. aureus*, *P. aeruginosa*, *E. coli*, *K. pneumoniae*, *C. albicans* and *Rhizopus* were susceptible or sensitive at concentrations of 12.5 and 6.25 mg/ml of the extract. Fungi growth was inhibited, resulting in visually clear tubes after 24 hours of incubation.

Table 5 Selected composition of neem oil extracted using hexane

Constituents	%Composition
Hexadecenoic acid	16.22
Squalene	47.10
11-Octadecenoic acid	22.87

3.5 Gas Chromatography Mass Spectrometry (GCMS)

Only three major components were reported in this work since they have significant quantity. Identified compounds were shown in Table 4, with squalene as the most abundant and likely to occur. Squalene is a fatty acid found in oil and fats. It is an organic compound with the formula $(C_5H_8)_6$. It is also a natural antioxidant and has anti-tumor properties, protecting the skin from carcinogens. It has anti-inflammatory properties that can reduce redness and swelling.

4 Conclusion

The results showed that the hexane extract of neem oil proves to be a good fungal inhibitory substance as it had high inhibition toward *Rhizopus* and *C. Albicans*. Drugs for eradicating these fungal diseases can be formulated from neem oil, which can also be used to combat storage and field pests. Also, compounds isolated from neem can also be used to develop drugs. Results from the phytochemical screening confirmed the claims that the crude extracts from various parts of neem have had medicinal application since time immemorial.

Acknowledgments The authors appreciate the sponsorship of Covenant University.

Conflict of Interest The authors declare that they have no conflict of interest.

References

Adeniyi, B., & Ayepola, O. (2008). The phytochemical screening and antimicrobial activity of leaf extracts of *Eucalyptus camaldulensis* and *Eucalyptus torelliana* (Myrtaceae). Res J. Med Plant, 2(1), 34-38.

Akin-Osanaiya, B., Nok, A., Ibrahim, S., Inuwa, H., Onyike, E., Amlabu, E., & Haruna, E. (2013). Antimalarial effect of neem leaf and neem stem bark extracts on *Plasmodium berghei* infected in the pathology and treatment of malaria. Inter J Res Biochem Biophy, 3(1), 7-14.

Akinmoladun, A. C., Ibukun, E., Afor, E., Obuotor, E., & Farombi, E. (2007). Phytochemical constituent and antioxidant activity of extract from the leaves of *Ocimum gratissimum*. Sci Res Essay, 2(5), 163-166.

Anibijuwon, I. I., Oladejo, B. O., Adetitun, D. O., & Kolawole, O. M. (2012). Antimicrobial Activities of Vernonia amygdalina Against Oral Microbes. Global J Pharm, 6, 178-185.

Bhowmik, D., Chiranjib, Y. J., Tripathi, K., & Kumar, K. S. (2010). Herbal remedies of *Azadirachta indica* and its medicinal application. J Chem Pharm Res, 2(1), 62-72.

Bhutada, P. R., Jadhav, A. J., Pinjari, D. V., Nemade, P. R., & Jain, R. D. (2016). Solvent assisted extraction of oil from *Moringa oleifera* Lam. seeds. Ind Crops Prod, 82, 74-80.

Biswas, K., Chattopadhyay, I., Banerjee, R. K., & Bandyopadhyay, U. (2002). Biological activities and medicinal properties of neem (Azadirachta indica). Curr Sci-Bang, 82(11), 1336-1345.

Champagne, D. E., Koul, O., Isman, M. B., Scudder, G. G., & Towers, G. N. (1992). Biological activity of limonoids from the Rutales. Phytochem, 31(2), 377-394.

Chuakul, W., Saralamp, P., & Boonpleng, A. (2002). Medicinal plants used in the Kutchum district, Yasothon Province, Thailand.

Claustra, A. L., Madulid, R. S., Aguinaldo, A. M., Espeso, E. I., Guevara, B. Q., Nonato, M. G., . . . del Castillo-Solevilla, R. C. (2005). A Guidebook to Plant Screening: Phytochemical and Biological. University of Santo Tomas Publishing House, Espana, Manila.

Daniel, A. K., & Dishi, K. (2011). Crude phytochemicals in The foliage and stem-bark of *Azadirachta indica*, Grown in Yola, Adamawa State, Nigeria. Global J Sci Frontier Res, 11(1), 9-13.

Edeoga, H. O., Okwu, D., & Mbaebie, B. (2005). Phytochemical constituents of some Nigerian medicinal plants. Afri J bio technology, 4(7), 685-688.

Elizabeth Babatunde, D., Otusemade, G. O., Elizabeth Ojewumi, M., Agboola, O., Oyeniyi, E., & Deborah Akinlabu, K. (2019). Antimicrobial activity and phytochemical screening of neem leaves and lemon grass essential oil extracts. Inter J. Mech Eng and Tech, 10(3).

Khare, C. P. (2008). Indian medicinal plants: an illustrated dictionary: Springer Science & Business Media.

Koul, O., Isman, M. B., & Ketkar, C. (1990). Properties and uses of neem, *Azadirachta indica*. Can J Botany, 68(1), 1-11.

Kumar, P. S., Mishra, D., Ghosh, G., & Panda, C. S. (2010). Available online at www. scholarsre-searchlibrary. com. Ann Bio Res 1(3), 24-34.

Kumar, R., Sharma, S., & Devi, L. (2018). Investigation of Total Phenolic, Flavonoid Contents and Antioxidant Activity from Extracts of *Azadirachta indica* of Bundelkhand Region. Int. J. Life. Sci. Scienti. Res. eISSN, 2455(1716), 1716.

Maithani, A., Parcha, V., Pant, G., Dhulia, I., & Kumar, D. (2011). *Azadirachta indica* (neem) leaf: A review. J Pharm Res, 4(6), 1824-1827.

Martins, P. F., De Melo, M., & Silva, C. (2016). Techno-economic optimization of the subcritical fluid extraction of oil from *Moringa oleifera* seeds and subsequent production of a purified sterols fraction. J Supercrit Fluids, 107, 682-689.

Nguyen, H. N., Pag-asa, D. G., Maridable, J. B., Malaluan, R. M., Hinode, H., Salim, C., & Huynh, H. K. (2011). Extraction of oil from *Moringa oleifera* kernels using supercritical carbon dioxide with ethanol for pretreatment: Optimization of the extraction process. Chemical Engineering and Processing: Process Intensification, 50(11-12), 1207-1213.

Ojewumi, M.E., (2018). Alternative Solvent Ratios for *Moringa oleifera* Seed Oil Extract. International Journal of Mechanical Engineering and Technology, 9(12), 295-307.

Ojewumi, M.E., & Owolabi, R. (2012). The Effectiveness of the Extract of '*Hyptis Sauveolens*' Leave (A Specie of Effinrin) in Repelling Mosquito. Trans J Sci Tech, 2(8), 79-87.

Ojewumi, M.E, Banjo, M., Ogunbiyi, T., Ayoola, A., Awolu, O., & Ojewumi, E. (2017a). Analytical investigation of the extract of lemon grass leaves in repelling mosquito. Inter J Pharm Sci Res, 8(5), 1000-1008.

Ojewumi, M.E., Adedokun, S. O., Omodara, O. J., Oyeniyi, E. A., Taiwo, O. S., & Ojewumi, E. O. (2017b). Phytochemical and Antimicrobial Activities of the Leaf Oil Extract of *Mentha Spicata* and its Efficacy in Repelling Mosquito. Inter J Pharm Res & Allied Sci, 6(4), 17-27.

Ojewumi, M. E., Banjo, M. G., Oresegun, M. O., Ogunbiyi, T. A., Ayoola, A. A., Awolu, O. O., & Ojewumi, E. O. (2017c). Analytical Investigation Of The Extract of Lemon Grass

Leaves in Repelling Mosquito. Inter J Pharm Sci and Res, 8(5), 2048-2055. doi: 10.13040/ijpsr.0975-8232.8(5).2048-55

Ojewumi, M.E., Adedokun, S. O., Ayoola, A. A., & Taiwo, O. S. (2018a). Evaluation of the oil Extract from *Mentha spicata* and its Chemical Constituents. 74(11/1), 68-89. DOI: 10.21506/j.ponte.2018.11.7. PONTE.

Ojewumi, M.E., Adeyemi, A.O., & Ojewumi, E.O. (2018b). Oil Extract From Local Leaves - An Alternative to Synthetic Mosquito Repellants. Pharmacophore, 9(2), 1-6.

Ojewumi, M.E., Ogele, P.C., Oyekunle, D.T., Omoleye, J.A., Taiwo, S.O, & Obafemi, Y.D. (2019a). Co-digestion of cow dung with organic kitchen waste to produce biogas using *Pseudomonas aeruginosa*. 3rd International Conference on Science and Sustainable Development (ICSSD 2019) IOP Conf. Series: Journal of Physics: Conf. Series 1299 (2019) 012011. doi:https://doi.org/10.1088/1742-6596/1299/1/012011

Ojewumi, M. E., Oyekunle, D., Amaefule, C., Omoleye, J., & Ogunbiyi, A. (2019b). Investigation into Alternative Energy Sources from Waste Citrus Peel (Orange): Approach to Environmental Protection. International Conference on Engineering for Sustainable World Journal of Physics: Conference Series 1378 (2019) 022066. doi:https://doi.org/10.1088/1742-6596/1378/2/022066

Ojewumi, M.E., Oyekunle, D., Ekanem, G., Obanla, O., & Owolabi, O. (2019c). Extraction of oil from selected plants using Response Surface Methodology [RSM]. International Conference on Engineering for Sustainable World Journal of Physics: Conference Series 1378 (2019) 042019. doi:https://doi.org/10.1088/1742-6596/1378/4/042019

Ojewumi, M.E., Oyekunle, D., Emetere, M., & Olanipekun, O. (2019d). Optimization of Oil from *Moringa oleifera* seed using Soxhlet Extraction method. Korean J Food & Health Conver, 5(3), 11-25.

Oseni, L., & Akwetey, G. (2012). An in-vivo evaluation of antiplasmodial activity of aqueous and ethanolic leaf extracts of *Azadirachta indica* in *Plasmodium berghei* infected balb/c mice.

Ucheya, R., Ucheya, U., & Amiegheme, F. (2011). Is a combine therapy of aqueous extract of *Azadirachta indica* leaf (Neem leaf) and chloroquine sulphate toxic to the histology of the rabbit cerebellum? Annals of med and health sci res, 1(2), 203-214.

Vaidya, A. D., & Devasagayam, T. P. (2007). Recent advances in Indian herbal drug research guest editor: thomas paul asir devasagayam current status of herbal drugs in India: an overview. J Cin biochem Nutri, 41(1), 1-11.

Yusoff, M. M., Gordon, M. H., Ezeh, O., & Niranjan, K. (2016). Aqueous enzymatic extraction of *Moringa oleifera* oil. Food Chem, 211, 400-408.

Proximate Analyses of Watermelon and Pineapple Wastes as Substrates for Single-Cell Protein Production

A. O. Salami, O. C. Nwinyi, E. F. Ahuekwe, and A. O. Adeyemi

1 Introduction

The exponential rise of the worldwide population, particularly in developing nations, has resulted in global issues such as food insecurity (protein deficiency) (Abarshi et al., 2017; Henchion et al., 2017). Thus, the production of biologically derived proteins from the fermentation of wastes derived from agricultural production using microorganisms has been the subject of debate for many researchers over the last decades (Saheed et al., 2016; Umesh et al., 2017). The search for new sources of dietary proteins with improved nutraceutical characteristics has led to the research on single-cell proteins (SCP) derived from microorganisms (Chee et al., 2019).

Single-cell protein (SCP) refers to the protein derived from pure cultures of microorganisms that can be used as a supplement for animal feed or human food (Canedo et al., 2016; Pruksasri et al., 2019). In most plant foods, there are limited sources of carbohydrates, fats, vitamins, nucleic acids, minerals, and essential amino acids. However, these essential nutrients can be derived from single-cell proteins, with some cells having a protein content of nearly 60–82% of their dry weight (Mondal et al., 2012). SCP derived from microorganisms can be used as chemical and pharmaceutical additives, protein supplements for human and animal feed, food aroma carriers, vitamin carriers, and emulsifying agents (Ukaegbu-Obi, 2016).

A. O. Salami (✉) · O. C. Nwinyi (✉) · E. F. Ahuekwe
Department of Biological Sciences, College of Science and Technology, Covenant University, Ota, Nigeria
e-mail: abimbola.salamipgs@stu.cu.edu.ng; obinna.nwinyi@covenantuniversity.edu.ng

A. O. Adeyemi
Departmentof Biochemistry, College of Science and Technology, Covenant University, Ota, Nigeria

© The Author(s), under exclusive license to Springer Nature Switzerland AG 2022

A. O. Ayeni et al. (eds.), *Bioenergy and Biochemical Processing Technologies*, Green Energy and Technology, https://doi.org/10.1007/978-3-030-96721-5_21

Nigeria has a large production of agricultural wastes due to its extensive agricultural activities and enormous landmass, and it is estimated that about 998 million tonnes of agricultural waste each year is produced (Obi et al., 2016; Jekayinfa et al., 2020). Improper waste disposal and management can pose a public health risk as well as environmental issues, such as diseases and air pollution (Yazid et al., 2017).

According to Mensah and Twumasi (2016), residues of orange peels, sugarcane waste, cassava waste, wheat straw, rice husk, coconut waste, sawdust, sugar beet pulp, paper mill waste, maize cobs, and mango waste are some of the common substrates that could be used to produce SCP. However, the extent of use of these wastes for SCP production will depend on the chemical composition of fruit wastes.

In this study, we selected the use of watermelon (*Citrullus lanatus*) and pineapple fruit (*Ananas comosus*) wastes for single-cell production because it has been reported to be rich in carbohydrates, sugar, as well as other basic nutrients that could support the growth of yeasts (Rabiu et al., 2018; Tanamool et al., 2020). We carried out the proximate analyses of the selected wastes to confirm their exact nutrient compositions due to influences of spatial and temporal variations on the abiotic and biotic factors prevailing in the tropical soil environment.

2 Materials and Methods

2.1 Chemicals and Reagents

Pure analytical grade chemicals and reagents were used for this study. These include perchloric acid, anthrone reagent, copper sulfate ($CuSO_4$), potassium sulfate (K_2SO_4), concentrated sulfuric acid (H_2SO_4), 40% caustic soda (NaOH), 4% boric acid, bromocresol green, methyl red, 0.1 N hydrochloric acid (HCl), trichloroacetic acid reagent, and petroleum ether.

The water used for the experiment was distilled, having been obtained from the distillation apparatus (D 4000) (Fig. 1).

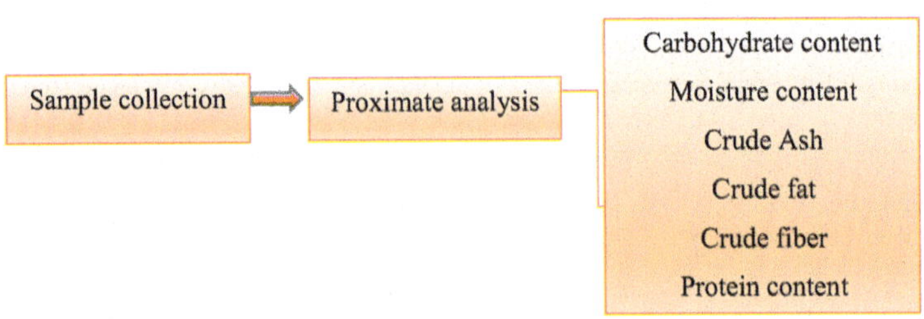

Fig. 1 A flow chart showing the stepwise analyses of the study

2.2 Experimental Methods

2.3 Sample Collection and Preparation

Watermelon and pineapple peels were collected immediately after cutting the fruits from the cafeteria of Covenant University, situated in the Ota district of Ogun State, Southwest Nigeria. The fruit peels were washed thoroughly with running tap water and then air-dried for 2 days, then completely dried in an oven (SM9053, Surgifriend Medicals, England) set at 60 °C. The oven-dried samples were then ground to powder using an electrical dry miller and sieved using a 2-mm mesh sieve. The ground samples were packed in Ziploc bags and stored at room temperature.

2.4 Proximate Analysis of the Sample

Twenty-five grams (25 g) of the dried pineapple and watermelon waste sample was used to carry out proximate analysis to determine carbohydrate content, crude fibre, ash content, crude fat, moisture content, and crude protein content as the percentage compositions of the substrate according to the methods described by the Association of Official Analytical Chemists (AOAC, 2016). Each analysis was performed in duplicates.

Estimation of Carbohydrate Content

The amount of carbohydrate in the fruit peels was calculated as the difference between all the other proximate parameters. Eq. 1 below was used to measure carbohydrate content:

$$\text{Carbohydrate}(\%) = (100 - \text{moisture}\% - \text{protein}\% - \text{fat}\% - \text{ash}\% - \text{fiber}\% \tag{1}$$

Estimation of Moisture Content

The dish was weighed (W_1), then the sample was added, and the total weight was taken (W_2). The sample was oven-dried for 5 hours at 125 °C, and the crucible weight and sample (W_3) were recorded. The percentage of moisture content was calculated using Eq. 2:

$$\text{Moisture}\left(\%\right) = \frac{W_3 - W_2}{W_2 - W_1} \text{x}\, 100 \tag{2}$$

Estimation of Crude Fibre

To measure the crude fibre content, 2 g of the samples was measured into a round-bottom flask with 50 mL of trichloroacetic acid reagent (TCA). The mixture was then boiled and vortexed. Weighted filter paper (W_1) that had been cooled to room temperature was used to filter the residue. Before the residue was used, it was cleaned in hot water four times and once in petroleum ether. The residue-containing filter was folded together and dried in an oven at 50 °C for 24 hours. After that, the sample was reweighed (W_2), ashed at 650 °C, cooled, and then reweighed (W_3). The percentage of crude fibre was calculated using Eq. 3:

$$\text{Crude fibre}\left(\%\right) = \frac{W_2 - W_3}{W_2 - W_1} \text{x}\, 100 \tag{3}$$

Estimation of Ash Content

Five grams of the samples were added into a clean dried crucible of weight (W_1), and the weight of the dish and its content was taken (W_2). The crucible that contained the sample was placed in the oven and heated until the sample was completely charred. Then it was further heated for 5 hours at 600 °C. Then the weight of the dish plus the content was reweighed (W_3). Eq. 4 below was used to calculate the percentage of ash content:

$$\text{Ash content}\left(\%\right) = \frac{W_2 - W_3}{W_2 - W_1} \text{x}\, 100 \tag{4}$$

Estimation of Crude Fat

Five grams of sample was weighed into a flask (W_1), and both the sample and flask weights were recorded (W_2). The samples were put in the Soxhlet extractor, and the flask was filled with 350-mL petroleum ether and left for 4–6 hours for extraction to occur. After extraction, the samples were dried and placed afterward in a desiccator to cool. The weight of the flask and sample after drying (W_3) was taken and recorded. Eq. 5 was used to calculate the percentage of fat content:

$$\text{Crude fat}\left(\%\right)=\frac{W_3-W_2}{W_1}\text{x}\,100 \tag{5}$$

Estimation of Protein Content

Two grams of the samples, 20 mL of concentrated sulfuric acid (H_2SO_4), 10 tablets of K_2SO_4, and 1 g of $CuSO_4$ were respectively weighed and added into the digestion tube, and the solution was digested for 5 hours at 420 °C until the solution appeared colorless. Two hundred milliliters of distilled water were added after the digested solution had cooled. One hundred milliliters of 0.1 N solution of hydrochloric acid (HCl) were pipetted into a conical flask, and 1 ml of mixed indicator (methyl red and bromocresol green) was added. The Kjeldahl flask that contained the digested sample was tilted, and 100 mL of 40% caustic soda (NaOH) was added then put in the distillation chamber (heating section). For ammonia collection, 30 mL of 4% boric acid was poured and placed beneath the distillation chamber, and then the solution changed from orange to green color. The solution in the conical flask was titrated in a burette until a color change from green to pink was observed, and the reading on the burette was recorded. Eqs. 6 and 7 were used to calculate the crude protein (CP):

$$\text{Crude protein}\left(\%\right)=\%\text{Nitrogen}\,\text{x}\,6.25 \tag{6}$$

$$\%\text{Nitrogen}=\frac{\left(A-B\right)x\,1.4007}{\text{Weight of sample}\left(\text{g}\right)}\text{x}\,100 \tag{7}$$

where A = normality of standard HCl * vol (mL) of standard HCl, B = normality of standard NaOH * vol (mL) of standard NaOH, and constant = 6.25.

3 Statistics

The results were presented as the mean with standard deviation (SD). The differences in the proximate properties of pineapple and watermelon were tested using the Student's t-test. Significance at $p < 0.05$ was used to assess significant differences between means of samples.

4 Results and Discussion

Fruit wastes are typically discarded as agricultural and home wastes. Thus, this study investigated the proximate properties of watermelon and pineapple wastes to establish a scientific foundation for their usage as substrates for SCP production.

Fig. 2 Proximate compositions of pineapple (*Ananas comosus*) peels

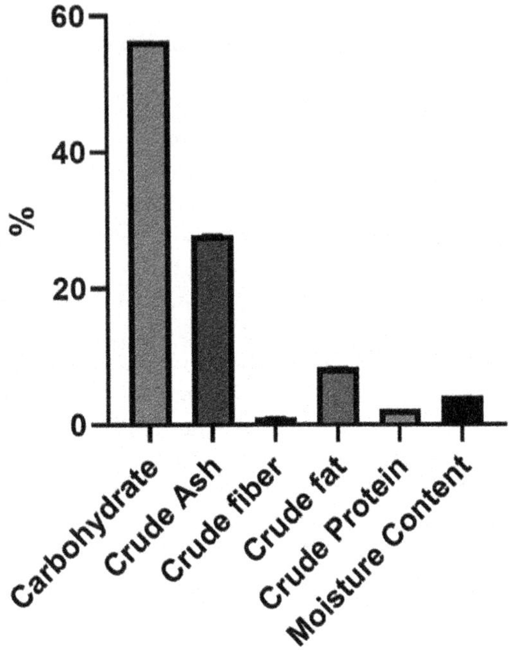

Fig. 3 Proximate compositions of watermelon (*Citrullus lanatus*) peels

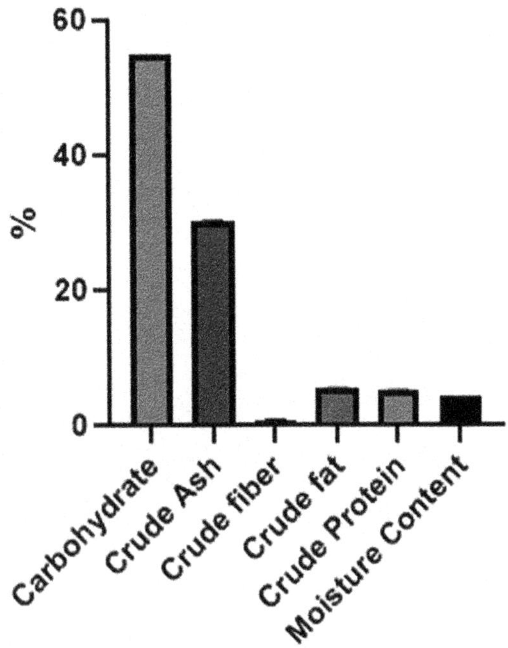

The result of the proximate analyses is presented in Figs. 2 and 3 for pineapple and watermelon dry peels, respectively.

From the results of the proximate analysis, it shows that the dried pineapple and watermelon peels contain variable ingredients with a major amount of carbohydrates and small amounts of protein, fat, fiber, and ash. The quantity of the total sugars reported in this study for pineapple peels was 56.38 ± 0.01% of dry weight. The result agrees with the observation of Zubairu et al. (2018). The value of carbohydrate obtained agrees with the data reported by Madika (2016) which was 59.65%. For watermelon peels, the total carbohydrate was 54.75% ± 0.05% of dry weight, and the result is in consonance with the observation of Umesh et al. (2019). Carbohydrate is the most favorable nutrient for the growth of microorganisms and the production of biomass. It has also been shown that pineapple and watermelon peels contain variable ingredients and may be used as carbon and energy sources for fungi growth in single-cell protein production.

For the protein content, pineapple peels were also found to contain 2.27 ± 0.00%, and this result correlated with the research carried out by Abarshi et al. (2017). This value was also lower than the protein content (5.11%) of pineapple peels observed in a study carried out by Feumba et al. (2016). The protein content of watermelon peels was recorded to be 5.15% ± 0.01%. This value was higher than the results obtained by Abdulazeez et al. (2020), which reported the protein content in the range of 0.55–0.80%; however, our values were lower when compared with the reports of Egbuonu (2015), which recorded the protein content of watermelon peels to be 7.04%. The findings from this study imply that pineapple and watermelon peels contain a good level of vital nutrients, like protein. The presence of protein content in fruit peels is most suitable for yeast fermentation to produce single-cell proteins (Dhanasekeran et al., 2011).

For the pineapple and watermelon peels, the ash content was reported to be 27.80 ± 0.05% and 30.24 ± 0.02%, respectively. Organic components or minerals are a natural substrate for microorganisms, and ash content showed the occurrence of organic matter in the fruit peels. This accounts for the potency of the fruit peels as a source of minerals for human health.

The crude fat results showed that the fruit peels contained fat contents (8.32 ± 0.06% for pineapple and 5.42 ± 0.03% for watermelon). The value for the pineapple peels was higher than the crude fat value reported by Madika (2016), which was 2.33%. The fat content values recorded for the watermelon peels were more than what Olayinka and Etejere (2018) reported in their study, which was 0.13% fat content. This reveals that cultivars and environmental and soil nutrient factors have an impact on the overall nutritional values of the fruits used in this study. These fruit peels can serve as a suitable substrate for the production of SCP since SCP must have low contents in fat.

The crude fiber obtained for the pineapple peels was 1.05 ± 0.04%. This is higher than the fiber value (0.60%) obtained in a study carried out by Hemalatha and Anbuselvi (2013) and twice the value of crude fiber obtained by Madika (2016). The analyzed watermelon peels had a fiber content of 0.78 ± 0.05%. The value was higher when compared with the works of Fila et al. (2013) and Olayinka and Etejere

(2018), which found a crude fiber content of 0.30% and 0.23%, respectively. Abarshi et al. (2017) recorded the fiber content of watermelon peels to be 2.42% higher than the values obtained in this study. Fiber is known to enhance digestion in animals and humans and also relieves constipation and hemorrhoids in humans (Abdulazeez et al., 2020). Therefore, the fruit peels can serve as substrates for single-cell protein production.

The moisture content recorded in this study was 4.18 ± 0.01% and 3.66 ± 0.01% for pineapple and watermelon peels, respectively. This value was lower than the data (8.65%) obtained by Bakri et al. (2020) on their proximate analysis of pineapple peels. The results for the moisture content of watermelon peels showed a lower value than the value (10.72%) reported by Iqbal (2015). The results obtained for the low moisture content showed that when pineapple and watermelon peels are stored after drying, they could serve a longer shelf life without microbial deterioration or chemical changes.

5 Conclusion

From this study, it is obvious that the selected fruit wastes contain the necessary nutritional content required for SCP production, which can be used as a good dietary supplement for good health.

Acknowledgments The authors appreciate Covenant University Research, Innovation, and Discovery (CURCID), for covering the cost of the publication.

References

Abarshi MM, Mada SB, Amin MI et al (2017) Effect of Nutrient Supplementation on Single Cell Protein Production from Watermelon and Pineapple Peels. Nig J Basic Appl Sci 25(1):130-136

Abdulazeez A, Abdullahi U, Audu, SS et al (2020) Proximate composition of rind and seed of watermelon (*Citrullus lanatus* L). Afr J Env Nat Sci Res 3(1):24-31

Bakri NF, Ishak Z, Jusoh AZ et al (2020) Quantification of Nutritional Composition and Some Antinutrient Factors of Banana Peels and Pineapple Skins. AFSJ 18(4):1-10.

Canedo MS, de Paula FG, Da Silva FA et al (2016) Protein enrichment of brewery spent grain from *Rhizopus oligosporus* by solid-state fermentation. Bioprocess Biosyst Eng 39(7):1105–1113

Chee JY, Lakshmanan M, Jeepery IF et al (2019) The potential application of *Cupriavidus necator* as polyhydroxyalkanoates producer and single-cell protein: A review on scientific, cultural and religious perspectives. Appl Food Biotechnol 6(1):19-34

Dhanasekeran D, Lawanya S, Saha S et al (2011) Production of Single Cell Protein from Pineapple Waste using Yeast. Innov Rom Food Biotechnol 8:26-32.

Egbuonu AC (2015) Comparative Investigation of the Proximate and Functional Properties of Watermelon (*Citrullus lanatus*) Rind and Seed. Res J Environ Toxicol 9:160-167.

Feumba DR, Ashwini RP, Ragu, SM (2016) Chemical composition of some selected fruit peels. Eur Food Sci Technol 4(4):12-21

Fila WA, Ifam EH, Johnson JT et al (2013). Comparative proximate compositions of watermelon *Citrullus lanatus*, Squash *Curcubita pepo'l* and Rambutan *Nephelium lappaceum*. Int J Sci Technol 2(1):81-88

Hemalatha R, Anbuselvi S (2013). Physicohemical constituents of pineapple pulp and waste. J Chem Pharm Res 5(2):240-242

Henchion M, Hayes M, Mullen AM et al (2017). Future Protein Supply and Demand: Strategies and Factors Influencing a Sustainable Equilibrium. Foods 6(7):53-59

Iqbal A (2015). Drying of Watermelon Rind and Development of Cakes from Rind Powder. Int J Novel Res Life Sci 2(1):14-21.

Jekayinfa SO, Orisaleye JI, Pecenka R (2020). An Assessment of Potential Resources for Biomass Energy in Nigeria: A Review. Resources 9: 92

Madika A (2016). Production of single cell protein from pineapple waste using Saccharomyces cerevisiae. Afr J Nat Sci 19:10–16

Mensah J, Twumasi P (2016). Use of pineapple waste for single-cell protein (SCP) production and the effect of substrate concentration on the yield. J Food Process Eng 40:12478

Mondal AK, Sengupta S, Bhowal J et al (2012). Utilization of fruit wastes in producing single-cell protein. Int J Environ Sci Technol 1(5):430 – 438

Obi FO, Ugwuishiwu BO, Nwakaire JN (2016). Agricultural waste concept, generation, utilization and management. Niger J Technol 35(4):957 – 964.

Olayinka BU, Etejere EO (2018). Proximate and Chemical Compositions of Watermelon (*Citrullus lanatus* (Thunb.) Matsum and Nakai cv Red and Cucumber (*Cucumis sativus* L. cv Pipino). Int Food Res J 25(3):1060-1066

Pruksasri S, Wollinger KK, Novalin S (2019). Transformation of rice bran into single-cell protein, extracted protein, soluble and insoluble dietary fiber, and minerals. J Sci Food Agric 99:5044–5049

Rabiu Z, Maigari FU, Lawan U et al (2018). Pineapple waste utilization as a sustainable means of waste management. In: Zakaria, Z.A. (Ed.), Sustainable Technologies for the Management of Agricultural Wastes. Springer Sci Rev 10:143–154

Saheed OK, Jamal P, Karim MI et al (2016). Utilization of fruit peels as carbon source for white-rot fungi biomass production under submerged state bioconversion. J King Saud Univ Sci 28:143–151

Tanamool V, Chantarangsee M, Soemphol W (2020). Simultaneous vinegar fermentation from a pineapple by-product using the co-inoculation of yeast and thermotolerant acetic acid bacteria and their physiochemical properties. Biotechnol J 3(10):1–11.

Ukaegbu-Obi KM (2016). Single Cell Protein: A Resort to Global Protein Challenge and Waste Management. J Microbiol 1(1):5

Umesh M, Priyanka K, Thazeem B et al (2017). Production of single-cell protein and Polyhydroxyalkanoate from *Carica papaya* waste. Arab J Sci Eng 42:2361–2369

Umesh M, Thazeem B, Preethi K (2019). Valorization of Pineapple Peels through Single Cell Protein Production Using *Saccharomyces cerevisiae* NCDC 364. Appl Food Biotechnol 6(4):255-263.

Yazid NA, Barrena R, Komilis D et al (2017). Solid-State Fermentation as a Novel Paradigm for Organic Waste Valorization: A Review. Sustainability 9:224

Zubairu A, Gimba A, Mamza W et al (2018). Proximate Analysis of Dry Watermelon (*Citrullus lanatus*) Rind and Seed Powder. J Sci Eng Res 5(3):473-478

Turbidity and Urine Turbidity: A Mini Review

C. C. Mbonu, O. Kilanko, M. B. Kilanko, and P. O. Babalola

1 Introduction

Turbidity, the measurement for impurity and the opposite phenomenon of clarity, is described as the reduced transparency of a liquid caused by the existence of undissolved matter (Kadaruddin & Zainuddin, 2020). The permissible volume of light through the liquid, or light that is not dispersed or absorbed but emitted through the liquid, provides a foundation for the analyses of various subject matters, such as liquid mass concentration and impurity identification (Institute & National Cancer Institute, 2020b). Turbidimetry (clarity measurement) or nephelometry can be used to determine the turbidity of a substance (a measure of cloudiness).

2 Theoretical Framework

Turbidimetry is the analysis of how much energy is reduced in an electromagnetic wave or photonic radiation after passing through a liquid to measure the turbidity or transparency of the liquid (Hamilton, 2014). The suspended particles in a liquid can be organic or inorganic (Mulyana & Hakim, 2018; Mahoney et al., 2019). Turbidity is calculated using two simple methodologies: turbidimetry, which determines the degree of light propagation, and nephelometry, which determines the degree of light scattering (Kitchener et al., 2017).

C. C. Mbonu · O. Kilanko (✉) · P. O. Babalola
Department of Mechanical Engineering, Covenant University, Ota, Ogun State, Nigeria
e-mail: oluwaseun.kilanko@covenantuniversity.edu.ng

M. B. Kilanko
General Hospital, Idiroko, Ipokia Local Government Area, Ota, Ogun State, Nigeria

© The Author(s), under exclusive license to Springer Nature Switzerland AG 2022

A. O. Ayeni et al. (eds.), *Bioenergy and Biochemical Processing Technologies*, Green Energy and Technology, https://doi.org/10.1007/978-3-030-96721-5_22

Gravimetric techniques can be used to determine the turbidity of a substance in addition to turbidimetry and nephelometry. Turbidimetry measurement involves a setup where the detector is positioned at a straight-on 180-degree angle relative to the incident light source. The tendency of suspended particles is to reflect light, which is detected using nephelometry by positioning a light detector at an angle relative to the incident light source since nephelometry measures light dispersion. This process was expressed through Rayleigh light scattering (Wiencek et al., 2020).

In the research carried out by Hannouche et al. (2011), it was observed that the relationship between turbidity and concentration is due to the amount of turbidity attainable to the concentration of the liquid. Based on the research, it is acceptable to conclude that the turbidity of a liquid is directly proportional to the concentration of the liquid. The concentration of a material is directly proportional to the amount of light absorbed or inversely proportional to the logarithm of the emitted light, according to Beer's law. This concept was to determine the absorption of light by a liquid sample (A) for a given absorptivity (a) and with a certain amount of concentration (c) (Kricka & Park, 2014). This is expressed mathematically as.

$$A = abc \tag{1}$$

where b is the distance of the light path.

According to the author, conformance to Beer's law is dependent on the following conditions:

- The concentration of solutes is within a specific limit.
- Between the molecule of interest and another solute or solvent molecule, there is no chemical reaction.
- As compared to solute absorbance, solvent absorption is negligible.
- On the substance of concern, incident radiation is monochromatic.
- There is an absence of optical interference.

2.1 Methods of Turbidimetry

Various optical methods have been employed in the detection and measurement of the turbidity of liquids. Most common methods such as turbidity tube, Jackson candle method, Secchi disk method, Baylis turbidimeter, turbidity rod, Formazin method, etc. serve as the underlining method being innovated to attain the present turbidimeter, this turbidity is capable of housing various technology with improved performance, efficiency, and accuracy of the turbidimeter.

Jackson Candle Method

Jackson's candle method is a conventional approach in the determination of the turbidity of a liquid sample. This method is a visual assessment method that employs the use of a metered calibrated flat-bottom glass tube, a clamp or stand, and a candle (Fig. 1). A stand or clamp is needed to hold or support the flat-bottom glass tube above the candle (Bratby, 2015; Dorea & Simpson, 2011). To determine the turbidity of a sample, the liquid sample is introduced to a metered flat-bottomed glass tube positioned above a burning candle. The sample is gradually introduced into the glass tube until the image of the candlelight fades away and the light produced by the candle is uniformly distributed across the sample. The observer determines the visibility of the candle. The turbidity of the liquid sample is observed with the eye (Johnson et al., 2006).

The observer looks into the calibrated flat-bottom glass tube as the sample is introduced into the tube gradually, as illustrated above. When the light transmitted by the candle becomes blurry, the observer stops the introduction of the sample into

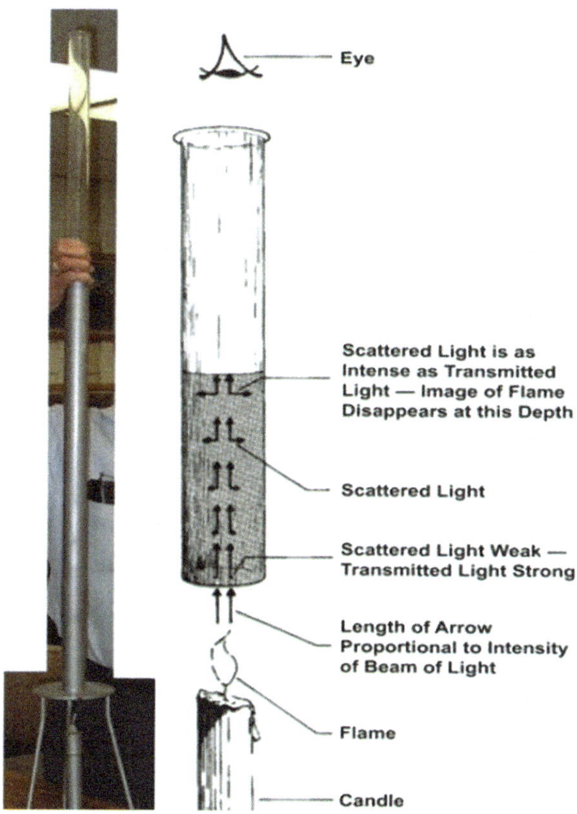

Fig. 1 Jackson candle method (Bratby, 2015)

the tube and records the light of the sample required to make the light blurry. The length recorded is used to determine the correspondent turbidity in Jackson turbidity unit (JTU) with the use of Jackson's turbidity table, as illustrated in Table 1 (Johnson et al., 2006). In Table 1, it is shown that the turbidity of the sample reduces with an increase in the length of the sample in centimeters. The Jackson candle turbidimeter is limited to a range of 25 JTU–1000 JTU. The Jackson method cannot account for turbidity less than 25 JTU (Kadaruddin & Zainuddin, 2020).

Secchi Disk Turbidimeter

The Secchi disk method involves the use of a 20-cm diameter disk, which is made up of two colors (black and white). The two colors are arranged alternately on the Secchi disk as shown in Fig. 2 (Bratby, 2015). It is made of heavy material, which allows the disk to sink when introduced to a liquid. At the midpoint of the disk, there is a provision that allows the attachment of a rope to the disk. To determine the turbidity of a sample, the Secchi disk is allowed to sink into the sample. Using the rope, the Secchi disk is gradually introduced into the sample until the disk is no longer visible to the observer. At this point, the rope is held stationary so that measurement can be conducted. When the Secchi disk is no longer visible, the point on the rope that intersects with the surface of the sample is noted (Boyd, 2020).

Table 1 Centimeter to Jackson turbidity unit (G. Johnson et al., 2006)

Light path	Turbidity	Light path	Turbidity	Light path	Turbidity
Cm	Units	Cm	Units	Cm	Units
2.3	1000	7.3	300	19.6	110
2.6	900	7.5	290	21.5	100
2.9	800	7.8	280	22.6	95
3.2	700	8.1	270	23.8	90
3.5	650	8.4	260	25.1	85
3.8	600	8.7	250	26.5	80
4.1	550	9.1	240	28.1	75
4.5	500	9.5	230	29.8	70
4.9	450	9.9	220	31.8	65
5.5	400	10.3	210	34.1	60
5.6	390	10.8	200	36.7	55
5.8	380	11.4	190	39.8	50
5.9	370	12	180	43.5	45
6.1	360	12.7	170	48.1	40
6.3	350	13.5	160	54	35
6.4	340	14.4	150	61.8	30
6.6	330	15.4	140	72.9	25
6.8	320	16.6	130		
7	310	18	120		

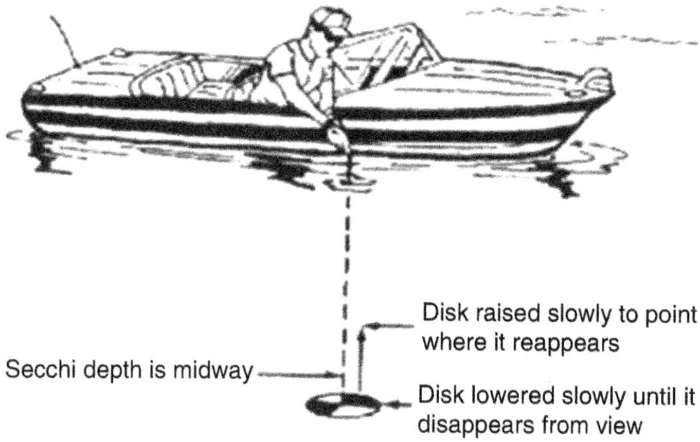

Fig. 2 Secchi disk method (Bratby, 2015)

The disk is removed from the sample, and the distance from the surface of the Secchi disk to the marked point on the rope is measured. The distance measured is called the Secchi depth. The conversion of the Secchi depth to turbidity is a complicated method because various factors affect the disappearance of the Secchi disk. These include light diffusion in the sample and the intensity of light reflected from the sun to the Secchi disk then to the eye of the observer. The diffusion of light in the sample is represented as diffusion attenuation coefficient denoted as K_d while the reflection from the sun to the surface of the Secchi disk then to the eye is represented as the beam attenuation coefficient denoted as C. In their study, Bower et al. (2020) illustrated that the Secchi depth (Z_{sd}) is inversely proportional to the sum of the beam attenuation coefficient (C) and the diffusion attenuation coefficient (K_d) as expressed in the equation below:

$$Z_{sd} = \frac{y}{(C + K_d)} \tag{2}$$

where y is a constant variable with a value, which is dependent on the reflectivity of the Secchi disk, the surrounding, and the sensitivity of the observer's eye. The Secchi disk's visibility or Secchi depth is also expressed by Boyd (2020) as

$$K_d = \frac{1.7}{Z_{sd}} \tag{3}$$

Turbidity Tube

The turbidity tube method of measuring the turbidity of a liquid sample involves the use of a clear tube, tube cap, viewing disk, and measuring device. The clear tube must be clean and very transparent to allow the optimum reflection of light into the tube. The tube is responsible for holding the sample and allowing the reflection of light during the duration of the analysis. The tube cap prevents the leakage of the sample from the tube, thereby keeping the liquid sample inside the clear tube. The viewing disk is similar to the Secchi disk, which has two white and two black quadrants on a circular disk, as illustrated in Fig. 3 below. The diameter of the viewing disk must fit properly into the internal diameter of the clear tube.

The measurement of the height of the liquid sample in the tube from the surface of the viewing disk to the maximum level of water can be obtained using two methods. The first method involves metering or calibration directly on the cylinder. Calibration starts from the level of the surface of the viewing disk upward. This ranges from 0 cm to up to 100 cm. The second method involves the use of an external measurement device to determine the height of the liquid sample. The starting point or 0 point of the external measuring device is placed on the same level as the surface of the viewing disk, while the rest of the external measurement device is aligned horizontally along the tube.

According to Mohammed (2015), to obtain the turbidity of a liquid sample using the turbidity tube method, the experiment must be conducted during daylight but must not have direct contact with sunrays. To achieve this, the observer must stand between the turbidity tube and the sun, thereby casting a shadow on the turbidity tube. To ensure proper results and attain the turbidity of the liquid sample, the following steps must be followed (Dorea & Simpson, 2011):

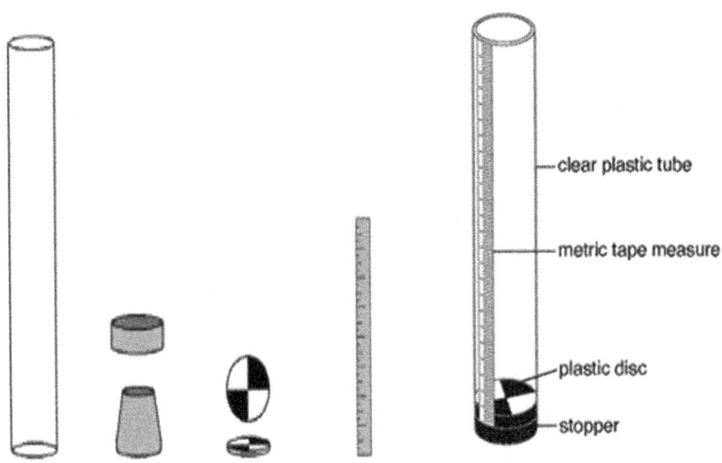

Fig. 3 Labeled representation of a turbidity tube (Myre & Shaw, 2006)

- A clean container is used to collect the sample to be introduced to the turbidity tube so that the integrity of the sample is secured.
- The observer is positioned 10–20 cm away from the top of the clear tube horizontally and must be observed directly above the clear tube such that the viewing disk is seen by the observer (Fig. 4).
- The liquid sample must be gently introduced into the tube. This continues until the viewing disk becomes blurry and difficult to see. At this point, the liquid sample is slowly introduced into the tube until the observer is no longer able to see the disk.
- The height of the liquid sample at which the observer is unable to see the viewing disk anymore is recorded and referred to length-to-turbidity conversion table (Table 2).

The turbidity tube method has certain elements and principles of operation of both the Jackson candle method of turbidimeter and the Secchi disk method of turbidimeter. It is important to note that the intermediate value of turbidity cannot be derived through linear interpolation due to the logarithmic nature or nonlinearity of the turbidity scale (Myre & Shaw, 2006). Hence, minimum and maximum values representing the upper and lower limits of the experiment are needed. The author used 85.4 cm–5 NTU as the lower limit and 7.3 cm–200 NTU as the upper limit. The maximum and minimum values preferred by the author, as highlighted in Table 2, were used to arrive at the intermediate values of turbidity using the equation below:

$$\text{Length of liquid}(\text{cm}) = 24.13\left(\text{turbidity}(NTU)\right)^{-0.662} \tag{4}$$

The summary of the Jackson candle turbidity, turbid meter, Secchi disk, and turbidity tube methods are listed in Table 3.

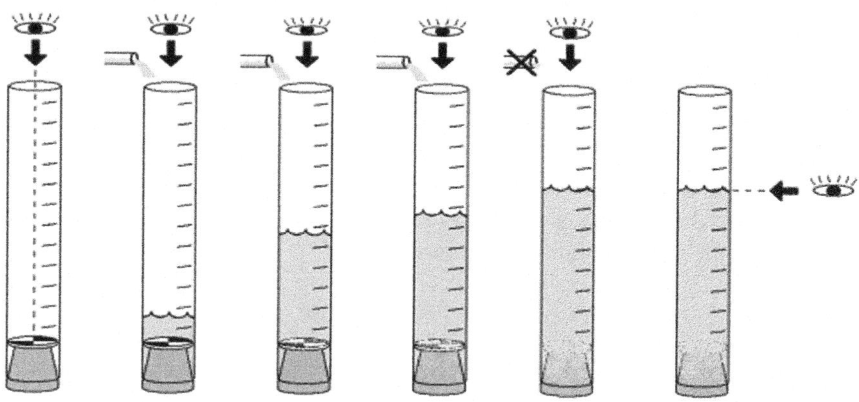

Fig. 4 Procedure for conducting turbidimetry using a turbidity tube (Mohammed, 2015)

Table 2 Length-to-turbidity conversion table (Myre & Shaw, 2006)

Length (cm)	NTU
6.7	240
7.3[a]	200[a]
8.9	150
11.5	100
17.9	50
20.4	40
25.5	30
33.1	21
35.6	19
38.2	17
40.7	15
43.3	14
45.8	13
48.3	12
50.9	11
53.4	10
85.4[a]	5[a]

[a]min and max values used in eq. 4

Baylis Method

Baylis method for determining the turbidity of a liquid sample is based on a color comparison between the sample and a standard reference liquid of the same class. The instrument for conducting turbidity measurement using the Baylis method involves the use of a closed galvanized iron box with two glass tubes on one side and a 250-watt light source with a reflector behind the light source. The glass tubes stand side by side with each other as the reference liquid is introduced in one, and the second glass makes the provision for the introduction of the liquid sample. A white opal glass plate supports the glass tubes at their lower ends, and blue cobalt glass plates surround them near the bottom section. To determine the turbidity of a liquid sample, the following steps are performed:

- A reference liquid with known turbidity is introduced into one of the glass tubes, while the liquid sample to be analyzed is introduced into the second glass tube.
- The light source is allowed to illuminate the interior part of the galvanized iron box. When the light ray strikes the blue cobalt glass plate, it causes blue illumination of the reference liquid and sample.
- If the resulting color obtained is not the same, the reference liquid is removed and a more suitable reference liquid is introduced to fit the color output of the liquid sample.
- When a fitting reference liquid is attained, the standard turbidity of the reference liquid is recorded as the turbidity of the sample.

Table 3 Summary of Jackson candle turbidity, turbid meter, Secchi disk, and turbidity tube (Mohammed, 2015)

Method	Advantage	Disadvantage
Jackson candle turbid meter	(historical method)	No longer a standard method
Water poured into tube		Cannot measure <25 JTU (25 NTU)
Reading taken when candle burning under tube can no longer be seen		
Turbid meter (nephelometer)	Extremely accurate	Expensive
Beam of light passed through water sample	Some are portable	Easily damaged
Amount of light scattered measured at a 90-degree angle	Can measure very low turbidity	Requires power source, requires calibration
Secchi disk	Low cost	Less accurate
Black-and-white disk lowered into water	Portable	Cannot be used in shallow water or swift currents
Maximum distance at which disk can be seen recorded	No consumables, easy to learn	Not applicable to small sample sizes
Turbidity tube (transparency tube)	Low cost	Less accurate
Jackson candle method combined with Secchi disk method	Portable, easy to learn, no consumables, suitable for all water sources	Cannot measure <5 NTU

2.2 *Light Scattering*

When light passes through a liquid sample, the scattering of light occurs at various angles due to suspended particles in the liquid as well as the shape and size of the particles. The intensity of light scattered depends on the suspended particles, which are attributed to the turbidity of the liquid. The optical techniques, which include visual turbidimetry, a detection method based on transmitted scattered light, and a detection method based on the transmission-scattering ratio, are used in all turbidity measuring methods (Cao et al., 2019). The intensity of the scattered light can be obtained by installing a photometer, which is a light detector, along the angle of detection. The detection angle is the angle created by the incident light beam's centerline, and the detectors receive the angle's centerline. The detection angle can have a great impact on particle detection from a size perspective as well as the instrument's turbidity range.

Light Scattering Theory

The principle of light scattering, which is due to the association of light with particles in a liquid, is used in nephelometry to address turbidity (Dasgupta & Wahed, 2013). As a result, nephelometry measures a liquid's turbidity by calculating the intensity of scattered light (Institute & National Cancer Institute, 2020a). As an initial light intensity (I_0) passes through a liquid with a given length (l) and turbidity coefficient (τ), the intensity of the light after passing through the liquid (I) is measured using the first equation below (Cao et al., 2019). The author also expressed the Rayleigh theorem, which describes the relationship between the intensity of light after scattering (I_s) and the quantity (n) and volume of particles (V^2) per unit volume of liquid, in the second equation below:

$$I = I_0 e^{-\tau l} \tag{5}$$

$$I_s = \frac{K l_0 n V^2}{\lambda^4} \tag{6}$$

where l_0 is the initial light wavelength.

As earlier mentioned, nephelometry is the detection of light scattered by suspended particles. The detection of the light scattered by suspended particles is performed through a nephelometer. This device measures the intensity of scattered light at an angle of 90 degrees to the direction of the propagated light (Boyd, 2020). Forward-scattering, side-scattering, and backscattering nephelometry are the three subcategories of nephelometry (Fig. 5). Side scattering involves the measurement of the intensity of scattered light at an angle of 90 degrees to the incident beam. Forward scattering involves the measurement of the intensity of scattered light within a range of 0 to 90 degrees, and backscattering involves the measurement of the intensity of scattered light within a range of 90 to 180 degrees (Kitchener et al., 2017).

The classification of particles in a liquid is given in ranges, according to Lawler (2016). The diameter (D) of the particles is used to identify the ranges of particles suspended in a liquid, with small particles being $D < 0.05$, large particles being $0.1 < D < 0.8$, and larger particles being $D > 0.8$. The light scattering theory is influenced by the diameter of the particle scattering light from a light source as a function of the wavelength of the electromagnetic wave (λ). Hence, the particle dimension with respect to light results in $D < 0.05\ \lambda$ for small particles, $0.1 < D < 0.8\ \lambda$ for large particles, and $D > 0.8\ \lambda$ for larger particles. This theory is known as the Mie scattering theory and is defined as follows:

$$\propto = \frac{\pi D}{\lambda} \tag{7}$$

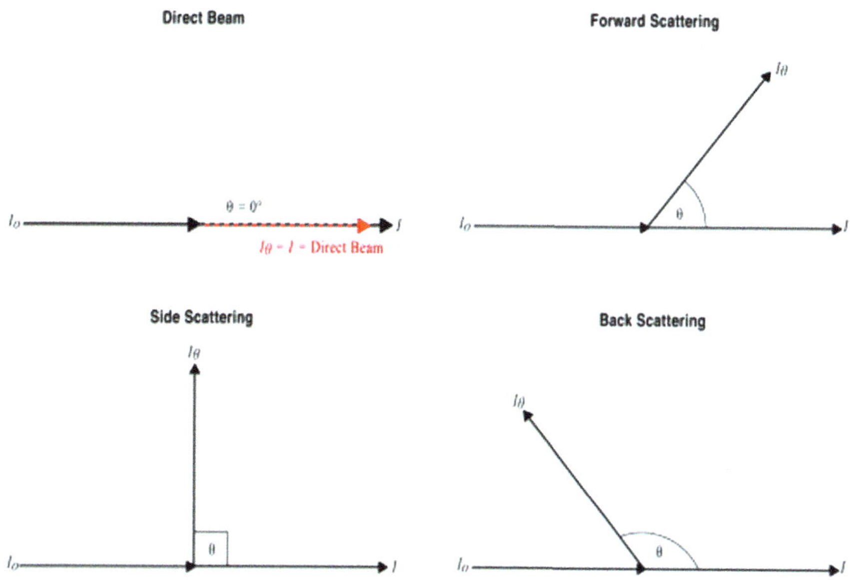

Fig. 5 Light scatter observation angles (Kitchener et al., 2017)

Rayleigh's scattering theory, formulated in 1871 for gases, is used for small particles. Rayleigh's scattering theorem for gases can be used to determine the distribution of small particles in a liquid due to their small size since Rayleigh's equation of scattering was designed to describe the angular distribution of a resultant light scattering. Rayleigh's equation is given as

$$\frac{i_o}{I_o} = \left(\left(\frac{n'}{n}\right) - 1\right)^2 \left(\frac{NV^2}{\lambda^4 r^2}\right)\left(1 + \cos^2\theta\right) \tag{8}$$

where i_o represents the dispersed light intensity at an angle, I_0 represents the initial light intensity, n' represents the particles' refractive index, n represents the liquid containing the particles' refractive index, r represents the distance of the liquid containing particles with volume V, and N represents the number of particles. A particle is referred to as a large particle with an asymmetric distribution of dispersed light if the size of the particle in a medium is within the range of $0.1 < D < 0.8$. The Mie scattering theory, which is expressed in the equation above, can be used to calculate large particles. It is worth noting that the larger the suspended particles are, the greater is the disruptive interference of light, resulting in more light scattering backward.

2.3 Urine and Turbidity

Various analyses can be conducted on urine to detect certain abnormalities, including metabolic imbalance or ailments (Libretexts, 2018). The examination of urine can be based on the urine's physical characteristics, microscopic makeup, or chemical composition. The physical characteristics of urine include color, specific gravity, odor, turbidity, and pH. The turbidity of urine is increased by the presence of cellular debris, cast, and, in some cases, crystal and other debris in the urine. Blood (both red and white blood cells), hemoglobin, cholesterol, albumin, leukocyte esterase, nitrites, ketones, bilirubin, and urobilinogen are all substances that are not expected to be found in urine, the presence of which can increase urine turbidity (Johnson, 2019; Jitta et al., 2018; Urinalysis, 2019).

A turbidity analysis of urine can give the following outcomes: clear urine, partially cloudy urine, cloudy urine, or opaque or flocculent urine (Libretexts, 2018). When the turbidity of a urine sample is high, it raises a red flag, requiring further investigation (Fan & Bai, 2020). Urine that has been freshly harvested but has an unclear and muddy look is referred to as turbid or cloudy urine.

Cystitis, diabetes, gonorrhea, chlamydia, kidney stones, pyelonephritis, urethritis, kidney failure, trichomonas, prostatitis, etc. are diseases associated with cloudy or turbid urine. Other causes of turbid urine include hydration, urinary tract infection, kidney infection and stones, sexually transmitted infection, vulvovaginitis, diet, and diabetes (Johnson, 2019). According to Gadalla et al. (2019), the urine of an individual who is healthy and properly hydrated and has not ingested any substance that can change the color of their urine is clear. Hence, a turbidity measurement of urine that does not report "clear" is a less desirable result.

Urine Turbidimetry

The conventional method for detecting whether urine is turbid or not is gross examination/inspection. Gross inspection involves the use of the eye to detect the transparency of urine. Generally, gross inspection is used to examine the physical properties of urine (Roxe, 1990). This method is similar in principle to Jackson's candle method as it focuses on the attenuation principle and turbidity tube method. The urine sample is either introduced into an illuminated environment and examines how much light the eye can see or place in front of a light source while a whited paper that has been printed on is placed between the light source and the transparent tube housing the urine sample. The turbidity of the urine sample is given by the visibility of the printed work through the urine sample. This method shows close similarities with the turbidity tube method.

Gross inspection usually provides inaccurate results as it is unable to measure the actual turbidity of urine. In recent times, however, laboratories have employed the use of more accurate methods. The use of photometers and spectrophotometers has now been a generally accepted method for calculating urine turbidity, and they yield

Fig. 6 Arduino turbidity sensor (Mithya et al., 2019)

more accurate and reliable results. Photometers measure the strength of light or the amount of light falling on a surface, while spectrophotometers calculate the strength of the light at specific wavelengths.

Mithya et al. (2019) proposed the use of an Arduino device to detect turbidity in liquid samples, which can be employed to determine the turbidity of urine. Arduino turbidity sensor can replicate a photometer to detect the number of suspended particles in urine by measuring transmitted and scattered light (Fig. 6). It consists of a light source and a detector, and the sample is placed between the sensor's probes. One probe houses the light source, while the other houses the detector. Here, turbidity is directly proportional to the intensity of light transmitted through the liquid (Chidera et al., 2020).

Photometry method involves the measurement of the intensity of light transmitted through a urine sample or scattered by the interaction of suspended particles in the urine sample. A personal health monitoring system can be developed from this principle. The device will adopt one of the principles of the method discussed above. A photometric device can be constructed and installed in the convenience to conduct real-time turbidity measurement. This can serve as a detection device or a monitoring device to conduct a follow-up on the response to treatment of ailment that results in deposition of particles in the urine.

3 Conclusion

The turbidity of turbid urine does not give the cause of the turbid state of the urine but the presence of suspended particles and a rough estimate of the amount of suspended particles present in the urine. Hence, turbidity of the urine sample should be carried regularly in the comfort of the user since turbidity measurement is a step to further investigation. Various modern devices approach for the detection of the turbidity liquid sample involves the measurement of light that passes through the sample and the measurement of the intensity of scattered light. Hence, developing a setup with the ability to describe the disturbance to the passage of light caused by suspended particles is needed. Internet of things (IoT) provides a means to measure the turbidity of a urine sample by measuring the intensity of light transmitted through the urine sample. Higher intensity depicts less amount suspended particles and turbidity while lower light intensity depicts a high amount of suspended particles and turbidity. This makes it possible to conduct turbidity measurement in the comfort of the user by developing compatible techniques to attain the turbidity of urine samples in real-time.

Acknowledgments The author acknowledges the management of Covenant University for the conference support given to this paper

References

Bowers, D. G., Roberts, E. M., Hoguane, A. M., Fall, K. A., Massey, G. M., & Friedrichs, C. T. (2020). Secchi disk measurements in turbid water. *Journal of Geophysical Research: Oceans*, 125, e2020JC016172. https://doi.org/10.1029/2020JC016172

Boyd, C. E. (2020) Suspended Solids, Color, Turbidity, and Light. In: Water Quality. *Springer, Cham*, pp. 199-133. https://doi.org/10.1007/978-3-030-23335-8_6

Bratby, J. (2015). Turbidity: Measurement of Filtrate and Supernatant Quality?. In Progress in *Filtration and Separation* (pp. 637-657). Academic Press.

Cao, P., Zhao, W., Liu, S., Shi, L., & Gao, H. (2019). Using a Digital Camera Combined with Fitting Algorithm and TS Fuzzy Neural Network to Determine the Turbidity in Water. IEEE Access, 7, 83589-83599.

Chidera, M., Kilanko, O., Azeta, J., Bolu, C.A. (2020). An LDR Based Colour Sensor for Urine Analysis: A Review. *International Journal of Emerging Trends in Engineering Research*, 8(10), 7704-7711. https://doi.org/10.30534/ijeter/2020/1618102020

Dasgupta, A., & Wahed, A. (2013). Clinical Chemistry, Immunology and Laboratory Quality Control: A Comprehensive Review for Board Preparation, Certification and Clinical Practice. Academic Press

Dorea, C. C., & Simpson, M. R. (2011). Turbidity tubes for drinking water quality assessments. *Journal of Water, Sanitation and Hygiene for Development*, 1(4), 233-241.

Fan, S.-L., & Bai, S. (2020). Urinalysis. In Contemporary Practice in Clinical Chemistry (pp. 665–680). https://doi.org/10.1016/b978-0-12-815499-1.00038-7

Gadalla, A. A. H., Friberg, I. M., Kift-Morgan, A., Zhang, J., Eberl, M., Topley, N., Weeks, I., Cuff, S., Wootton, M., Gal, M., Parekh, G., Davis, P., Gregory, C., Hood, K., Hughes, K., Butler, C., & Francis, N. A. (2019). Identification of clinical and urine biomarkers for uncomplicated urinary tract infection using machine learning algorithms. *Scientific Reports*, 9(1), 19694.

Hamilton, R. G. (2014). Methods (In Vitro and In Vivo): Nephelometry and Turbidimetry. In Encyclopedia of Medical Immunology (pp. 484–486). https://doi.org/10.1007/978-1-4614-9194-1_301

Hannouche, A., Chebbo, G., Ruban, G., Tassin, B., Lemaire, B. J., & Joannis, C. (2011). Relationship between turbidity and total suspended solids concentration within a combined sewer system. Water Science and Technology: *A Journal of the International Association on Water Pollution Research*, 64(12), 2445–2452.

Institute, N. C., & National Cancer Institute. (2020a). Nephelometry. In *Definitions*. https://doi.org/10.32388/qwg57a

Institute, N. C., & National Cancer Institute. (2020b). Turbidimetry. In Definitions. https://doi.org/10.32388/9ypp18

Johnson, G., Gervino, N., Gunderson, L., Hotka, L., & Klang, J. (2006). Turbidity TMDL Protocols and Submittal Requirements. http://dx.doi.org/

Johnson, J. (2019). Nine causes of cloudy urine. https://www.medicalnewstoday.com/articles/324443. Accessed 01 September 2021

Josephus Jitta, N., Veneman, S. E., Maatman, R., Aan de Stegge, W. B., & Veneman, T. F. (2018). Urine changing from clear to milky-white. *The Netherlands Journal of Medicine*, 76(8), 379–380.

Kadaruddin, K., & Zainuddin, M. (2020, April). A brief review on low-cost turbidimeter designs. *In IOP Conference Series: Earth and Environmental Science* (Vol. 476, No. 1, p. 012096). IOP Publishing.

Kitchener, B. G., Wainwright, J., & Parsons, A. J. (2017). A review of the principles of turbidity measurement. *Progress in Physical Geography*, 41(5), 620-642.

Lawler, D. M. (2016). Turbidity, Turbidimetry, and Nephelometry. In Reference Module in Chemistry, Molecular Sciences and Chemical Engineering. https://doi.org/10.1016/b978-0-12-409547-2.11006-6

Libretexts. (2018). 24.4A: Physical Characteristics of Urine. Libretexts. https://med.libretexts.org/Bookshelves/Anatomy_and_Physiology/Book%3A_Anatomy_and_Physiology_(Boundless)/24%3A__Urinary_System/24.4%3A_Urine/24.4A%3A_Physical_Characteristics_of_Urine. Accessed 01 September 2021.

Mahoney, E., Kun, J., Smieja, M., & Fang, Q. (2019). Point-of-care urinalysis with emerging sensing and imaging technologies. Society *Journal of The Electrochemical*, 167(3), 037518.

Mohammed, S. S. (2015). Effect of pH on the Turbidity Removal of Wastewater. *In OALib* (Vol. 02, Issue 12, pp. 1–9). https://doi.org/https://doi.org/10.4236/oalib.1102283.

Mithya, V., Prabha, N. D., Samlein, S., & Madhumitha, M. (2019). Smart toilets using turbidity sensor. *International journal of innovative technology and exploring engineering (IJITEE)*, 8(5S).

Mulyana, Y., & Hakim, D. L. (2018). Prototype of Water Turbidity Monitoring System. *In IOP Conference Series: Materials Science and Engineering* (Vol. 384, No. 1, p. 012052). IOP Publishing.

Myre, E., & Shaw, R. (2006). The turbidity tube: simple and accurate measurement of turbidity in the field. Department of Civil and Environmental Engineering, Michigan Technological University.

Roxe, D. M. (1990). Urinalysis. In Clinical Methods: The History, Physical, and Laboratory Examinations. 3rd edition. Butterworths.

Wiencek, J. R., Duh, S.-H., & Christenson, R. H. (2020). Proteins: analysis and interpretation in serum, urine, and cerebrospinal fluid. In Contemporary Practice in Clinical Chemistry (pp. 365–390). https://doi.org/10.1016/b978-0-12-815499-1.00022-3

Antibiotic Resistance Status of *Pseudomonas aeruginosa* in Clinical Isolates in Ogun State

H. U. Ohore, P. A. Akinduti, E. F. Ahuekwe, A. O. Salami, and G. I. Olasehinde

1 Introduction

Antibiotic-resistant bacterial infection among hospitalized patients is currently one of the most serious global concerns and results in high mortality and increased hospital costs. According to the World Health Organization (WHO) and the US Centers for Disease Control and Prevention (CDC), the rising trend in antibiotic resistance by several bacteria is an imminent threat to public health. *Pseudomonas aeruginosa*, a member of the ESKAPE group, causes a wide range of infections that are difficult to treat because of their natural resistance to a host of anti-microbial compounds (Ahmad et al., 2016).

The infections caused by *Pseudomonas aeruginosa* are associated with relatively high treatment failures due to its intrinsic and acquired resistance to commonly available antibiotics. This is particularly shown in patients hospitalized with burns (Mirzaei et al., 2020) and malignancies and those with cystic fibrosis (Botelho et al., 2019), as well as those with fulminant infections, such as sepsis and pneumonia, most of which are fatal with a high mortality rate. Mainly, these infections arise due to the development of drug resistance patterns, biofilm formation and virulence factor production. All these have contributed to the resistance of pathogens. At present, *P. aeruginosa* has been described as multi-drug resistant (MDR), being resistant to at least one agent in three or more anti-microbial classes, or pan-drug resistant, being resistant to all agents in all anti-microbial classes (Horcajada et al., 2019). This is due to the fact that they have developed or acquired resistance to a number of clinically useful antibiotics.

H. U. Ohore (✉) · P. A. Akinduti · E. F. Ahuekwe · A. O. Salami · G. I. Olasehinde (✉)
Department of Biological Sciences, College of Science and Technology, Covenant University, Ota, Nigeria
e-mail: grace.olasehinde@covenantuniversity.edu.ng

© The Author(s), under exclusive license to Springer Nature Switzerland AG 2022

A. O. Ayeni et al. (eds.), *Bioenergy and Biochemical Processing Technologies*, Green Energy and Technology, https://doi.org/10.1007/978-3-030-96721-5_23

Pseudomonas aeruginosa is a Gram-negative bacilli that has a mortality rate of up to 61% from nosocomial infections (Mirzaei et al., 2020), and its resistance has been increasing. Extended-spectrum beta-lactamases (ESBLs) are enzymes that hydrolyze and induce resistance to cephalosporin monobactams (Ahmad et al., 2016). Beta-lactamase production is the most frequently utilized resistance mechanism (Ogunrinola et al. 2020; Shaikh et al., 2015). It is a cause of resistance in Gram-negative bacteria, particularly among *P. aeruginosa*, *Klebsiella pneumoniae* and *Escherichia coli* (Pérez et al., 2019).

Since resistance is driven by antibiotic use, the drug options of MDR Gram-negative bacteria have become less effective as they become more available for clinical use. A number of studies in Nigeria have reported antibiotic resistance in *P. aeruginosa* in various locations (Adesoji et al., 2015; Jombo et al., 2008; Ozumba, 2003; Ogbolu et al., 2008; Olayinka et al., 2009; Akinduti et al. 2021 & Smith et al., 2012). However, there is a paucity of information on the multi-drug-resistant *P. aeruginosa* in Ogun State, especially in clinical samples. The primary aim of this study is therefore to determine the prevalence of beta-lactamase producing *P. aeruginosa* in hospitals in Ogun State as a way to generate local data necessary for planning and advocacy, particularly for empiric therapy, antibiotic stewardship and infection control.

2 Methodology

2.1 Study Site

This study was carried out at the following hospitals: Ota General Hospital, Ota; Federal Medical Center, Abeokuta; Medicare, Ota; and Covenant University Health Center, Ota, Ogun State.

2.2 Ethical Approval

Ethical approval was obtained from the Covenant Health Research Ethics Committee (CHREC), with protocol number CHREC/055/2020 Covenant University, Idi-Iroko, Ota, Ogun State.

2.3 Study Subjects

Extra-intestinal samples (urine, wound swabs, ear swab and high vaginal swab samples) were collected from the study sites over the course of 3 months, from 1 September 2020 to 30 November 2020. Both male and female samples were collected from all ages.

2.4 Sample Size

The sample size obtained for this study totalled 150 isolates (77 urine, 42 high vaginal swabs, 17 ear swabs, 13 wound swabs and two blood cultures).

2.5 Specimen Collection, Culturing and Bacterial Isolation

Using swab sticks and universal bottles, a total of 150 extra-intestinal isolates were recovered from urine, wound and skin from the selected hospitals. The swab stick was used on skin by rubbing on affected areas, and mid-stream urine samples were recovered in universal bottles. The swabs were placed in Amies transport media. All swabs and universal bottles were properly labelled and transported to the Covenant University Microbiology laboratory for further analysis.

The samples obtained were inoculated on cetrimide agar and nutrient agar aseptically and aerobically incubated for 24 hours at 37 °C. Positive growth on cetrimide agar, indicated by a yellow green to blue colour or fluorescence, suggested positive for *Pseudomonas* growth (cetrimide is a selective medium for *P. aeruginosa*), while no growth suggested negative for *P. aeruginosa*. The suspected positive isolates were inoculated on nutrient agar and stored in nutrient broth for further analysis. Based on colony morphology, colour, Gram staining and biochemical tests such as catalase and oxidase, presumptive bacterial identification was achieved. Pure culture bacterial isolates were identified based on their morphology and biochemical characteristics.

2.6 Antibiotic Susceptibility Testing

The susceptibility pattern of the *P. aeruginosa* isolates was determined using the modified Kirby-Bauer disc diffusion system. The antibiotics used were cefuroxime, ciprofloxacin, ceftazidime, penicillin, ofloxacin, amoxicillin, gentamicin and streptomycin (Oxoid, UK). The susceptibility test was carried out on Mueller-Hinton agar (HiMedia, India). Plates were incubated at 37 °C for 18–24 hours and examined for zones of inhibition following the incubation period. The diameters of the zones were measured and recorded accordingly. The following Clinical and Laboratory Standard Institute (CLSI) guidelines were used for the interpretation of the zone of inhibition for each antibiotic: resistant (R), intermediate (I) and sensitive (S).

2.7 Phenotypic Detection of β-Lactamase

The β-lactamase assay by plate method was done following the method of Krishnamurthy et al. (2013). A loopful culture of the test organism was spot inoculated on Mueller Hinton agar (MHA) containing 1% starch and penicillin (10,000 units). After the incubation of plates at 37 °C for 18–24 hours, the plates were flooded with freshly prepared phosphate-buffered saline (PBS) containing potassium iodide. The appearance of the clear, colourless zone around the bacterial growth, an indication of β-lactamase production, was present in 27 isolates.

3 Statistical Analysis

Statistical analysis was performed using chi-square from the Statistical Package for Social Sciences (SPSS) version 20. A p value of ≤ 0.05 was considered statistically significant.

4 Results and Discussion

The prevalence of *Pseudomonas aeruginosa* infection in this study was 18%. The results presented here are from a subset of 27 (18%) participants from whose samples *Pseudomonas aeruginosa* was isolated.

The antibiotic profile of the organisms isolated in this study is presented in Table 2. Isolates were most susceptible to gentamicin (74.1%), ciprofloxacin (59.3%) and ofloxacin (59.3%). The isolates were resistant to all other antibiotics assessed in this study. Over half (51.8%) of the isolates were susceptible to at least two antibiotics, while only six (22.2%) of the isolates were resistant to all antibiotics used in this study. All isolates in this study were multi-drug resistant, i.e. resistant to more than one class of antibiotics (Fig. 1).

Antibiotic resistance has become such a challenge to health professionals that constant monitoring is required to stay abreast of the evolving changes in resistance patterns, gene encoding for resistance and the antibiotics that are no longer effective to ensure the control of the antibiotic resistance epidemic. *Pseudomonas aeruginosa* has frequently found a means to overcome the bactericidal effects of various antibiotics, such as beta-lactams, making the treatment of infections caused by it difficult, especially in immunocompromised individuals.

Table 1 shows the prevalence of *P. aeruginosa* based on patient demographics and the source of collection. In this study, 27 isolates out of 150 were positive for *P. aeruginosa*. There was a higher prevalence of *P. aeruginosa* in ear and would swabs, as opposed to the other sample sites. The differences in prevalence could mean that ear and wound swabs are more likely to be infected by *P. aeruginosa*, as

Table 2 Antibiotic profile of 27 *P. aeruginosa* isolates obtained in this study

Antibiotics	Susceptible		Resistant		Intermediate	
	Number (*n*)	Percentage (%)	Number (*n*)	Percentage (%)	Number (*n*)	Percentage (%)
CAZ	0	0.0	27	100.0	0	0.0
CPX	0	0.0	27	100.0	0	0.0
GEN	201	74.1	6	22.2	1	3.7
CPR	16	59.3	11	40.7	0	0.0
OFL	16	59.3	11	40.7	0	0.0
STR	8	29.6	19	70.4	0	0.0
PEN	0	0.0	27	100.0	0	0.0
AMP	0	0.0	27	100.0	0	0.0

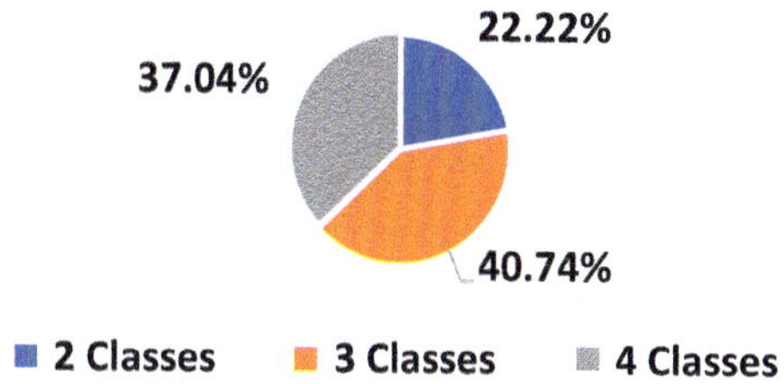

Fig. 1 Multi-drug resistance pattern of *P. aeruginosa* isolates obtained in this study

compared to samples from other parts of the body, such as blood. This was the same in studies by, for example, Jombo et al. (2010), Ogbolu et al. (2008) and Raja and Singh (2007). There was also a higher prevalence of *P. aeruginosa* among the ages of 0–10 and 21–30. This was similar to previous studies done in Nigeria (Ullah et al., 2018 & Ajibade et al., 2019) and other parts of the world (Sambrano et al., 2021). The higher incidence of *P. aeruginosa* in patients aged 0–10 may be due to their weaker immune system and recreational activities; also, during this stage of life, children are likely to be exposed to nosocomial infections. Young adults can also be exposed to this pathogen during hospital visits or recreational activities.

The antibiotic susceptibility pattern of *P. aeruginosa* isolates in this study is consistent with previously published studies. Isolates in this study were most susceptible to four of the eight antibiotics screened: gentamicin (74.10%), ofloxacin (59.3%), ciprofloxacin (59%) and streptomycin (44%)—in that order. Absolute resistance (100%) was shown to ceftazidime, cefuroxime and ampicillin (Table 2). *P. aeruginosa* showed 96.30% resistance to penicillin. The research of Ahmad et al. (2016), conducted in Pakistan, showed a contrast in susceptibility to gentamicin as they recorded a resistance level of 74.25%. Kaur and Singh (2018) found ceftazidime to have the least effect on *Pseudomonas*, with 39.1% susceptibility. Gentamicin and ciprofloxacin showed average susceptibility of 55.7% and 57.2%, respectively. This is in direct contrast to the study carried out by Olowokeere et al. (2018) in the North Central, where *P. aeruginosa* was resistant to the mentioned antibiotics.

This study agrees with that of Shaikh et al. (2015), which found 100% resistance to ampicillin and 91.46% to ceftazidime. The authors did record a higher level of

Table 1 Prevalence of *Pseudomonas aeruginosa* according to the demographic data of the patients recruited for the study

Demographics	Source	Number (n)	Percentage (%)		X2
	Blood	2	7.41	7.630	0.106
	Ear	10	37.04		
	HVS	3	11.11		
	Urine	5	18.52		
	Wound	7	25.93		
Gender	Female	20	74.07	6.259	0.012
	Male	7	25.93		
Age group	0–10	7	25.90	6.519	0.164
	11–20	1	3.70		
	21–30	9	33.30		
	31–40	5	18.50		
	>40	5	18.50		

$^{*}P < 0.05$ is significant ($N = 27$)

ciprofloxacin resistance (74.47%) than was found in this study. Conversely, gentamicin resistance was recorded at 95.74%. This trend of resistance was also observed by Zubair and Iregbu (2018). This is in direct contrast to the study carried out by Olowo-Okere et al. (2018) in the North Central, where *P. aeruginosa* showed resistance to gentamicin and ciprofloxacin.

Where this study found a 40% resistance to the fluoroquinolones Ciprofloxacin and Ofloxacin, the results of Braide et al. (2018) showed no resistance at all. The disparity in the results in Imo State (Braide et al., 2018) versus our findings in Ogun State show that anti-microbial resistance levels vary with location. To highlight this, we found a total resistance to ceftazidime, whereas Braide et al. (2018) recorded 4%. Back in 2007, ceftazidime showed susceptibility of 79.4% in Lagos (Aibinu et al., 2007), and in 2003, ceftazidime had an 88.5% susceptibility in Enugu (Ozumba, 2003). The current surge in resistance could be caused by over-prescription. The findings of gentamicin, ofloxacin, and ciprofloxacin by the same authors were almost evenly distributed.

A systematic review of the prevalence of ESBLs in Nigeria was conducted by Tanko et al. (2020), and *P. aeruginosa* has been confirmed to produce ESBLs. Nasarawa State had the highest levels of ESBLs: 53.6% in the North Central, Bauchi; 82.3% in the North East, Kano; 41.2% in the North West, Ebonyi; 74.3% in the South East, Akwa-Ibom; 47.1% in the South-South and Oyo; and 76.9% in the South West. Overall, on average, there is a higher prevalence of ESBLs in southern Nigeria than in northern Nigeria. Other organisms frequently screened for ESBLs in Nigeria are *Escherichia coli* and *Klebsiella pneumoniae*.

The literature shows that ceftazidime is almost ineffective against pseudomonal infections. Based on the findings of this study, gentamicin shows the most likelihood of treating pseudomonal infections. The fluoroquinolones could also be a treatment option based on their average susceptibility.

5 Conclusion

The high level of *Pseudomonal* resistance recorded is likely due to the abuse of antibiotic usage and the acquisition of resistance elements in hospital and community environments. From this study, gentamicin shows promise, but second- and third-generation cephalosporins are quickly becoming ineffective. This requires urgent attention and an urgent search for more antibiotic agents with effectiveness against *Pseudomonas aeruginosa.*

Acknowledgement I would like to acknowledge all the authors who contributed to this work. I also acknowledge Covenant University Center for Research, Innovation and Development (CUCRID), Ota, Ogun State, Nigeria, for the support for publication.

References

Adesoji AT, Ogunjobi AA, Olatoye IO (2015). Molecular characterization of selected multidrug resistant *Pseudomonas* from water distribution systems in southwestern Nigeria. Ann Clin Microbiol 14(1): 1–11

Ahmad M, Hassan M, Khalid A et al (2016). Prevalence of Extended Spectrum β-Lactamase and Antimicrobial Susceptibility Pattern of Clinical Isolates of *Pseudomonas* from Patients of Khyber Pakhtunkhwa, Pakistan. BioMed Res Int 2016: 6068429

Aibinu I, Nwanneka T, Odugbemi T (2007). Occurrence of ESBL and MBL in Clinical Isolates of *Pseudomonas aeruginosa* From Lagos, Nigeria. J Am Sci 3(4):81-85

Ajibade O, Oladipo EK, Aina KT et al (2019). Incidence of *Pseudomonas aeruginosa* Resistance in Clinical Isolates from Selected Hospitals in Oyo State, Nigeria. Appl Microbiol 5(2):164

Akinduti P, Obafemi YD, Isibor PO, Ishola R, Ahuekwe FE, Ayodele OA, Oduleye OS, Oziegbe O, Onagbesan OM (2021) Antibacterial kinetics and phylogenetic analysis of Aloe vera plants. Open Access Maced J Med Sci 9(A):946-954

Botelho J, Grosso F, Peixe L (2019). Antibiotic resistance in *Pseudomonas aeruginosa* – Mechanisms, epidemiology and evolution. Drug Resist Updat 44(2019):1100640.

Braide W, Madu LC, Adeleye SA et al (2018). Prevalence of Extended Spectrum Beta Lactamase producing *Escherichia coli* and *Pseudomonas aeruginosa* isolated from clinical samples. Int J Sci 4(2): 89–93.

Horcajada JP, Montero M, Oliver A et al (2019). Epidemiology and treatment of multidrug-resistant and extensively drug-resistant *Pseudomonas aeruginosa* infections. Clin Microbiol Rev 32(4): e00031-19.

Jombo G, Akpan S, Epoke J et al (2010). Multidrug resistant *Pseudomonas aeruginosa* infections complicating surgical wounds and the potential challenges in managing post–operative wound infections: University of Calabar Teaching Hospital experience. Asian Pac J Trop Med 3(6): 479–482.

Jombo GTA, Jonah P, Ayeni JA (2008). Multidrug resistant *Pseudomonas aeruginosa* in contemporary medical practice: findings from urinary isolates at a Nigerian University Teaching Hospital. Niger J Physiol Sci 23(1–2):105-109

Kaur A, Singh S (2018). Prevalence of Extended Spectrum Betalactamase (ESBL) and Metallobetalactamase (MBL) Producing *Pseudomonas aeruginosa* and *Acinetobacter baumannii* Isolated from Various Clinical Samples . J Pathog 2018: 1–7.

Krishnamurthy V, Vijaykumar GS, Kumar S et al (2013). Phenotypic and genotypic methods for detection of extended spectrum β lactamase producing *Escherichia coli* and *Klebsiella pneumoniae* isolated from ventilator associated pneumonia. J Clin Diagn Res 7(9): 1975-1978.

Mirzaei B, Bazgir ZN, Goli HR et al (2020). Prevalence of multi-drug resistant (MDR) and extensively drug-resistant (XDR) phenotypes of *Pseudomonas aeruginosa* and *Acinetobacter baumannii* isolated in clinical samples from Northeast of Iran. BMC Res Notes 13(1): 380

Ogbolu DO, Ogunledun A, Adebiyi OE et al (2008). Antibiotic susceptibility patterns of *Pseudomonas aeruginosa* to available antipseudomonal drugs in Ibadan, Nigeria. Afr J Med Med Sci 37(4): 339–344.

Ogunrinola GA, Oyewale GO, Oshamika OO, Olasehinde GI (2020) The human microbiome and its impacts on health. Int J Microbiol. 2020:8045646

Olayinka AT, Olayinka BO, Onile BA (2009). Antibiotic susceptibility and plasmid pattern of *Pseudomonas aeruginosa* from the surgical unit of a university teaching hospital in North central Nigeria. Int J Med Med Sci 1(3): 79–83.

Olowo-Okere A, Ibrahim YKE, Olayinka BO. (2018). Molecular characterisation of extended-spectrum β-lactamase-producing Gram-negative bacterial isolates from surgical wounds of patients at a hospital in North Central Nigeria. J Glob Antimicrob Resist 14: 85–89.

Ozumba UC (2003). Antibiotic sensitivity of isolates of *Pseudomonas aeruginosa* in Enugu, Nigeria. African J Clin Exp Microbiol 4(1): 48–51.

Pérez A, Gato E, Pérez-Llarena J et al (2019). High incidence of MDR and XDR *Pseudomonas aeruginosa* isolates obtained from patients with ventilator-associated pneumonia in Greece, Italy and Spain as part of the MagicBullet clinical trial. J Antimicrob Chemother 74(5): 1244–1252.

Raja NS, Singh NN (2007). Antimicrobial susceptibility pattern of clinical isolates of *Pseudomonas aeruginosa* in a tertiary care hospital. J Microbiol Immunol Infect 40(1): 45–49.

Sambrano H, Castillo JC, Ramos CW et al (2021). Prevalence of antibiotic resistance and virulent factors in nosocomial clinical isolates of *Pseudomonas aeruginosa* from Panamá. Braz J Infect Dis 25(1): 101038.

Shaikh S, Fatima J, Shakil S et al (2015). Prevalence of multidrug resistant and extended spectrum beta-lactamase producing *Pseudomonas aeruginosa* in a tertiary care hospital. Saudi J Biol Sci 22(1):62–64.

Smith S, Ganiyu O, John R et al (2012). Antimicrobial resistance and molecular typing of *Pseudomonas aeruginosa* isolated from surgical wounds in Lagos, Nigeria. Acta Med Iran 50(6):433–438.

Tanko N, Bolaji RO, Olayinka AT et al (2020). A systematic review on the prevalence of extended-spectrum beta lactamase-producing Gram-negative bacteria in Nigeria. J Glob Antimicrob Resist 22: 488–496

Ullah W, Qasim M, Rahman H et al (2018). Incidence of *Pseudomonas aeruginosa* resistance in clinical isolates from selected hospitals in Oyo state, Nigeria. J Chin Med Assoc 27(4):176–182.

Zubair K, Iregbu K (2018). Resistance Pattern and Detection of Metallo-beta-lactamase Genes in Clinical Isolates of *Pseudomonas aeruginosa* in a Central Nigeria Tertiary Hospital. Niger J Clin Pract 21: 176–182.

Microbial Quality of Watermelons Sold in Ota

N. O. Fasuyi

1 Introduction

Fruit and vegetables are heralded as life giving and life maintaining because they are packed with substances that promote health, prevent diseases, and boost longevity (FAO 2020). A diet that adequately represents them foretells a life with minimal trips to the doctor. Fruits have an abundance of sugars, phytochemicals, fiber, water, and proteins (Erhirhie et al. 2020). The widely recommended daily intake of fruit and vegetables by regulatory agencies for each is two servings and three servings, respectively (Miller et al. 2016; Wallace et al. 2019), an equivalent to the World Health Organization's approximation of 400 grams of fruit and vegetables to be consumed daily (Frank et al. 2019; FAO 2020). The consumption of fruit and vegetables in Africa is at a much slower rate when compared to that in other continents. Africa brings up the rear. This low consumption rate is probably due to population explosion and slow economic expansion (Balali et al. 2020). Despite this, the production of fruit and vegetables has increased, with street-vended fresh fruit and vegetables accounting for 86% of the total market produce, thus necessitating crops free from microbial contamination (Balali et al. 2020). Generally, fruits are consumed whole in Nigeria, but many are cut into smaller sizes and sold in open markets. Watermelons and pineapples are the most commonly cut ready-to-eat (RTE) fruits in Nigeria (Oyedele et al. 2020). Recent years have seen an increase in cut RTE fruits as these do not have to be washed, peeled, and cut before ingestion. These fruits are cheaper than whole fruits, convenient to eat on the go, nutritious, and are more accessible (Orji et al. 2016; Osuntokun 2018).

N. O. Fasuyi (✉)
Department of Biological Sciences, Microbiology Programme, Covenant University, Ota, Ogun State, Nigeria

© The Author(s), under exclusive license to Springer Nature Switzerland AG 2022
A. O. Ayeni et al. (eds.), *Bioenergy and Biochemical Processing Technologies*, Green Energy and Technology, https://doi.org/10.1007/978-3-030-96721-5_24

Fruits are generally highly perishable and are prone to contamination by pathogenic microorganisms (FAO 2020). Their sugars, vitamins, minerals, and other organic substances supply substrate for microorganisms and contribute to the density and variety of the microbial population found in food (Mailafia et al. 2017). Microbiological contaminants of priority concern are parasites, bacteria, viruses, and prions (Alegbeleye et al. 2018). Bacterial pathogens are the most frequently implicated contaminants in fresh produce. Contamination by foodborne organisms will differ based on varying conditions and the means of contamination. Microorganisms get into fruits through various channels, including sewage, organic fertilizers, air, manure, soil, raw/irrigation water, livestock, and insects (Rajwar et al. 2016; Jideani et al. 2017). Postharvest contamination can arise from handling, equipment, environment, transportation, storage, and insects (Machado-Moreira et al. 2019). When the fruits are injured with cuts and punctures, organisms on the skin may be internalized within the fruit and replicate under favorable conditions (Alegbeleye et al. 2018; Edeghor et al. 2019). The increasing rate of foodborne illnesses from fruits has cast a shadow on the safety of fruits in the market. Foodborne illnesses refer to ill health, which is a direct result of the consumption of infected or poisoned food. When two or more individuals present with similar symptoms after eating the same food, this means a foodborne illness outbreak has occurred (Kearney 2018).

Ready-to-eat fruits sold by street vendors are frequently prepared in areas with limited sanitary provisions, thus facilitating contamination from handling practices (Augustin et al. 2020). Exposure to environmental pollution such as market refuse, wastewater, crawling and flying insects, and exhaust fumes can enable pathogenic microbes in the atmosphere and immediate environment, which can settle on the fruits' surface (Erhirhie et al. 2020). There are records of outbreaks of fruits implicated in foodborne illnesses in developed countries. Africa's compromised public health system has made African countries the recipients of the bulk of foodborne diseases (Faour-Klingbeil and Todd 2019). The implicated pathogens include *Bacillus cereus*, *Staphylococcus aureus*, *Pseudomonas aeruginosa*, *Listeria monocytogenes*, *Escherichia coli*, *Campylobacter jejuni*, *Klebsiella pneumoniae*, *Serratia marcescens*, and *Enterobacter aerogenes*. Thus, this study was carried out to identify some of the pathogens present in ready-to-eat watermelons sold in Ota, Ogun State, Nigeria.

2 Materials and Methods

2.1 Sample Collection

Two samples each of freshly cut watermelons tied up in transparent polythene and displayed for sale were purchased from four different roadside vendors along Ota road (Winners, Iyana, Oja-Ota, and Oju-Ore), Ogun State, Nigeria. The samples were stored in sterile Ziploc bags and transferred to the Microbiology Teaching Laboratory, Covenant University, Ota, Nigeria, for microbiological analysis.

2.2 *Isolation of Microorganisms*

The samples from each vendor were homogenized in a high-power blender, and 1 ml of each homogenate was diluted in 9 ml of sterile distilled water up to fivefold dilutions. Nutrient agar (NA), MacConkey agar (MCA), eosin methylene blue agar (EMB), and mannitol salt agar (MSA) were prepared and sterilized following the manufacturer's instructions. An aliquot of 1 ml (equivalent to 1 g) from the 10^{-5} dilutions was pipetted into sterile Petri dishes, and cooled molten agar was mixed in. The plates were swirled gently for an even distribution of the inoculum. When fully set, they were inverted and incubated for 24 hours at 37 °C. The colonies that were observed the following day were subjected to subculturing for pure isolates and for the monitoring of phenotypic characteristics.

2.3 *Determination of Microbial Load*

The pure colonies obtained from the pour plate method were enumerated using a colony counter (JP SELECTA 4905000) and expressed in the CFU/mL standard form.

3 Results and Discussion

3.1 *Microbial Load of Watermelon Samples*

The number of distinct colonies observed and enumerated varied among the watermelon samples from different vendors. Table 1 contains the details of the average plate count. The bacterial load was in descending order, from the highest to the lowest, from vendor 4 to vendor 1, respectively. Among the four vendors, vendor 4 had the highest count at 2.7×10^8 CFU/mL, followed by vendor 3 with 2.3×10^7 CFU/mL and vendor 2 with 3.0×10^7 CFU/mL; finally, vendor 1 had the least count at

Table 1 Total aerobic plate count of watermelons sampled in Ota, Ogun State

Mean bacterial load (CFU/mL)				
	Vendor 1 (Winners)	Vendor 2 (Iyana)	Vendor 3 (Oja-Ota)	Vendor 4 (Oju-Ore)
	1.6×10^7	3.0×10^7	2.3×10^7	2.7×10^8

1.6×10^7 CFU/mL. The least bacterial count was beyond the acceptable levels of 10^3 CFU/mL.

3.2 *Isolated Organisms from Watermelon Samples*

The phenotypic summary of the organisms isolated from the watermelon samples is given below. The features considered include Gram's reaction, shape, colony, and color. Green metallic sheen on EMB agar was indicative of *Escherichia coli*, yellow round colonies on MSA indicated *Staphylococcus* sp., and white colonies on MSA signaled *Micrococcus* sp. White colonies on MCA were presumptive of the presence of *Proteus* and pink mucoid colonies on EMB of *Klebsiella* sp. (Table 2).

Fruits contain microbial contaminants more than one would want to believe (Machado-Moreira et al. 2019; Balali et al. 2020). Previous research has highlighted the variety of bacterial pathogens that have been recovered from watermelons and other fruits. It is recommended that 1 g of fruit should not contain more than 10^3 of microorganisms CFU/g (FAO 1992). The high values found by many researchers within the 10^4–10^7 range all converge to the point that fruit samples are home to more contaminants and likely pathogens than is generally regarded as safe (Daniel 2014; Auta et al. 2017; Musa et al. 2019). The present research found the microbial load of the watermelons to be more than twice the recommended ranges stated by the Food and Agriculture Organization (FAO). It can thus be interpreted that the microbial quantities in whole fruits would be astronomical, many of which may be contaminants risky to public health. This would also mean that the high titers of inoculum in fruits would mean that a higher chance of foodborne infections when consumed as a factor for pathogenicity is the size of the pathogen inoculum. The general safety of the fruits is largely dependent on the pre- and postharvest practices adhered to by food growers, as well as the conditions the fruits are exposed to. It is easy for the fruits to become contaminated after harvest, before they eventually reach the customers despite thorough hygiene practices by the farmers. Coupled with the internalization ability of some pathogens like *Salmonella*, which becomes buried in the pulp of fruits and replicates, the enforcement of food safety in the retail sector is critical.

Table 2 Isolated organisms from watermelon samples

S/N	Phenotypic summary	Gram's reaction	Probable organism
1	Green metallic sheen on EMB agar	– Pink rods	*Escherichia coli*
2	Pink mucoid colonies on EMB agar	– pink rods	*Klebsiella spp.*
3	Yellow round colonies on MSA	+ purple cocci	*Staphylococcus aureus*
4	Pink round colonies on MSA	+ purple cocci	*Staphylococcus spp.*
5	White colonies on MCA	– Pink rods	*Proteus spp.*
6	White colonies on MSA	+ purple tiny cocci	*Micrococci*

Five genera of bacteria were identified in this study: *Escherichia coli*, *Klebsiella*, *Staphylococcus*, *Proteus*, and *Micrococcus*. *Escherichia coli* is the most well-known food pathogen of utmost importance. This coliform is used to confirm the absence of fecal contamination. If present, it can be assumed that a part of the production process was conducted under unsanitary conditions (Ekici and Dümen 2019). It is frequently implicated in diarrheic infections, with complications such as sepsis and hemolytic uremic syndrome (HUS). Pathogenic strains of *E. coli*, such as 0157:H7, have been implicated in foodborne outbreaks from the ingestion of contaminated meat, poultry, and fruit and vegetables with a record of fatalities (Ekici and Dümen 2019).

Klebsiella pneumoniae, a nosocomial pathogen, is quickly becoming a concern in food safety as it has been recovered from raw and cooked meat and vegetables, as well as fruit juices (Riley 2020). Its multidrug-resistant status compels the urgent need for safe practices. The organism could cause liver abscess when it colonizes the gastrointestinal tract (Hartantyo et al. 2020).

With an estimated 241,000 illnesses per year in the United States, foodborne diseases caused by *Staphylococcus aureus* have become a problem (Kadariya and Smith 2014). It is dominantly transmitted via poor handwashing practices by food handlers. Its symptoms include abdominal cramps, vomiting, and diarrhea. Clinically recognized as a cause of urinary tract infections as well as a commensal of the digestive tract, *Proteus* sp. has been implicated in gastrointestinal infections, Crohn's disease, and appendicitis (Hamilton et al. 2018). These swarming/swimming bacteria are equipped with antibiotic resistance, hemolysins, enterotoxins, mucosal adherence, biofilm formation, and other defense mechanisms that enable the successful execution of infections. The first report of food poisoning by *P. mirabilis* occurred in 2008 and was linked to stewed pork balls (Wang et al. 2010). Between 1994 and 2005, 124 events of *Proteus* food infections were recorded (Wang et al. 2005), with *P. mirabilis* having the highest number (61) of cases. Although not frequently found, *Proteus* is rapidly gaining a reputation as a foodborne pathogen.

These organisms recovered in this study are risky to public health. They could have gained access to the fruits through poor preharvest and postharvest strategies, such as the use of contaminated irrigation water or slicing fruits with unclean knives and using dirty utensils (Machado-Moreira et al. 2019). The general optimal temperature of these organisms is 37 °C; hence, they can multiply uninhibited and even reach dangerous quantities within the fruits. The high water activity of watermelons also creates suitable conditions for this phenomenon to occur. *Staphylococcus aureus* and *Escherichia coli* may be involved with food intoxication due to their ability to produce toxins. *Staphylococcus aureus* produces the staphylococcal heat-stable enterotoxins A–J, which interact with the immune system, leading to the low effectiveness of proteolytic enzymes and mast cells (Kadariya and Smith 2014). *E. coli* has a variety of toxins, including the Shiga toxin and verotoxin, which are responsible for the diarrheic features of *E. coli* infections (Kim et al. 2020). In rare cases, *Proteus* infections can result in gastroenteritis. This is more frequent in *Klebsiella* infections, but *Micrococcus* is not thought to be a true pathogen.

4 Conclusion

The results of this study show that the fresh-cut watermelons assessed are of poor quality, having numbers totally unacceptable by established standards. It reflects that hygiene levels in the markets are poor and should be reviewed. Measures to ensure that the levels of microbial populations fall within the set limits and that food handlers and food preparation areas are well sanitized need to be implemented within the short possible time to help reduce public health concerns on ready-to-eat fruits.

Acknowledgments The author wishes to thank Covenant University for supporting this publication.

References

Alegbeleye OO, Singleton I, Sant'Ana AS (2018) Sources and contamination routes of microbial pathogens to fresh produce during field cultivation: A review. Food Microbiol 73:177. https://doi.org/10.1016/J.FM.2018.01.003

Augustin J-C, Kooh P, Bayeux T, Guillier L, Meyer T, Jourdan-Da Silva N, Villena I, Sanaa M, Cerf O (2020) Contribution of Foods and Poor Food-Handling Practices to the Burden of Foodborne Infectious Diseases in France. Foods 9:1644. https://doi.org/10.3390/foods9111644

Auta KI, Madinatu B, Yayock HC, Solomon B (2017) Microbial Quality Assessment of Sliced Pineapple and. IJRDO-Journal Biol Sci 3:13–29

Balali GI, Yar DD, Afua Dela VG, Adjei-Kusi P (2020) Microbial Contamination, an Increasing Threat to the Consumption of Fresh Fruits and Vegetables in Today's World. Int J Microbiol 2020: https://doi.org/10.1155/2020/3029295

Daniel AA, Danfulani S 2, Peter G 4 &, Ajewole AE (2014) MICROBIOLOGICAL QUALITY OF SLICED FRESH FRUITS SOLD IN BIDA, NIGERIA. 3:178–180

Edeghor U, John GE, Origbu C (2019) Bacteriological profile of cut fruits sold in Calabar Metropolis. World News Nat Sci 23:43–55

Ekici G, Dümen E (2019) Escherichia coli and Food Safety. Universe Escherichia coli. https://doi.org/10.5772/INTECHOPEN.82375

Erhirhie EO, Omoirri MA, Chikodiri SC, Ujam TN, Kesiena EE, Oseyomon JO (2020) Evaluation of Microbial Quality of Vegetables and Fruits in Nigeria: A Review. Int J Nutr Sci 5:99–108

FAO (2020) FRUIT AND VEGETABLES-YOUR DIETARY ESSENTIALS. Rome

FAO (1992) Manuals of food quality control

Faour-Klingbeil D, C. D. Todd E (2019) Prevention and Control of Foodborne Diseases in Middle-East North African Countries: Review of National Control Systems. Int J Environ Res Public Health 17:70. https://doi.org/10.3390/ijerph17010070

Frank SM, Webster J, McKenzie B, Geldsetzer P, Manne-Goehler J, Andall-Brereton G, Houehanou C, Houinato D, Gurung MS, Bicaba BW, McClure RW, Supiyev A, Zhumadilov Z, Stokes A, Labadarios D, Sibai AM, Norov B, Aryal KK, Karki KB, Kagaruki GB, Mayige MT, Martins JS, Atun R, Bärnighausen T, Vollmer S, Jaacks LM (2019) Consumption of Fruits and Vegetables Among Individuals 15 Years and Older in 28 Low- and Middle-Income Countries. J Nutr 149:1252–1259.

Hamilton AL, Kamm MA, Ng SC, Morrison M (2018) Proteus spp. as putative gastrointestinal pathogens. Clin Microbiol Rev 31: https://doi.org/10.1128/CMR.00085-17/ASSET/4DDC8A7C-AECE-4EC9-9555-5998F7BA2F24/ASSETS/GRAPHIC/ZCM0031826350003.JPEG

Hartantyo SHP, Chau ML, Koh TH, Yap M, Yi T, Cao DYH, Gutiérrez RA, Ng LC (2020) Foodborne Klebsiella pneumoniae: Virulence Potential, Antibiotic Resistance, and Risks to Food Safety. J Food Prot 83:1096–1103. https://doi.org/10.4315/JFP-19-520

Jideani AIO, Anyasi TA, Mchau GRA, O.Udoro E, Onipe OO (2017) Processing and Preservation of Fresh-Cut Fruit and Vegetable Products. Postharvest Handl https://doi.org/10.5772/INTECHOPEN.69763

Kadariya J, Smith TC, Thapaliya D (2014) Staphylococcus aureus and staphylococcal food-borne disease: an ongoing challenge in public health. Biomed Res Int 2014: https://doi.org/10.1155/2014/827965

Kearney GD (2018) Introduction to Foodborne Illness Outbreak Investigations. Environ Public Heal Pract Guid https://doi.org/10.2105/9780875532943CH13

Kim JS, Lee MS, Kim JH (2020) Recent Updates on Outbreaks of Shiga Toxin-Producing Escherichia coli and Its Potential Reservoirs. Front Cell Infect Microbiol 10:273. https://doi.org/10.3389/FCIMB.2020.00273/BIBTEX

Machado-Moreira B, Richards K, Brennan F, Abram F, Burgess CM (2019) Microbial Contamination of Fresh Produce: What, Where, and How? Compr Rev Food Sci Food Saf 18:1727–1750

Mailafia S, Okoh GR, Olabode HOK, Osanupin R (2017) Isolation and identification of fungi associated with spoilt fruits vended in Gwagwalada market, Abuja, Nigeria. Vet World 10:393–397. https://doi.org/10.14202/vetworld.2017.393-397

Miller V, Yusuf S, Chow CK, Dehghan M, Corsi DJ, Lock K, Popkin B, Rangarajan S, Khatib R, Lear SA, Mony P, Kaur M, Mohan V, Vijayakumar K, Gupta R, Kruger A, Tsolekile L, Mohammadifard N, Rahman O, Rosengren A, Avezum A, Orlandini A, Ismail N, Lopez-Jaramillo P, Yusufali A, Karsidag K, Iqbal R, Chifamba J, Oakley SM, Ariffin F, Zatonska K, Poirier P, Wei L, Jian B, Hui C, Xu L, Xiulin B, Teo K, Mente A (2016) Availability, affordability, and consumption of fruits and vegetables in 18 countries across income levels: findings from the Prospective Urban Rural Epidemiology (PURE) study. Lancet Glob Heal 4:e695–e703. https://doi.org/10.1016/S2214-109X(16)30186-3

Musa F, Alhassan S, Nasir N (2019) Assessment of Bacterial Quality of Some Fruits Sold in Selected Markets Within Kaduna Metropolis. Sci World J 14:2019

Orji JO, Orinya Chinedu I, Okonkwo Eucharia O, Uzoh Chukwuma V, Ekuma Uchechukwu O, Ibiam Gideon Ama, Onuh Euslar N (2016) The Microbial Contamination of Ready-To-Eat Vended Fruits in Abakpa Main Market, Abakaliki, Ebonyi State, Nigeria. J Pharm Biol Sci 11:71–80

Osuntokun O (2018) Resistance profiling of bacterial isolate from cut and sliced ready- to-eat polyethylene packed watermelon (citrullus lanatus) sold in akoko communities. J Bacteriol Infect Dis 0:

Oyedele OA, Kuzamani KY, Adetunji MC, Osopale BA, Makinde OM, Onyebuenyi OE, Ogunmola OM, Mozea OC, Ayeni KI, Ezeokoli OT, Oyinloye AM, Ngoma L, Mwanza M, Ezekiel CN (2020) Bacteriological assessment of tropical retail fresh-cut, ready-to-eat fruits in south-western Nigeria. Sci African 9:e00505. https://doi.org/10.1016/j.sciaf.2020.e00505

Rajwar A, Srivastava P, Sahgal M (2016) Microbiology of Fresh Produce: Route of Contamination, Detection Methods, and Remedy. http://dx.doi.org/101080/1040839820138411 19 56:2383–2390. https://doi.org/10.1080/10408398.2013.841119

Riley LW (2020) Extraintestinal Foodborne Pathogens. https://doi.org/101146/annurev-food-032519-051618 11:275–294. https://doi.org/10.1146/ANNUREV-FOOD-032519-051618

Wallace TC, Bailey RL, Blumberg JB, Burton-Freeman B, Chen CO, Crowe-White KM, Drewnowski A, Hooshmand S, Johnson E, Lewis R, Murray R, Shapses SA, Wang DD (2019) Fruits, vegetables, and health: A comprehensive narrative, umbrella review of the science and recommendations for enhanced public policy to improve intake. https://doi.org/101080/1040839820191632258 60:2174–2211.

Wang S, Duan H, Zhang W, Li J-W (2005) Analysis of bacterial foodborne disease outbreaks in China between 1994 and 2005. https://doi.org/10.1111/j.1574-695X.2007.00305.x

Wang Y, Zhang S, Yu J, Zhang H, Yuan Z, Sun Y, Zhang L, Zhu Y, Song H (2010) An outbreak of Proteus mirabilis food poisoning associated with eating stewed pork balls in brown sauce, Beijing. Food Control 21:302–305. https://doi.org/10.1016/J.FOODCONT.2009.06.009

underrepresentation of MM, systemic infections, and the impact of mucormycosis as a re-emerging fungal infection associated with COVID-19 in Africa.

2 Prevalence of Mucormycosis

Mucormycosis is acquired by sporangiospores inhalation from air, occasionally through ingestion of fungal spore contaminated food or traumatic inoculation, and it has an often fatal prognosis (Alqarihi et al. 2020). Mucormycosis is accompanied with general symptoms such as swelling of one side of the face, headache, congestion of the nasal cavity, black lesions on nasal bridge or upper palate of the mouth which can escalate quickly and become chronic, vessel thrombosis, tissue necrosis, and fever (CDC 2021). This often results to the need for surgical removal of infected tissues (Skiada et al. 2018) and survivors often require reconstructive surgeries to mitigate the disfiguring defects (Augustine et al. 2017). *Lichtheimia spp.*, *Mucor spp.*, and *Rhizopus spp.* have been associated with approximately 70% of the mucormycosis reported cases and are the visibly common mucormycosis etiologic agents (Skiada et al. 2020).

Despite mucormycosis angioinasiveness, reports of systemic infection associated mucormycosis have been reported. Systemic spread of mucormycosis are usually observed primarily in the lungs (pulmonary mucormycosis) which could further disseminate to the nervous system, sinus, liver, kidney, heart, orbital space, and large intestine (Chen et al. 2019). Currently, mucormycosis epidemic in India reported cerebral and orbital area as common site of infection following entry via the lungs and other organs (Prakash and Chakrabarti 2019).

2.1 Epidemiology of Mucormycosis

Globally, MM reports remain scanty and estimation of its prevalence varies due to prevalence in different high-risk populations. Most of the cases discovered post-mortem are rhinocerebral and pulmonary mucormycosis (Ramadorai et al. 2019).

Epidemiology in Africa

Cases of mucormycosis have been recorded in some parts of Africa but COVID-19 associated mucormycosis has not been properly documented. Mucormycosis has also not been properly documented in Nigeria but Yusuf and Onyiriuka (2012) recorded a case of subcutaneous mucormycosis based on physical analysis and light microscopy. Oladeji et al. (2013) also recorded rhinocerebral MM in a 40-year-old female with uncontrolled diabetes mellitus, four-week history of facial rash, complete nasal bridge collapse, nasopalatal fistula with black eschars on the mucosa, ulceration of the mid face, and bilateral visual loss. Based on available data, there has been no case of CAM in Nigeria. Onyango et al. (2002) reported rhinocerebral mucormycosis from a male patient, a Kenyan resident, with non-insulin dependent diabetes, with characteristic symptom of abdominal pain, left facial palsy, vomiting, and dyspnea; a 39-year-old female with tuberculosis (TB)/human immunodeficiency virus (HIV) co-infection was suspected to have rare renal disseminated MM in South Africa, although it was not confirmed by fungal culture or molecular analysis (PCR) (Khaba et al. 2021). Cases of *Rhizomucor spp.* mediated MM were reported in child cancer patients with fatality in Egypt (El-Mahallawy et al. 2016), with diabetic ketoacidosis as the major risk factor associated with MM prevalence in male patients (Daoud et al. 2014). Egypt recorded her first case of COVID-19 associated mucormycosis on June 1, 2021, at Mansoura University Hospital (Egypt Independent 2021).

2.2 *Systemic Mucormycosis*

Rhinocerebral Mucormycosis

A common form of mucormycosis; it affects the nasal cavity and spreads quickly to the brain. Mostly found among people with diabetic ketoacidosis and diabetes mellitus (Jeong et al. 2019) and kidney transplant patients (CDC 2021). Rhinocerebral mucormycosis spreads and extends from the nasal area to the brain (Sahoo et al. 2017). Common non-ophthalmic symptoms include facial swelling, bone destruction, fever, facial pain, headache, nasal discharge, facial nerve palsy, epistaxis, hemiplegia, tooth ache, sinusitis, nasal ulceration, palatal eschar, altered mental status, and facial numbness (Prakash and Chakrabarti 2019), ophthalmic manifestations include ophthalmoplegia eye pain, proptosis, decreased vision, chemosis, orbital cellulitis, necrosis, ptosis, and periorbital discolouration (Vaughn et al. 2018).

Pulmonary Mucormycosis

Also a common type of mucormycosis usually found with people with hematological malignancy, diabetes mellitus, stem cell transplant, organ transplant, kidney disease (Feng and Sun 2018), and pulmonary tuberculosis (Prakash et al. 2019). Symptoms include high fever, persistent cough, pleuritic chest pain, hemoptysis, and dyspnea; diagnosis is difficult because of non-specificity of imaging studies.

Cutaneous Mucormycosis

Frequently seen in trauma or breach of skin epithelial usually after skin burns, surgery, and any type of skin trauma (CDC 2021). Trauma predisposes people to this type of mucormycosis and risk factors such as injection in the intramuscular part of the body in local health centers, motor vehicle accident, surgery, contaminated dressings, burns, natural disasters, and so on (Prakash and Chakrabarti 2019). It is a localized infection with deep permeation in and around the subcutaneous tissue, it is also classified as a disseminated infection, based on the extent of the invasion. Cutaneous mucormycosis is confined to the cutaneous and subcutaneous tissue and does not infect adjacent sites, or permeation of tissue deep extension involving muscle invasion, bone invasion as well as tendons, it has a clinical presentation of necrotic tissue having a black appearance composed of dried blood and exudate (Paduraru et al. 2016).

Gastrointestinal Mucormycosis

Gastrointestinal (GI) mucormycosis is more present in young children especially children born premature and frail looking babies, with immunosuppressed immune system because of surgery, antibiotics, or other medications that cause immunosuppresion (CDC 2021) and common among malnourished patients or patients undergoing dialysis. Gastrointestinal mucormycosis is very difficult disease to diagnose after death (Kaur et al. 2018). It also occurs in immunocompromised patients especially those who have just undergone solid organ transplants (Wotiye et al. 2020) and in individuals with underlying diseases such as diabetes mellitus, and people who use antibiotics indiscriminately are at risk of this type of mucormycosis (Kaur et al. 2018). Sites of infection include large intestine, stomach, small intestine, and oesophagus (Dioverti et al. 2015), and symptoms often include abdominal pain, gastrointestinal bleed, abdominal distension, and diarrhea (Prakash and Chakrabarti 2019) (Tables 1 and 2).

Renal Mucormycosis

Individuals who underwent kidney transplant or immunosuppression have been reported with systemic mucormycosis invading the kidney, but most recently a unique case was also reported in an apparently healthy person (Saneesh et al. 2021). Presenting flank pain, anuria, and fever (Bhadauria et al. 2018). Computer tomography (CT) scans, ultrasound, and polymerase chain reaction (PCR) can be employed for early diagnosis of this systemic mucormycosis that invades the kidney. CT scans show infarction in the parenchyma, renal thrombosis with thickening of the renal pelvis (Shanmugam et al. 2019).

Table 1 Prevalence of mucormycosis in selected countries

Area of study	Duration	Incidence of cases	Reference
San Francisco Bay, Area	1992–1993	1.7 per 1 million individuals (500 cases per year)	Rees et al. (1998)
Mucormycosis-related hospitalizations	January 2005–June 2014	0.12 per 10,000 discharges during, later increased to 0.16 per 10,000 discharges	Kontoyiannis et al. (2016)
France	1997–2006	0.7 cases per million- 1.2 per million	Lantermier et al. (2012)
Belgium	2000–2009	0.019 /10,000 patient-days- 0.148/10,000 patient-days	Saegeman et al. (2010)
India	10-year period 5-year period 18-month period	12.9 cases/year 35.6 cases/year eighteen 50 cases/ year	Chakrabarti et al. (2001). Chakrabarti et al. (2006) Chakrabarti et al. (2009)

Table 2 Recent report of mucormycosis in Africa

	Country	Mucormycosis cases per 100,000	Duration	Reference
Africa	Senegal	0.2	2012–2014	Badiane et al. (2015)
	Cameroon	0.2	2014–2017	Mandengue and Denning (2018)
	Nigeria	–	–	–

Disseminated Mucormycosis

It occurs by spreading via the bloodstream to other organs of the body, e.g., brain, heart, lung, nervous system, skin, and spleen (CDC 2021). The most common site of dissemination is the lungs (Fadhel et al. 2019). Patients who receive organ transplants and patients with leukemia, myeloma, and lymphoma are more at risk of this type of mucormycosis (Jeong et al. 2019).

2.3 *Underrepresentation and Neglect of Mucormycosis in Africa*

Generally, fungal infections have low representation and are often neglected in Africa, compared with other microorganisms over the years (Rodrigues and Albuquerque 2018). With relative death and morbidity associated with mycoses in sub-Saharan Africa, it is noteworthy that they are not recognized as neglected tropical diseases (Govender et al. 2011). Cryptococcal meningitis, a deadly infectious disease responsible for approximately 1.5 million deaths (WHO 2018) received 0.5% of global funding (Chapman et al. 2017), lesser than malaria, tuberculosis, diarrheal diseases; cryptococcal meningitidis (Chapman et al. 2017) kills 20 times more than Neisseria meningitidis while it remains neglected in several countries. Mortality and morbidity caused by fungal infections are often common among people in underdeveloped or developing countries (Schwartz et al. 2017), and the mycoses epidemiology is difficult to predict and often shows multiple resistance to antifungal drugs which could pose a threat to human health, e.g., *Candida auris* (Lamoth and Kontoyiannis 2018). Reliable and cost-effective diagnostic tools are limited, and therapeutic options are poorly effective, toxic, and expensive (Pinalto and Alspaugh 2016), available ones are not readily available where needed or even not registered in areas where needed (GAFFI 2018). Amphotericin B discovered in 1958 (Carolus et al. 2020) and fluconazole (Richardson et al. 1990) were previous antifungals available before echinocandins, which is the most recent antifungal agent approved for mycoses therapy was approved in 2002 (Denning 2002). Over 20 antibiotics have been approved since then. Similarities shared by fungi with their host hinder the development of antifungal compounds (Rodrigues and Nosanchuk 2020).

There is a huge need to investigate and increase the surveillance of fungal infections and therapeutic intervention to MM. As shown by the recent cases of COVID-19 associated mucormycosis (CAM) and COVID-19 associated pulmonary aspergillosis (CAPA), that fungal infections can complicate primary infections (COVID-19) and also be very fatal, hence the urgent need to develop novel therapeutic, diagnostic, and preventive tools and antifungal agents (Ramaswani et al. 2021).

2.4 *Systemic Pathogenesis and Tissue Pathology*

Pathogenesis of Mucormycosis

Inhaled spores' ability to germinate, form hyphae in the invaded host is crucial in the establishment of infection in mucormycosis (Ibrahim et al. 2012). Mucormycosis is characteristically incapable of penetrating unbroken skin unless the skin is broken through burns, cuts, etc. allowing spores, contaminated adhesive tape and the use of contaminated therapeutics for immunocompromised patients (Hartnett et al. 2019). Prevalence is common during natural disasters, i.e., contaminated soil or water (e.g., Indonesia tsunami 2004) (Andresen et al. 2005), and after devastating tornadoes in Joplin, Missouri in June 2011 (Neblett Fanfair et al. 2012) saw outbreaks of mucormycosis. Mucormycosis flourish in iron-rich environments as iron (Fe) is necessary for fungal cell growth, metabolism, and development, and elevated iron levels in the serum predisposes patients to mucormycosis (Adrianaki et al. 2018).

Factors That Aid Pathogenesis

Host Defense and Mucormycosis Pathogenesis

Studies have shown that immunocompromised individuals or impaired phagocytes were at higher risk of mucormycosis (Ibraham et al. 2012; Inglesfield et al. 2018; Hassan and Voight 2019). Fungal spores when inhaled could germinate and hyphae is formed, which is necessary for establishment of infection, inhalation of mucor spores by immunocompetent individuals rarely have resultant effect on the development of mucormycosis (He et al. 2021). Individuals who have their immune system suppressed because of use of corticosteroid and have DKA die of progressive pulmonary and blood transmitted infection (Baldin and Ibrahim 2017). Diabetic ketoacidosis characterized with hyperglycemia and low blood pH have shown to have dysfunctional phagocytes, impaired chemotaxis put one at risk of MM (Palermo et al. 2020). The mechanisms by which phagocytes lead to severe infection such as diabetes mellitus and diabetes ketoacidosis and its relationship with corticosteroids are not yet known so phagocyte dysfunction does not really throw light on the high incidence of mucormycosis in individuals with DKA (Inglesfield et al. 2018).

Iron Uptake and Mucormycosis Pathogenesis

Mucorales fungi are virulent and it enables them to cause disease in their host. These virulence factors include the ability to acquire iron as essential element for growth and functioning of their cell from their host using different processes (Stanford and Voight 2020). It grows well in serum or environment rich in iron and at acidic environment, e.g., rhizopus (Morales-Franco et al. 2021). Studies have shown that the iron bound to proteins in humans to avoid toxicity of free iron in human (Caza and Kronstad 2013; Eid et al. 2017; Yiannikourides and Latunde-Dada 2019) and the unbound iron is present in the serum of patient with DKA predisposes to mucormycosis (Adam et al. 2016). MM acquire their iron from their host through siderophores. Siderophores are organic chelators having a low molecular weight with distinct affinity for Fe (III), and microbial cells iron uptake is mediated by siderophores (Morales-Franco et al. 2021); example is rhizopus known to secrete rhizoferrin of the polycarboxylate family (Carroll et al. 2017). The mechanism of rhizoferrin is yet to be deciphered, the uncertainty is whether it transports iron by extracellular release or internalization of the siderophore before the release of iron in the cytoplasm of the fungi, but the certain part of this mechanism of iron uptake with the use of siderophores is the inefficiency of rhizoferrin in obtaining iron from serum (Challa 2019). Hence, fungi siderophore contribution of its virulence in human is minimal if not non-significant (Fatima et al. 2017). Fungal adaptation to use xenosiderophores is a more structured instrument for acquiring iron from their host (Gerwein et al. 2018).

Another technique fungi acquire iron from host is the use of heme oxygenase (Kornitzer and Roy 2020). Studies revealed that this heme oxygenase may enable species of Rhizopus, e.g., *Rhizopus oryzae*, to obtain iron from human heamoglobin which gives an explanation to their angioinvasiveness (Ibrahim et al. 2012; Petrikkos and Tsioutis 2018; Jose et al. 2021). Ibrahim et al. (2010) reported pathogenesis of *Rhizopus oryzae* producing numerical reduction of FTR1, reduced biomass on heme supplemented media, thus FTR1 could as well be acting as a permease for cytoplasmic membrane which help the uptake of heme intracellularly, whereby Fe^{3+} is released through intracellular degradation with heme oxygenases (Ibrahim et al. 2012).

Host-Pathogen Interactions and Mucormycosis Pathogenesis

The major characteristic of mucormycosis is its angioinvasiveness, which is its ability to penetrate the vessels especially blood vessels and it often leads to blood clot formation (vessel thrombosis), subsequently, infected tissue in the body (tissue necrosis) (Huang et al. 2021). The necrosis of infected tissues prevents leukocytes and antifungals from the focal infection. This angioinvasion likely links the ability of mucormycosis to spread to other organs causing disseminated mucormycosis. It is therefore essential to understand the mechanisms by which mucormycosis angioinvasion occurs to give insight to approaches in prevention and treatment of mucormycosis. Other considered but not established factors that aid pathogenesis of mucormycosis; Rhizopus species, e.g., *Rhizopus microsporus* (Jennenssen et al. 2005) and *Rhizopus chinensis* (White et al. 2002) through a symbiotic relationship with bacterium of genus Burkholderia intracellularly produce mycotoxin known as rhizoxin (an antimitotic macrocyclic polyketide metabolite) (Lackner and Hertweck 2011) but it has not been confirmed as causal mechanism of MM pathogenesis.

2.5 Risk Factor for Transmission and Spread

The infection can be obtained through spore inhalation into the sinus and lungs, or penetration into broken skin either by forced trauma, cuts, or injury, or eating food contaminated with fungal spores as fungi colonize quickly in foods rich in carbohydrates (Luo et al. 2021). These ingested spores can quickly develop and produce hyphae which leads to death of blood tissue due to blood vessel clot, as a result of the blood vessels invasion by the hyphae (CDC 2021). However, rare nosocomial outbreaks of infections are increasingly reported from many hospitals (Sanchez-Gil et al. 2017) through contaminated bandages (Prakash and Chakrabarti 2019), hospital served ready-to-eat food, contaminated hospital linens (Sundermann et al. 2019), ventilators, medical equipment, e.g., wooden sticks, catheter, elastoplast adhesive dressings, and wooden tongue depressors contaminated with spores (Prakash and Chakrabarti 2019). The most common site is traumatized or open skin (CDC 2021). Associated risk factors to this infection are diabetes with or without ketoacidosis, diabetes mellitus, renal failure, tumor, neutropenia, drug associated immunosuppression therapy, and organ transplant. There are reported cases with HCA mucormycosis from adhesive bandages, hospital linen, and wooden tongue depressors (Sharma et al. 2021). More recently, use of corticosteroids is an associated risk factor with CAM which makes COVID-19 patients susceptible of due to weak immune systems caused by decreased levels of T cells (CD4 + T and CD8 + T) (Sen et al. 2021).

2.6 Control and Prevention

It can be controlled with persistent antifungal drug therapy with the use of amphotericin B, oral drugs such as posaconazole and isavuconazole, or surgery; neutrophils can also be supplied with sufficient oxygen through a hyperbaric chamber to supply enough oxygen to kill fungi; Dexamethasone, a drug popularly known as "mawu mawu" (weight enhancer) or 'yodi' (buttocks enhancer) in the south western part of Nigeria, western Africa, due to its ability to cause one to have increased appetite subsequently increase their fat serves and body weight (Michel and Cabanac 2016); Glucocorticoids, primary stress hormones that reduce inflammation and help maintain physiological processes (Ramamoorthy and Cidlowski 2016), and they have anti-inflammatory and immunosuppressive ability (Youssef et al. 2016). Glucocorticoid's use has been accompanied with an array of side effects including diabetes, osteoporosis, glaucoma andavascular necrosis among others (Youssef et al. 2016). It can also be prevented by avoiding indiscriminate use of Corticoids or drugs that can induce immunosuppression, avoidance of oxygen masks sharing (Banerjee et al. 2021), and proper border control (Afroz et al. 2021).

2.7 COVID-19 Associated Mucormycosis

Worldwide occurence of COVID-19 brought about an increase in the mucormycosis incidence, with quite a number of incidence temporally associated with COVID-19, birthing the name COVID-19 associated mucormycosis (CAM) (Pal et al. 2021). Corticosteroids continous use in COVID-19 therapy causes immunosuppression which was identified as one of the major factors that bring about mucormycosis as a secondary infection (Singh et al. 2021). Oxygen treatment in ICU and use of humidifiers increase the risk of black fungus (Banerjee et al. 2021) among hospitalized individuals. Cases of CAM remain very low as pertinent to the Delta variant (Banerjee et al. 2021). The diagnosis of CAPA relies on early prognosis and consideration of consistent radiology scans, prevailing risk factors, and fungal spore demonstration in microscopy and tissue culture (Koehler et al. 2020). In IPA cases, the galactomannan in bronchoalveolar lavage is a useful marker (Hsu et al. 2010). IPA shares similarities with pulmonary mucormycosis in their risk factors, radiology scans, and in their clinical manifestations, but not with CAM, making its diagnosis more difficult (Garg et al. 2021). Poor prognosis, limited clinical symptom, and difficulty in causal fungi isolation add to issues of poor diagnosis of MM, but diagnosis rely more on microscopy as most cases are diagnosed with post mortem findings (Garg et al. 2021).

3 Global Impact of Mucormycosis on COVID-19

Reports of fungal secondary infection associated with COVID-19 (IPA, CAPA) have been perviously recorded (Koehler et al. 2020) but most recently a severe fungal secondary infection on the rise is COVID-19 associated mucormycosis (Pal et al. 2021). Increasing reports of cases of CAM has been recorded mostly among individuals with underlying health complications, such as DKA, diabetes mellitus (DM), or excessive and indiscriminate use of steriods (Walsh et al. 2012). The huge upsurge of CAM cases recorded in India calls for global attention and surveillance of individuals with COVID-19, especially people at risk. India has recorded the highest number of people with CAM with an estimate of over 4000 people (Fig. 1). Studies carried out on CAM in India have shown that three quarter of those affected by the disease are male, a huge number of patients who were diagnosed CAM had been treated with gluco-corticosteroids and have underlying health issues, those with diabetes mellitus showing the greatest susceptibility (Sen et al. 2021), with rhino-orbital mucormycosis being the most frequent type of mucormycosis associated with this disease in India (PAHO 2021). There are increasing concern for fatalities that accompany CAM, making use of dexamethasone in treating COVID-19 cases remains unachievable; John Hopkins School of Public Health describes it as a "nightmare in the making" and "the signaling of a global problem." Mucormycosis has been declared an epidemic in India due to report of over 41,000 cases and 3554 as of 16 July 2021. The use of amphotericin B, which targets a sterol found in fungi cell wall known as ergosterol, must be increased, but its administration is limited by infusion-related toxicity (Laniado-Laborin and Cabrales-Vargas 2009) often mediated by cytokine release from monocytes (Mihu et al. 2014).

3.1 Potential Impact of COVID-19 Associated Mucormycosis (CAM) in Africa

Over 19 million adults (age 20–79) living with diabetes and 60% of subjects in Africa living with diabetes and do not know (IDF 2019), over 20 million cases of cancer and 26 million HIV positive people in Africa (continent with highest incidence of HIV) are the potentially affected, with adverse impact of CAM in Africa which is very devastating. Low precipitations and elevated temperature during dry season are identified as possible factors for increased incidence due to clustering

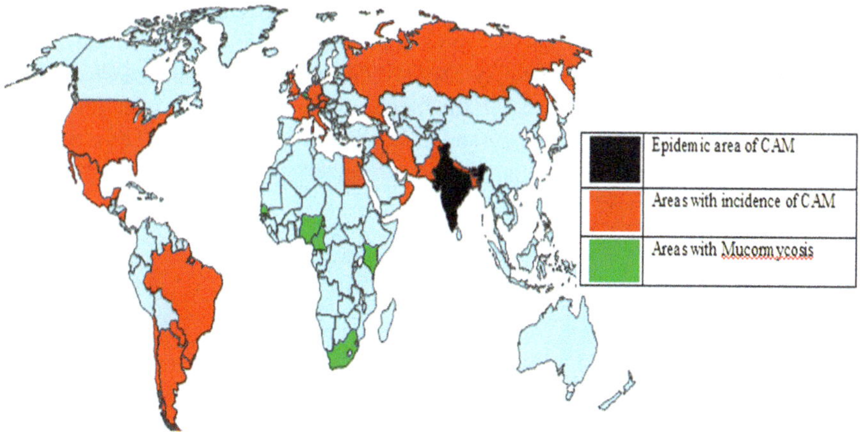

Fig. 1 Worldwide incidence of CAM and mucormycosis

observed during these periods (Stemler et al. 2020). As diabetes has been recognized as a major risk factor linked with CAM, countries such as South Africa share a huge case burden of diabetic patients with over 7.8 million living with HIV, and has also been an epicenter of COVID-19 in Africa with over 2 million cases (WHO 2021). The reduced use of immunosuppressants should be considered and border laws must be enforced to prevent the movement of people with a high CAM (CDC 2021).

4 Interventions

The COVID-19 global epidemic has recorded a total mortality of over four million people worldwide. The development of vaccine has reduced the impact globally, but the distribution of this vaccine is limited in developing countries. There is still need for supportive care in management of COVID-19 in developing countries (Garg et al. 2021). Glucocorticoids, remdesivir drugs proven to increase survival rate of COVID-19, compared with others, are very much less expensive, readily available, and have shown to increase survival rate and reduce mortality rate of COVID-19 infected individuals (Sterne et al. 2020). More funding is needed for holistic epidemiological study on mucormycosis (Prakash and Chakrabarti 2019). The incidence rate and worldwide occurrence of mucormycosis is unknown because it is often a secondary infection and opportunistic (Dallalzadeh et al. 2021) which often lead to late prognosis (Bellazreg et al. 2015). Most cases are identified post-mortem (Garg et al. 2021), and lack of diagnostic tools lead to non-reportability of the disease (Skiada et al. 2020). Most figures of incidence are based on estimates from a large population-based studies (Roden et al. 2005), which do not present epidemiological interventions. There is a wide gap of research studies, medical intervention, and risk factor, peculiar to these countries (Prakash et al. 2019). High incidence of diabetes mellitus, post-tuberculosis, chronic renal failure and long-hospitalization (Jeong et al. 2019), high poverty rate, and poor health care system (Prakash and Chakrabarti 2019) are risk factors for increase in the incidence of mucormycosis in developing countries, while immunosuppression and recipients of organ transplant (Kontoyiannis et al. 2016) are commonly associated MM risk factors in developed countries.

5 Conclusion

Mucormycosis is a disease widely associated with developing country due to poor detection and diagnostic facilities required for diagnosis and effective therapeutic management. Population at risk needs to be monitored with adequate awareness for prevention. There is a need to regulate the administration of dexamethasone, its dosage, and its duration of use when treating COVID-19 in patient with low risk of complications. There is a need to establish effective surveillance for public health and diagnosis of MM; regulation of dexamethasone, the use and application during COVID-19 infection, and detection of MM should be included in the diagnosis and treatment of COVID-19. There is a need for public surveillance of MM. There is also a need for further research to determine the role of environmental factors in unprecedented increase of CAM cases in India and developing countries, and possible outbreak of MM and CAM; also, research on the difference or similarities between virulence properties of MM and CAM strains, and the possible link of immunosuppression mediated by each SARS-CoV-2 variant especially (Delta or Delta plus variant).

Acknowledgement The authors sincerely thank and acknowledge the management of the Covenant University Centre for Research, Innovation and Discovery (CUCRID), Ota, Nigeria, for financial support in the publication of this review study.

References

Adam S, Grecian S, Syed AA. (2016). Hereditary haemochromatosis presenting with diabetic ketoacidosis. QJM-Int J Med. 109(11),747-748.

Afroz F, Barai L, Rahim MA, Kanta SS, Hossain MD. (2021). Post-COVID Pulmonary Mucormycosis: first case report from Bangladesh. Bangladesh j. med. sci. 32(2):156-160

Alqarihi A, Gebremariam T, Gu Y, Swidergall M, Alkhazraji S, Soliman SS, Bruno VM, Edwards Jr JE, Filler SG, Uppuluri P, Ibrahim AS (2020). GRP78 and integrins play different roles in host cell invasion during mucormycosis. Mbio, 11(3):01087-20.

Andresen D, Donaldson A, Choo L, Knox A, Klaassen M, Ursic C, Vonthethoff L, Krilis S, Konecny P. (2005). Multifocal cutaneous mucormycosis complicating polymicrobial wound infections in a tsunami survivor from Sri Lanka. Lancet. 365(9462):876-878

Andrianaki AM, Kyrmizi I, Thanopoulou K, Baldin C, Drakos E, Soliman SS, Shetty AC, McCracken C, Akoumianaki T, Stylianou K, Ioannou P. (2018). Iron restriction inside macrophages regulates pulmonary host defense against Rhizopus species. Nat. Commun. 9(1):1-17

Augustine HF, White C, Bain J (2017). Aggressive combined medical and surgical management of mucormycosis results in disease eradication in 2 pediatric patients. Plast. Surg. 25(3):211-217.

Badiane AS, Ndiaye D, Denning DW (2015). Burden of fungal infections in Senegal. Mycoses, 58:63-69.

Baldin C, Ibrahim AS. (2017). Molecular mechanisms of mucormycosis—the bitter and the sweet. PLoS pathog. 13(8):1006408.

Banerjee I, Robinson J, Asim M, Sathian B, Banerjee I. (2021). Mucormycosis and COVID-19 an epidemic in a pandemic? Nepal J. Epidemiology. 11(2):1034.

Bellazreg F, Hattab Z, Meksi S, Mansouri S, Hachfi W, Kaabia N, Said MB, Letaief A. (2015). Outcome of mucormycosis after treatment: report of five cases. New microbes new infect. 6:49-52.

Bhadauria D, Etta P, Chelappan A, Gurjar M, Kaul A, Sharma RK, Gupta A, Prasad N, Marak RS, Jain M, Srivastava A (2018). Isolated bilateral renal mucormycosis in apparently immunocompetent patients—a case series from India and review of the literature. Clin. Kidney J. 11(6):769-776.

Carolus H, Pierson S, Lagrou K, Van Dijck P. (2020) Amphotericin B and Other Polyenes—Discovery, Clinical Use, Mode of Action and Drug Resistance. J Fungi. 6(4):321.

Carroll CS, Grieve CL, Murugathasan I, Bennet AJ, Czekster CM, Liu H, Naismith J, Moore MM. (2017). The rhizoferrin biosynthetic gene in the fungal pathogen Rhizopus delemar is a novel member of the NIS gene family. Int. J. Biochem. Cell. Bio. 89:136-146.

Caza M, Kronstad J. (2013). Shared and distinct mechanisms of iron acquisition by bacterial and fungal pathogens of humans. Front. Cell. Infect. Microbiol. 3:80.

Centers for Disease Control and Prevention www.cdc.gov/fungal/diseases/mucormycosis January 14, 2021 Accessed July 25th 2021

Chakrabarti A, Das A, Sharma A, Panda N, Das S, Gupta KL, Sakhuja V (2001) Ten Years' Experience in Zygomycosis at a Tertiary Care Centre in India. J. Infect. 42:261–266

Chakrabarti A, Das A, Mandal J, Shivaprakash MR, George VK, Tarai B, Rao P, Panda N, Verma SC, Sakhuja V (2006). The rising trend of invasive zygomycosis in patients with uncontrolled diabetes mellitus. Med. Mycol. 44:335–342.

Chakrabarti A, Chatterjee SS, Das A, Panda N, Shivaprakash MR, Kaur A, Varma SC, Singhi S, Bhansali A, Sakhuja V (2009). Invasive zygomycosis in India: Experience in a tertiary care hospital. Postgrad. Med. J. 85:573–581

Challa S. (2019). Mucormycosis: Pathogenesis and pathology. Curr. Fungal Infect. Rep. 13(1):11-20.

Chapman N, Doubell A, Oversteegen L, Chowdhary V, Rugarabamu G, Zanetti R, Ong M, Borri J. (2017). Neglected disease research and development: reflecting on a decade of global investment. Policy cures research.

Chen Q, Chen K, Qian S, Wu S, Xu L, Huang X, Shi P, Wang K, Wang M, Wang X. (2019) Disseminated mucormycosis with cerebellum involvement due to Rhizomucor pusillus in a patient with multiple myeloma and secondary myelodysplastic syndrome: A case report. Experimental and therapeutic medicine. 18(5):4076-80.

Dallalzadeh LO, Ozzello DJ, Liu CY, Kikkawa DO, Korn BS. (2021). Secondary infection with rhino-orbital cerebral mucormycosis associated with COVID-19. Orbit. 25:1-4.

Daoud A, Elbendary A, Elfishawi M, Rabea M, Alfishawy M, (2014). Diabetic ketoacidosis with two life threatening infections: mucormycosis, and bilateral emphysematous pyelonepritis, preciptating erythema nodosum leprosum as the initial presentation of diabetes. Diabetes Metab J. 5(433):2.

Denning DW. (2002). Echinocandins: a new class of antifungal. J. Antimicrob. Chemother. 49(6):889-891.

Dioverti MV, Cawcutt KA, Abidi M, Sohail MR, Walker RC, Osmon DR (2015). Gastrointestinal mucormycosis in immunocompromised hosts. Mycoses. 58(12):714-718.

Eid R, Arab NT, Greenwood MT. (2017). Iron mediated toxicity and programmed cell death: A review and a re-examination of existing paradigms. Biochim Biophys Acta Mol Cell Res. 1864(2):399-430.

El-Mahallawy HA, Khedr R, Taha H, Shalaby L, Mostafa A (2016). Investigation and management of a Rhizomucor outbreak in a pediatric cancer hospital in Egypt. Pediatr. Blood Cancer. 63(1):171-173.

Fadhel M, Patel SV, Liu E, Fune L, Wasserman EJ, Asif A (2019). Disseminated pulmonary with isolated muscular mucormycosis in an acute myeloid leukemia patient: a case report and literature review. Am. J. Med. Case Rep. 20:1210.

Fatima N, Javaid K, Lahmo K, Banday S, Sharma P, Masoodi, L. (2017). Siderophore in fungal physiology and virulence. Int. J. Res. Phytochem. Pharmacol. 6(5):1073-1080.

Feng J, Sun X (2018). Characteristics of pulmonary mucormycosis and predictive risk factors for the outcome. Infection. 46(4):503-512.

GAFFI. Global Fund for Fungal Infections. 2018. Available from: http://www.gaffi.org. Accessed 23 July 2021.

Garg D, Muthu V, Sehgal IS, Ramachandran R, Kaur H, Bhalla A, Puri GD, Chakrabarti A, Agarwal R. (2021). Coronavirus disease (Covid-19) associated mucormycosis (CAM): case report and systematic review of literature. Mycopathologia. 186(2):289–298.

Gerwien F, Skrahina V, Kasper L, Hube B, Brunke S. (2018). Metals in fungal virulence. FEMS Microbiol. Rev. 42(1):50.

Govender NP, Chiller TM, Poonsamy B, Frean JA. (2011) Neglected fungal diseases in sub-Saharan Africa: A call to action. Curr Fungal Infect Rep. 5(4):224–32.

Hartnett KP, Jackson BR, Perkins KM, Glowicz J, Kerins JL, Black SR, Lockhart SR, Christensen BE, Beer KD. (2019). A guide to investigating suspected outbreaks of mucormycosis in healthcare. J. Fungus. 5(3):69

Hassan MIA, Voigt K. (2019). Pathogenicity patterns of mucormycosis: epidemiology, interaction with immune cells and virulence factors. Med. Mycol. J. 57(2):245-S256.

He J, Sheng G, Yue H, Zhang F, Zhang HL. (2021). Isolated pulmonary mucormycosis in an immunocompetent patient: a case report and systematic review of the literature, BMC Pulm. Med. 21(1):1-8.

Hsu LY, Ding Y, Phua J, Koh LP, Chan DS, Khoo KL, Tambyah, PA. (2010). Galactomannan testing of bronchoalveolar lavage fluid is useful for diagnosis of invasive pulmonary aspergillosis in hematology patients. BMJ Infect. Dis. 10(1):1-6. http://www.fda.gov/cder/approval/index.html Accessed 25 July 2021.

Huang H, Xie L, Zheng Z, Yu H, Tu L, Cui C, and Yu J. (2021). Mucormycosis-induced upper gastrointestinal ulcer perforation in immunocompetent patients: a report of two cases. BMC gastroenterol. 21(311):1-7.

Ibrahim AS, Gebremariam T, Lin L, Luo G, Husseiny MI, Skory CD, Fu Y, French SW, Edwards Jr JE, Spellberg B. (2010). The high affinity iron permease is a key virulence factor required for Rhizopus oryzae pathogenesis. Mol. Microbiol. 77(3):587-604

Ibrahim AS, Spellberg B, Walsh TJ, Kontoyiannis, DP (2012). Pathogenesis of mucormycosis. Clin. Infect. Dis. 54(1):16-22.

Inglesfield S, Jasiulewicz A, Hopwood M., Tyrrell J, Youlden G, Mazon-Moya M, Millington OR, Mostowy S, Jabbari S, Voelz K. (2018). Robust phagocyte recruitment controls the opportunistic fungal pathogen Mucor circinelloides in innate granulomas in vivo. MBio. 9(2):10-17.

International Diabetes Federation Idf diabetes atlas. 2019. https://diabetesatlas.org/en/resources/ www.idf.org Accessed 25 July 2021.

Jennessen J, Nielsen KF, Houbraken J, Lyhne EK, Schnürer J, Frisvad JC, Samson RA. (2005). Secondary metabolite and mycotoxin production by the Rhizopus microsporus group. J. Agric. Food Chem. 53(5):1833-1840.

Jeong W, Keighley C, Wolfe R, Lee WL, Slavin MA, Kong DCM, Chen, S. A. (2019). The epidemiology and clinical manifestations of mucormycosis: a systematic review and meta-analysis of case reports. Clin. Microbiol. Infect. 25(1):26-34.

Jose A, Singh S, Roychoudhury A, Kholakiya Y, Arya S, Roychoudhury S. (2021). Current Understanding in the Pathophysiology of SARS-CoV-2-Associated Rhino-Orbito-Cerebral Mucormycosis: A Comprehensive Review. J. Oral Maxillofac. Surg. 20:373-380.

Kaur H, Ghosh A, Rudramurthy SM, Chakrabarti A (2018). Gastrointestinal mucormycosis in apparently immunocompetent hosts—A review. Mycoses. 61(12):898-908.

Khaba MC, Nevondo LM, Moroatshehla SM, Makhado NA (2021). Disseminated mucormycosis presenting as a renal mass in a human immunodeficiency virus-infected patient: A case report. S Afr. J. Infect D. 36(1):4.

Koehler P, Cornely OA, Böttiger BW, Dusse F, Eichenauer DA, Fuchs F, Hallek M, Jung N, Klein F, Persigehl T, Rybniker J. (2020). COVID-19 associated pulmonary aspergillosis. Mycoses. 63(6):528-534.

Kontoyiannis DP, Yang H, Song J, Kelkar SS, Yang X, Azie N, Harrington R, Fan A, Lee E, Spalding J. R. (2016). Prevalence, clinical and economic burden of mucormycosis-related hospitalizations in the United States: a retrospective study. BMC Infect. Dis. 16:730

Kornitzer D, Roy U. (2020). Pathways of heme utilization in fungi. Biochim Biophys Acta Mol Cell Res. 1867(11):118817

Lackner G, Hertweck C. (2011). Impact of endofungal bacteria on infection biology, food safety, and drug development. PLoS pathog. 7(6):1002096.

Lamoth F, Kontoyiannis DP. (2018). The Candida auris alert: facts and perspectives. J. Infect. Dis. 217(4):516-520.

Laniado-Laborín R, Cabrales-Vargas MN. (2009). Amphotericin B: side effects and toxicity. Rev. Iberoam. Micol. 26(4):223-227.

Lanternier F, Dannaoui E, Morizot G, Elie C, Garcia-Hermoso D, Huerre M, Bitar D, Dromer F, & Lortholary O (2012). The French Mycosis Study Group. A Global Analysis of Mucormycosis in France: The RetroZygo Study (2005–2007). Clin. Infect. Dis. 54:35–43.

Li H, Liu SM, Yu XH, Tang SL, Tang CK (2020). Coronavirus disease 2019 (COVID-19): current status and future perspectives. Int. J. Antimicrob. Agents 55(5):105951.

Luo Y, Li J, Zhou H, Yu B, He J, Wu A, Huang Z, Zheng P, Mao X, Yu J, Li H. (2021). The Nutritional Significance of Intestinal Fungi: Alteration of Dietary Carbohydrate Composition Triggers Colonic Fungal Community Shifts in a Pig Model. Appl. Environ. Microbiol. 87(10):38-21.

Maital S, Barzani E (2020). The global economic impact of COVID-19: A summary of research. Samuel Neaman Institute for National Policy Research, 2020, 1-12.

Mandengue CE, Denning DW (2018). The burden of serious fungal infections in Cameroon. J. Fungus. 4(2):44.

Michel C, Cabanac M. (2016). Effects of Dexamethasone on the Body Weight Set-Point of Rats. Physiol. Behav. 60:817-821.

Mihu MR, Pattabhi R, Nosanchuk JD. (2014). The impact of antifungals on toll-like receptors. Front. Microbiol. 5:99.

Morales-Franco B, Nava-Villalba M, Medina-Guerrero EO, Sánchez-Nuño YA, Davila-Villa P, Anaya-Ambriz EJ, Charles-Niño CL. (2021). Host-Pathogen Molecular Factors Contribute to the Pathogenesis of Rhizopus spp. in Diabetes Mellitus. Curr. Trop. Med. Rep. 8:6–17

Neblett Fanfair R, Benedict K, Bos J, Bennett SD, Lo YC, Adebanjo T, Etienne K, Deak E, Derado G, Shieh WJ, Drew C. (2012). Necrotizing cutaneous mucormycosis after a tornado in Joplin, Missouri, in 2011. N. Engl. J. Med. 367(23):2214-2225.

Oladeji SM, Amusa YB, Olabanji JK, Adisa AO (2013). Rhinocerebral mucormycosis in a diabetic case report. J West Afr. Coll. Surg. 3(1):93-102

Onyango JF, Kayima JK, Owen WO (2002). Rhinocerebral mucormycosis: case report. East Afr. Med. J. 79(7):390-393.

Paduraru M, Moreno-Sanz C, Gallardo JMO. (2016). Primary cutaneous mucormycosis in an immunocompetent patient. Case Reports, 2016, p. bcr2016214982.

Pal R, Singh B, Bhadada SK, Banerjee M, Bhogal RS, Hage N, Kumar A. (2021). COVID-19-associated mucormycosis: An updated systematic review of literature. Mycoses. 00:1– 8

Palermo NE, Sadhu AR, McDonnell ME. (2020). Diabetic ketoacidosis in COVID-19: unique concerns and considerations. J. Clin. Endocrinol. Metab. 105(8):2819-2829.

Pan American Health Organization / World Health Organization. Epidemiological Alert: COVID-19 associated Mucormycosis. June 2021, Washington, D.C.: PAHO/WHO; 2021 Accessed 30 July 2021.

Petrikkos G, Tsioutis C. (2018). Recent advances in the pathogenesis of mucormycoses. Clin. ther. 40(6):894-902.

Pianalto KM., Alspaugh JA. (2016). New horizons in antifungal therapy. J. Fungus. 2(4):26.

Prakash H, Chakrabarti A (2019). Global epidemiology of mucormycosis. J. Fungus. 5(1):26.

Prakash H, Chakrabarti A (2021). Epidemiology of mucormycosis in India. Microorganisms. 9(3):523.

Prakash H, Ghosh AK, Rudramurthy SM, Singh P, Xess I, Savio J, Pamidimukkala U, Jillwin J, Varma S, Das A, Panda NK (2019). A prospective multicenter study on mucormycosis in India: Epidemiology, diagnosis, and treatment. Med. Mycol. J. 57(4):395-402.

Ramadorai A, Ravi P, Narayanan V (2019). Rhinocerebral mucormycosis: A prospective analysis of an effective treatment protocol Ann. Maxillofac. Surg. 9(1):192.

Ramamoorthy S, Cidlowski JA. (2016). Corticosteroids: mechanisms of action in health and disease. Rheum. Dis. Clin. 42(1):15-31.

Ramaswami A, Sahu AK, Kumar A, Suresh S, Nair A, Gupta D, Chouhan R, Bhat R, Mathew R, Aggarwal P, Nayer J. (2021). COVID-19 associated Mucormycosis Presenting to the Emergency Department–An Observational Study of 70 Patients. QJM: An International Journal of Medicine.

Rees JR, Pinner RW, Hajjeh RA, Brandt ME, Reingold AL. (1998). The epidemiological features of invasive mycotic infections in the San Francisco Bay area, 1992-1993: results of population-based laboratory active surveillance. Clin. Infect. Dis. 27(5):1138–1147.

Richardson K, Cooper K, Marriott MS, Tarbit MH., Troke F, Whittle PJ. (1990). Discovery of fluconazole, a novel antifungal agent. Rev. Infect. Dis. 12(3):267-271.

Roden MM, Zaoutis TE, Buchanan WL, Knudsen TA, Sarkisova TA, Schaufele RL, Sein M, Sein T, Chiou CC, Chu JH, Kontoyiannis, DP. (2005). Epidemiology and outcome of zygomycosis: a review of 929 reported cases. Clin. Infect. Dis. 41(5):634-653.

Rodrigues ML, Albuquerque PC (2018). Searching for a change: The need for increased support for public health and research on fungal diseases. PLOS Negl. Trop. Dis. 12(6):0006479.

Rodrigues ML, Nosanchuk JD. (2020). Fungal diseases as neglected pathogens: A wake-up call to public health officials. PLOS Negl. Trop. Dis. 14(2):0007964.

Saegeman V, Maertens J, Meersseman W, Spriet I, Verbeken E, Lagrou K (2010). Increasing incidence of mucormycosis in University Hospital, Belgium. Emerg. Infect. Dis. 16(9):1456.

Sahoo NK, Kulkarni V, Bhandari AK, Kumar A. (2017). Mucormycosis of the frontal sinus: A rare case report and review. Ann. Maxillofac. Surg. 7(1):120.

Sánchez-Gil J, Guirao-Arrabal E, Parra-García GD, del Pilar Luzón-García M, Bautista-Marín MF, Barayobre-Barayobre M, Fontalba-Navas A. (2017). Nosocomial rhinocerebral mucormycosis: two cases with a temporal relationship. Mycopathologia. 182(9):933-935.

Saneesh PS, Yelamanchi R, Pilllai S. (2021). Isolated renal mucormycosis presenting with bilateral renal artery thrombosis: a case report. Afr. J. Urol. 27(1):1-4.

Schwartz IS, Kenyon C, Lehloenya R, Claasens S, Spengane Z, Prozesky H, Burton R, Parker A, Wasserman S, Meintjes G, Mendelson M. (2017) AIDS-related endemic mycoses in Western Cape, South Africa, and clinical mimics: a cross-sectional study of adults with advanced HIV and recent-onset, widespread skin lesions. In Open forum infectious diseases US: Oxford University Press. 4(4):186.

Sen M, Lahane S, Lahane TP, Parekh R, Honavar SG. (2021). Mucor in a viral land: a tale of two pathogens. Indian J. Ophthalmol. 69(2):244.

Shanmugam S, Rajeshwari B, Niamath S, Ghosh M (2019). Isolated Renal Zygomycosis, A Rare and Lethal Cause of Massive Renal Infarction-Report of Two Cases. J. Clin. Diagnostic Res. 13(5):1-3.

Sharma S, Grover M, Bhargava S, Samdani S, Kataria T. (2021). Post coronavirus disease mucormycosis: a deadly addition to the pandemic spectrum. J Laryngol Otol. 135(5):442-447.

Singh AK, Singh R, Joshi SR, Misra A. (2021) Mucormycosis in COVID-19: a systematic review of cases reported worldwide and in India. Diabetes Metab Syndr. 15(4):102146.

Skiada A, Lass-Floerl C, Klimko N, Ibrahim A, Roilides E, Petrikkos G (2018). Challenges in the diagnosis and treatment of mucormycosis. Med. mycol. 56(1):93-101.

Skiada A, Pavleas I, Drogari-Apiranthitou M. (2020). Epidemiology and diagnosis of mucormycosis: an update. J. Fungus. 6(4):265.

Stanford FA, Voigt K. (2020). Iron Assimilation during Emerging Infections Caused by Opportunistic Fungi with emphasis on Mucorales and the Development of Antifungal Resistance. Genes. 11(11):1296.

Stemler J, Hamed K, Salmanton-García J, Rezaei-Matehkolaei A, Gräfe SK, Sal E, Zarrouk M, Seidel D, Abdelaziz Khedr R, Ben-Ami R, Ben-Chetrit E. (2020). Mucormycosis in the Middle East and North Africa: Analysis of the FungiScope® registry and cases from the literature. Mycoses. 63(10):1060-1068.

Sterne JA, Murthy S, Diaz JV, Slutsky AS, Villar J, Angus DC, Annane D, Azevedo LCP, Berwanger O, Cavalcanti AB, Dequin PF. (2020). Association between administration of systemic corticosteroids and mortality among critically ill patients with COVID-19: a meta-analysis. Jama. 324(13):1330-1341.

Sukaina M (2021). Re-emergence of mucormycosis in COVID-19 recovered patients transiting from silent threat to an epidemic in India. J. Glob. Health. 5:2021067.

Sundermann AJ, Clancy CJ, Pasculle AW, Liu G, Cumbie RB, Driscoll E, Ayres A, Donahue L, Pergam SA, Abbo L, Andes DR. (2019). How clean is the linen at my hospital? The Mucorales on unclean linen discovery study of large United States transplant and cancer centers. Clin. Infect. Dis. 68(5):850-853.

Vaughan C, Bartolo A, Vallabh N, Leong SC (2018). A meta-analysis of survival factors in rhino-orbital-cerebral mucormycosis—has anything changed in the past 20 years? Clin Otolaryngol. 43(6):1454-1464.

Walsh TJ, Gamaletsou MN, McGinnis MR, Hayden RT, Kontoyiannis DP. (2012). Early clinical and laboratory diagnosis of invasive pulmonary, extrapulmonary, and disseminated mucormycosis (zygomycosis). Clin. Infect. Dis. 54(1):55-60.

White JD, Blakemore PR, Green NJ, Hauser EB, Holoboski MA, Keown LE, Nylund Kolz CS, Phillips BW. (2002). Total Synthesis of Rhizoxin D, a Potent Antimitotic Agent from the Fungus Rhizopus chinensis. J. Org. Chem. 67(22):7750-7760.

WHO. Coronavirus disease (COVID-19): Dexamethasone https://www.who.int/news-room/q-a-detail/coronavirus-disease-covid-19-dexamethasone Accessed August 3, 2021.

WHO. Fact sheet: Tuberculosis. http://www.who.int/mediacentre/factsheets/fs104/en/. 2018. Accessed 25 July 2021.

Wotiye AB, Poornachandra KS, Ayele BA. (2020). Invasive intestinal mucormycosis in a 40-year old immunocompetent patient-a rarely reported clinical phenomenon: a case report. BMC Gastroenterol. 20(61):1-6.

www.egyptindependent.com/mansoura-university-hospital-reports-black-fungus-cases/. Accessed 25 July 2021

www.scopus.com/search/form.uri?display=basic#basic Accessed August 5, 2021.

Yiannikourides A, Latunde-Dada GO. (2019). A short review of iron metabolism and pathophysiology of iron disorders. Medicines. 6(3):85.

Youssef J, Novosad SA., Winthrop KL. (2016). Infection risk and safety of corticosteroid use. Rheum. Dis. Clin. 42(1):157-176.

Yusuf EO, Onyiriuka AN (2012). Subcutaneous Mucormycosis in an Immunocompetent Nigerian Child. Turk. J. Med. Sci. 39(2):309-311.

Evaluation of Preservative and Shelf-Life Quality of Probiotic-Lactobacilli Fortified Nigerian Fermented Condiments

Yemisi D. Obafemi, Solomon U. Oranusi, Kolawole O. Ajanaku, and Paul A. Akinduti

1 Introduction

Fermented foods are crucial to socio-economic advancement in Nigeria as well as in addressing a major dietary need of people because of their nutritional benefits, digestibility, accessibility, and greater safety. However, they have not achieved their full societal and commercial value due to short shelf-life, and the non-optimization of their health-promoting characteristics (Okechukwu et al. 2019; Olasupo and Okorie 2019). Contamination of fermented foods during processing, transportation, and storage by food-borne pathogens such as *Listeria monocytogenes*, *Staphylococcus aureus*, *Escherichia coli*, *Salmonella enterica*, and spoilage microorganisms such as *Bacillus subtilis* pose a major food safety risk, low shelf-life quality, and public rejection of products in Nigeria (Okorie et al. 2017; Ezeokoli et al. 2018; Olanbiwoninu and Odunfa 2018; Adewumi et al. 2019). The food safety and shelf-life quality of many food products were prolonged by the use of chemical food preservatives. However, prolonged consumption of chemical preservatives usually poses very serious adverse impacts on consumers' health (Ugwuanyi and Okpara 2019). The increasing awareness of food quality-related issues by consumers has resulted in an urgent need for research and development of safe, cost-effective, and specific natural substances, which are health-promoting and yield novel bio-preservatives food industries (Atere et al. 2020; Owusu-Kwarteng et al. 2020). Probiotic strains (live microorganisms which when taken in adequate amount confer health benefits to the host) and other fermented food strains are of

Y. D. Obafemi · S. U. Oranusi (✉) · P. A. Akinduti
Department of Biological Sciences, Covenant University, Ota, Ogun State, Nigeria
e-mail: solomon.oranusi@covenantuniversity.edu.ng

K. O. Ajanaku
Department of Chemistry, Covenant University, Ota, Ogun State, Nigeria

© The Author(s), under exclusive license to Springer Nature Switzerland AG 2022

A. O. Ayeni et al. (eds.), *Bioenergy and Biochemical Processing Technologies*, Green Energy and Technology, https://doi.org/10.1007/978-3-030-96721-5_26

considerable economic significance because of their use in food fermentation processes and their ability to produce antimicrobial substances which could interfere with the growth of pathogenic microorganisms, yielding safe and healthy fermented foods (Uzodinma et al. 2020). This study evaluates the preservative potential and shelf-life analysis of probiotic-fortified Nigerian fermented condiments for improved food safety by using genotyped probiotic *Lactobacillus* strains as starter cultures for co-fermentation of the condiments.

2 Materials and Methods

2.1 *Identification Lactobacillus Species*

Identified *Lactobacillus* species (already genotyped) was recovered in MRS broth and further serially diluted to 0.5-McFarland standard turbidity inoculum suspensions in Difco™ Lactobacilli MRS Agar (BD BBL, USA) anaerobically incubated at 37 °C for 48 hours as described by Chukwu et al. (2019). Pure *Lactobacillus* strains were maintained at 4 °C for further analysis.

2.2 *Probiotic Assessment of Lactobacillus Isolates*

The probiotic activities of selected *Lactobacillus* isolates were carried at different pH (2, 4, 3, 5.5), bile (0.1%, 0.3%, 0.7%, 1.0%), phenol (0.1%, 0.4%, 0.5%, 1.0%), and cholesterol (200 mg/dl). The control experiment was carried out at pH = 7 and was not supplemented with bile, phenol, and cholesterol reagents. Then, 100 μL of each of the *Lactobacillus* cultures were inoculated into the supplemented media and incubated at 37 °C for 24 hours. Survived viable colonies were enumerated and survival rate was calculated using the equation below as described (Ahire et al. 2021).

$$\text{Survival rate}(\%) = \frac{\text{Number of viable cells survived}(\text{CFU / ml})}{\text{Number of initial viable cells}(\text{CFU / ml})} \times 100$$

2.3 *Co-Fermentation of Raw Seeds with Probiotic Lactobacilli Starter*

The raw condiment seeds were prepared as described (Ifesan et al. 2017; Ojewumi et al. 2021). About 1 kg of each of the condiment types (Iru, Ugba, and Ogiri) were boiled for 6–8 hours in a pressure cooker (Master Chef, Australia) and de-hulled.

The de-hulled Iru and Ogiri seeds were washed severally and soaked in clean water overnight while only the ugba seeds were sliced with a knife to a thickness of 0.2–0.4 cm and drained thoroughly to remove excess water. Fifty grams (50 g) of each of the seeds were transferred into three air-tight clean containers. Then, 10% (v/w) inoculum containing probiotic *Lactobacillus* species consortium were aseptically added to one set of the condiments and 10% (v/w) inoculum containing probiotic *Lactobacillus* species consortium supplemented with 1% (v/v) glacial acetic acid were also added to another set of the condiments while no inoculum was added to set of control condiments. All the seeds were allowed to ferment in the air-tight container for 72 hours while monitoring the pH and *Lactobacillus* count 12 hourly. After the fermentation, the fermented condiments were properly stored awaiting shelf-life analysis.

2.4 Shelf-Life Analysis of Probiotic Fortified Fermented Condiments

The fermented condiments were stored in three parts at room temperature 25 ± 2 °C for 3 weeks. The first parts (control) were preserved with sterile distilled water. The second part (pLAB) was co-fermented with probiotic lactobacilli as the starter and the third part (pLAB_GAC) was also co-fermented with probiotic lactobacilli and glacial acetic acid (1%v/v) starter culture. Shelf-life assessments were evaluated weekly as described (Shukla et al. 2018). The pHs of fermented condiments were determined using calibrated digital pH meter (EUTECH Instruments, Singapore) placed directly in homogenized samples (10-g of samples with 50 mL distilled water for 60 seconds). Total aerobic bacteria, fungal, and lactobacilli counts were determined by using the pour plate method and expressed as LogCFU/g.

3 Results and Discussion

3.1 Identification of Lactobacillus Species

Ten strains of probiotic lactobacilli were previously obtained from Nigerian fermented condiments. Creamy-colored isolates with smooth edges on Difco™ Lactobacilli MRS Agar incubated at 30–37 °C for 24–48 hours. Selected isolates were Gram-positive bacilli, catalase-negative, oxidase-negative, facultative anaerobes non-spore-forming, and non-motile microorganisms. Gas production was observed with the fermentation of D-glucose, D-galactose, D-glucose, D-fructose, D-mannitol, D-sorbitol (Table 1). These Nigerian fermented condiments were normally locally produced with a spontaneous fermentation process which often results in unsafe and poor shelf-life quality. Genotyped *Lactobacillus* strains with

Table 1 Occurrence of *Lactobacillus* species ($N = 185$) in fermented condiments

	Iru	Ugba	Ogiri
Presumptive isolates	*n*(%)	*n*(%)	*n*(%)
Lactobacillus plantarum ($N = 69$)	26.0(14.1)	38.0(20.5)	5.0(2.7)
Lactobacillus pentosus ($N = 5$)	3.0(1.6)	2.0(1.1)	0.0(0.0)
Lactobacillus paracasei ($N = 46$)	3.0(1.6)	1.0(0.5)	42.0(22.7)
Lactobacillus ruteri ($N = 28$)	4.0(2.7)	16.0(8.7)	8.0(4.2)
Lactobacillus fermentum ($N = 35$)	6.0(3.2)	24.0(2.2)	5.0(2.7)
Lactobacillus brevis ($N = 2$)	1.0(0.5)	1.0(0.5)	0.0(0.0)

Key: *n* number of positive isolates, % percentage occurrence

phenotypic probiotics including *Lactobacillus plajomi*, *Lactobacillus plantarum*, *Lactobacillus fermentum*, and *Lactobacillus paracasei* from wide varieties form alkaline fermented foods (Dosumu et al. 2012; Afolabi et al. 2016; Braide et al. 2018; Chen et al. 2019; Goel et al. 2020) and were potential strains that could enhance the fermentation of condiments. However, *Bacillus* species have been widely reported as the most predominant bacterial species in fermented condiments (Nwanekwu and Dike 2018; Adewumi et al. 2019; Nwagu et al. 2020; Musliu et al. 2021).

3.2 Probiotic Assessment of *Lactobacillus* Isolates

Selected *Lactobacillus* isolates from fermented condiments significantly exhibited high tolerance to the extreme to a moderately acidic environment (ranging from 2.0 to 5.5) for at least 6 hours which signifies the potentials of isolates to survive in the GI tract (pH = 3) with food passage before digestion for about 3 hours. The isolates also significantly exhibited high tolerance in 1.0% bile after 6 hours of incubation. Extreme bile concentration often decreases the survivability of bacteria in the gastric environment which was not observed in these isolates. There was also significant tolerance (≥97% survival at $P < 0.05$) to 1.0% phenol concentration after 24 hours of incubation. There was also high assimilation of cholesterol after 24 hours of incubation. The isolates significantly assimilated the cholesterol at $P < 0.05$ with a survival rate ≥ 90% after 24 hours of incubation (Fig. 1). These probiotic strains could survive the human gut with characteristic growth patterns in low pH, bile salts, phenol, and cholesterol assimilation. Selected strains tolerated up to pH = 2.0, 0.7% bile salts, 1.0% phenol with ≥70% survival rate, and ≥ 90% ability to assimilate cholesterol (200 mg/dl) within 24 hours in agreement with previously conducted studies (Khalil et al. 2018; Mulaw et al. 2019; He et al. 2021).

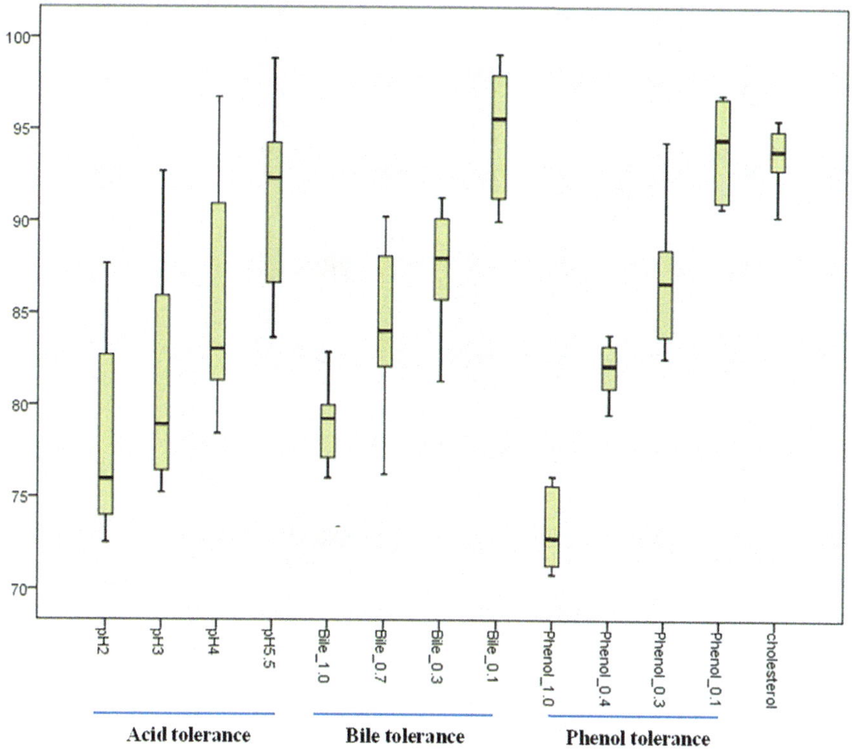

Fig. 1 Probiotic assessment of isolated *Lactobacillus* species from various condiments

3.3 Spontaneous Co-Fermentation of Raw Seeds with Probiotic Lactobacilli Starter

The result obtained in this study revealed fermented condiments supplemented with probiotic lactobacilli and glacial acetic acid (1%v/v) starter culture (pLAB_GAC) exhibited higher acidity (pH = 4.5–8.2) when compared with fermented condiments supplemented with only probiotic lactobacilli (pLAB) (pH = 5.7–9.2). The control fermented condiments exhibited ammoniacal fermentation (pH = 6.8–9.9) (Fig. 2). The pLAB_GAC fermented condiments also exhibited significantly higher *Lactobacillus* count when monitored for the 72-hour fermentation period than the pLAB and control fermented condiments (Fig. 3). The results obtained in this study evaluated the fermentation potentials of probiotic lactobacilli (pLAB) and probiotic lactobacilli supplemented with glacial acetic acid (1%v/v) (pLAB_GAC). There was a lower alkaline 72 h-fermentation environment (pH = 4.5–8.2) in the pLAB and pLAB_GAC compared with spontaneous ammoniacal fermentation. This result agrees with previously reported studies (Fakoya et al. 2018; Duru and Maduka 2021). There was also a significant increase in the proliferation of the probiotic *Lactobacillus* population during the 72 h-fermentation in the pLAB and

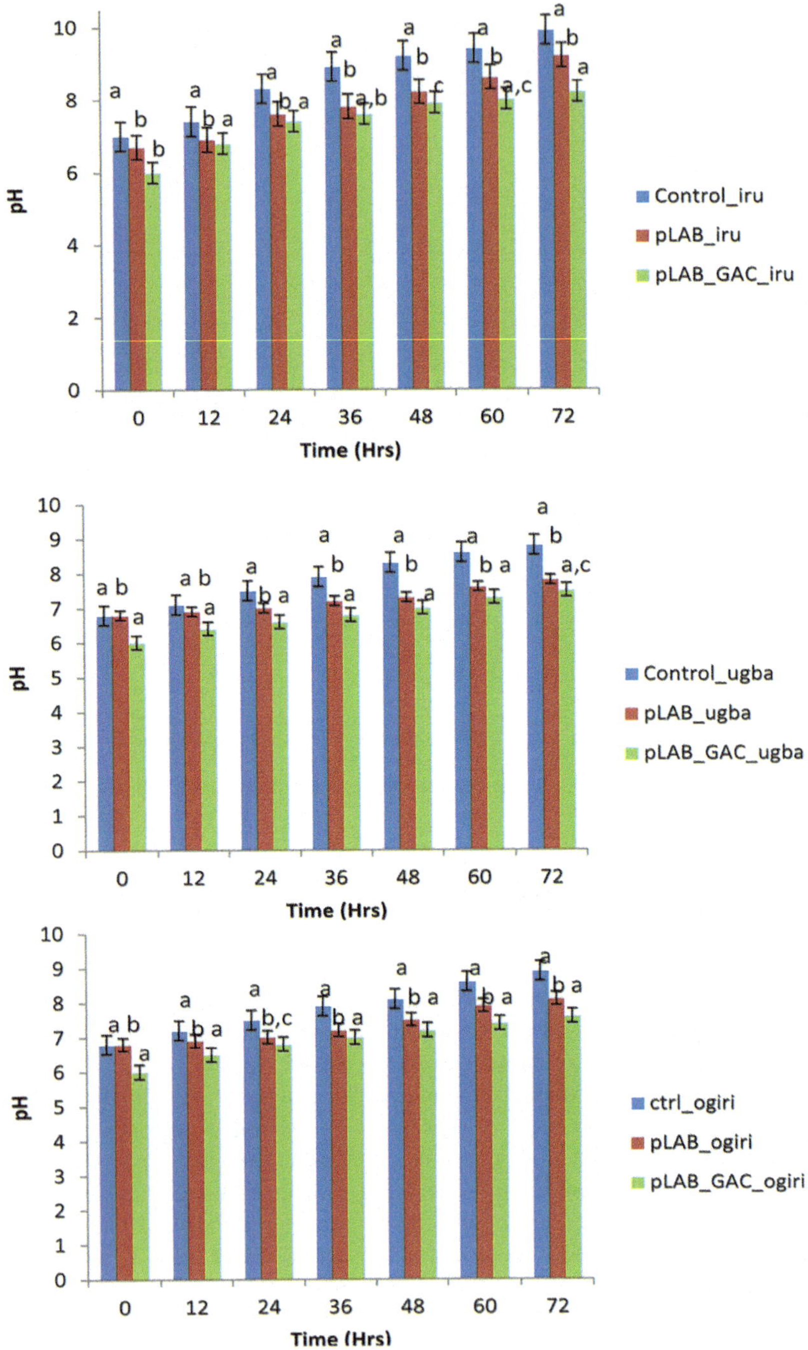

Fig. 2 Comparative pH changes during the fermentation of condiments

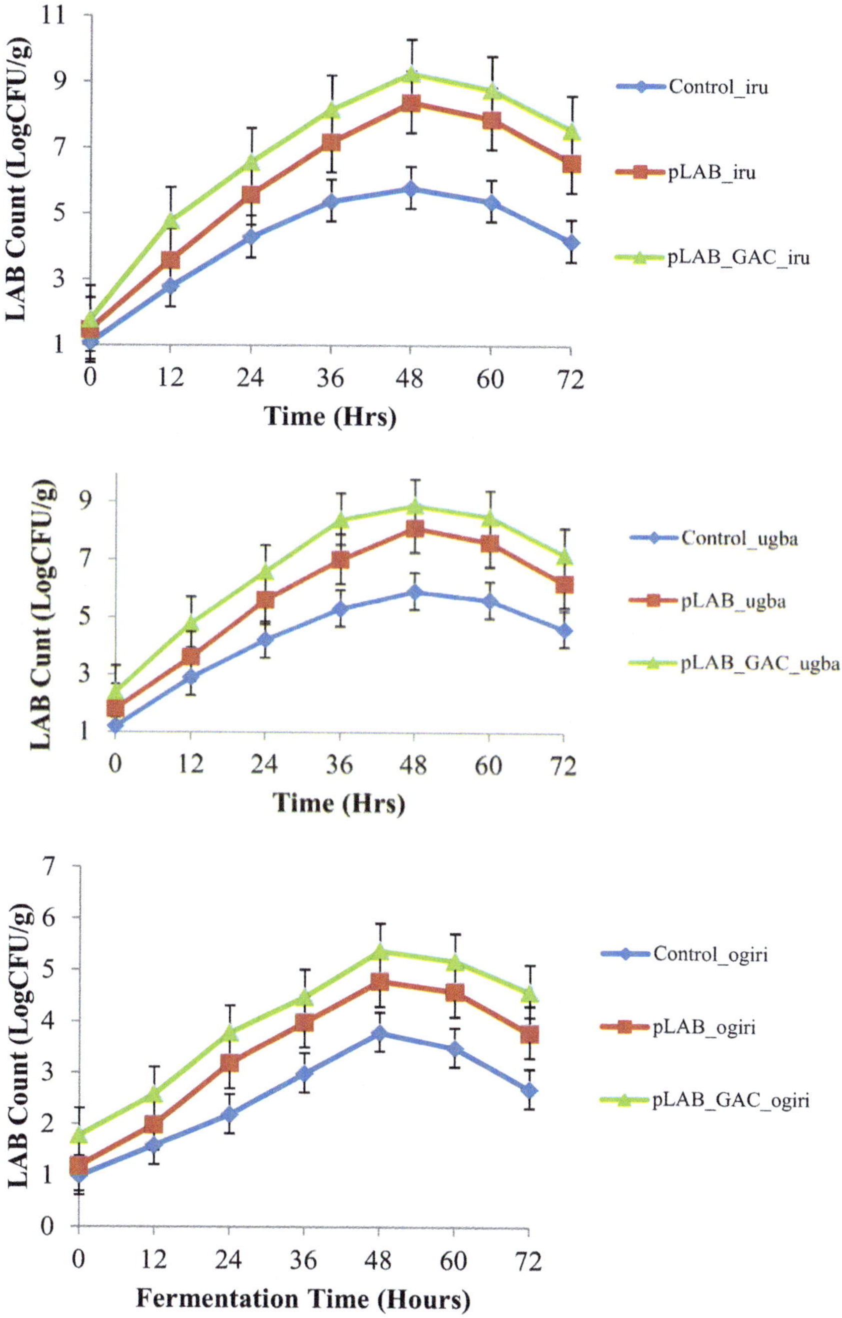

Fig. 3 Growth of probiotic lactobacilli culture during fermentation of condiments

pLAB_GAC as compared with the spontaneous fermentation (Adebayo 2018). Previous report by Ojewunmi et al. (2018) recorded increase in pH from 5.9 to 8.2 during aerobic fermentation and reduction in pH during anaerobic fermentation of legume-based seed condiments by consortium of *Bacillus* species. Co-fermentation of lebbeck seeds with *Lactobacillus plantarum* reportedly lead to reduction of all anti-nutritional components which significantly improves odor and texture after fermentation (Fakoya et al. 2018). The impacts of starter cultures, namely, *Lactobacillus plantarum*, *Lactobacillus pentosus* and *Lactobacillus paraplantarum* used for fermentation of pickled gherkins (alkaline fermented vegetable from Korea), with significant changes in the pH, total mesophilic aerobic bacteria, lactic acid bacteria and fungi population. It was observed that total mesophilic aerobic bacteria and fungi count in pickle sample containing *Lb. plantarum*-49 strain were significantly reduced and the amount of biogenic amines produced were considerably below the toxic level. It was reported that pickle sample fermented with *Lb. plantarum*-49 strain had best acceptable fermentation profile and shelf-life quality which could be of great biotechnology importance to the food industry (Alan 2019). The probiotic *L. plantarum* ATCC 14917 was applied in fermentation of pomegranate juice, resulting in production of low alcoholic functional beverage. This strain was also recommended for production of functional food with health and nutritional benefits (Mantzourani et al. 2019).

3.4 Shelf-Life Analysis Over 3-Week Storage Time

The pLAB (probiotic *Lactobacillus* fortified) and control fermented condiments showed higher aerobic bacteria and visible fungal growth from week 1 storage through to week 3, however, there was also lower lactobacilli presence when compared with the pLAB_GAC (probiotic *Lactobacillus* with glacial acetic acid). There was also a faster rate of spoilage as physically observed in color and pungent odor in the control fermented condiments when compared with the pLAB_GAC fermented condiments. After 3 weeks of storage, comparative shelf-life studies evaluated revealed the pLAB-GAC fermented condiments as the best-fermented condiments with improved quality and shelf-life (Table 2). The shelf-life result also exhibited moderately alkaline pH, high *Lactobacillus* count, no physical spoilage signs, and lower aerobic bacterial and fungal population in pLAB_GAC fermented condiments from week 1 through week 3 storage periods. The pLAB and spontaneous (control) fermented condiments exhibited visible moldy growths on Ugba with reduced lactobacilli population and pungent odor in Ogiri from week 1 storage through to week 3. The result in this study is in agreement with previous reports (Adebayo 2018). Higher alkaline pH, lower lactobacilli count, with faster visible spoilage signs, ammoniacal, pungent odor changes, the higher aerobic bacterial and fungal population in pLAB, and spontaneous (control) fermented condiments were observed during the 3-week shelf-life (Fig. 4). This result disproves common practices of traditional preservation by salting which were explained by the previous

Table 2 Shelf-life analysis of fermented condiments fortified with probiotic lactobacilli starter over 3 weeks storage

Storage period	Fermented condiments	Physical properties		Mean Microbial Load ± SD		
		Color	Odor	TBC	TFC	TLC
Week _1	Control_Iru	Brown	Pungent	7.6 ± 1.2	5.2 ± 0.9	4.2 ± 0.6
	pLAB_Iru	Brown	Normal	5.4 ± 0.9	4.2 ± 0.6	5.5 ± 1.0
	pLAB_GAC_Iru	Brown	Normal	4.2 ± 0.8	3.0 ± 0.4	6.5 ± 1.2
	Control_Ugba	Brown	Normal	5.4 ± 0.9	6.5 ± 1.2	4.6 ± 0.6
	pLAB_Ugba	Brown	Normal	4.8 ± 0.6	4.6 ± 0.6	5.6 ± 1.0
	pLAB_GAC_Ugba	Brown	Normal	2.4 ± 0.4	2.0 ± 0.4	6.8 ± 1.2
	Control_Ogiri	White	Ammoniacal	8.7 ± 2.0	6.8 ± 1.6	4.2 ± 0.8
	pLAB_Ogiri	White	Pungent	5.7 ± 1.1	4.6 ± 0.6	4.8 ± 0.6
	pLAB_GAC_Ogiri	White	Normal	3.2 ± 0.4	2.6 ± 0.4	5.2 ± 1.2
Week_2	Control_Iru	Brown	Pungent	8.5 ± 1.8	5.2 ± 0.7	4.2 ± 0.4
	pLAB_Iru	Brown	Pungent	6.4 ± 1.0	4.2 ± 0.6	5.5 ± 0.9
	pLAB_GAC_Iru	Brown	Normal	4.8 ± 0.6	3.0 ± 0.4	6.5 ± 1.3
	Control_Ugba	White (mouldy)	Normal	5.2 ± 0.9	7.5 ± 1.2	4.8 ± 0.6
	pLAB_Ugba	Brown	Normal	3.2 ± 0.5	4.8 ± 0.8	5.9 ± 1.1
	pLAB_GAC_Ugba	Brown	Normal	2.8 ± 0.3	2.0 ± 0.1	6.8 ± 1.3
	Control_Ogiri	Brown	Ammoniacal	6.7 ± 1.3	7.8 ± 1.6	4.6 ± 0.6
	pLAB_Ogiri	Yellow	Pungent	4.7 ± 0.7	5.6 ± 1.0	5.2 ± 1.2
	pLAB_GAC_Ogiri	White	Normal	3.2 ± 0.4	2.0 ± 0.1	6.2 ± 1.1
Week_3	Control_Iru	Brown(marshy)	Normal	9.6 ± 2.4	5.2 ± 0.9	4.2 ± 0.6
	pLAB_Iru	Brown	Normal	2.3 ± 1.3	4.2 ± 0.6	5.5 ± 0.9
	pLAB_GAC_Iru	Dark brown	Normal	1.0 ± 0.2	2.0 ± 0.1	6.5 ± 1.3
	Control_Ugba	White (mouldy)	Normal	6.0 ± 1.2	8.8 ± 1.9	5.2 ± 0.8
	pLAB_Ugba	Brown	Normal	3.0 ± 0.1	5.8 ± 0.9	6.9 ± 1.3
	pLAB_GAC_Ugba	Brown	Normal	2.9 ± 0.2	2.4 ± 0.1	7.8 ± 1.5
	Control_Ogiri	Black	Ammoniacal	7.9 ± 1.4	6.8 ± 1.4	4.9 ± 0.8
	pLAB_Ogiri	Brown	Pungent	5.4 ± 0.7	4.6 ± 0.2	5.6 ± 0.8
	pLAB_GAC_Ogiri	White	Normal	3.9 ± 0.4	2.8 ± 0.3	6.9 ± 1.1

Key: *TBC* total bacteria count (LogCFU × 10^6/g), *TFC* total fungal count (LogCFU X10^3/g), *TLC* total Lactobacilli count (LogCFU × 10^5/g

researchers due to the public health impact associated with consumption of salts (Balogun and Omoloso 2017; Ojewumi 2018). This study revealed that fermented foods were preserved naturally by fortifying the food with probiotic lactobacilli starter culture and thus the pLAB-GAC fermented condiments had improved safety

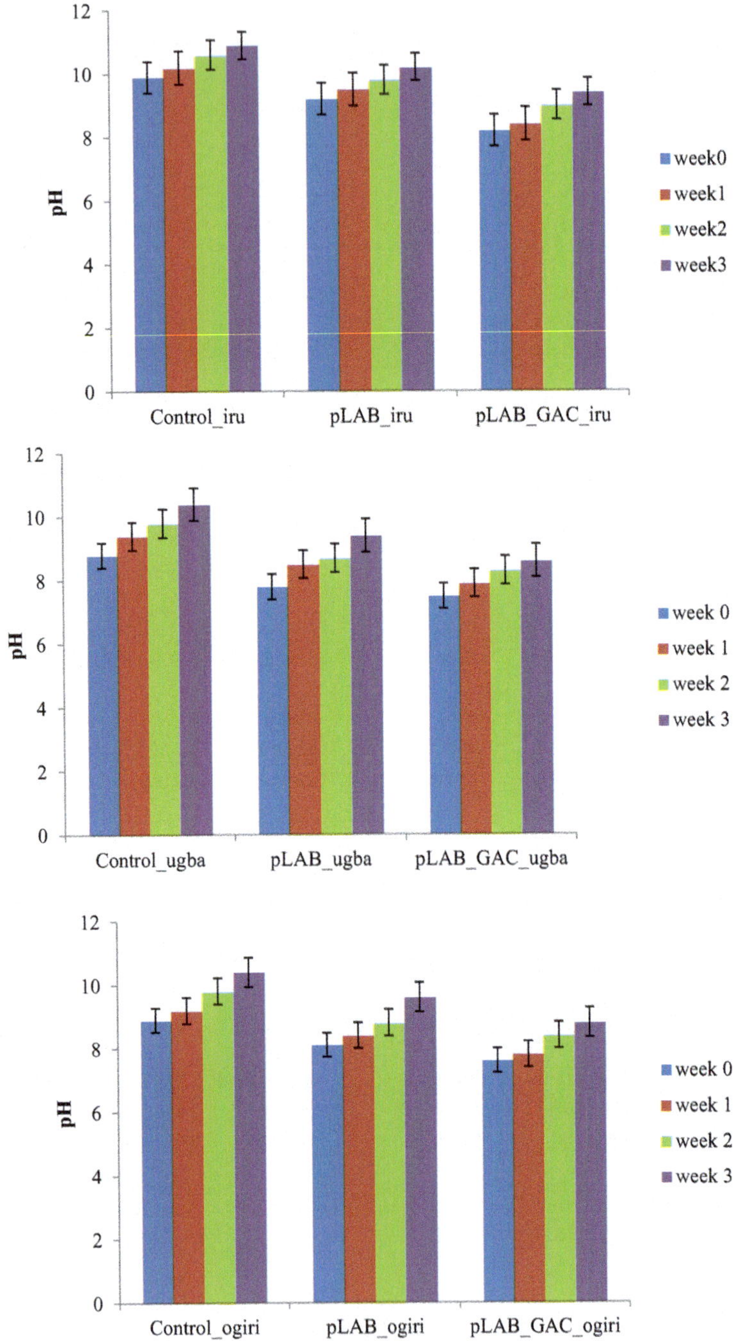

Fig. 4 pH changes during 3-week storage of the fermented condiments

and shelf-life quality. The use of health-promoting starter cultures is crucial in carrying out the industrial fermentation of foods for desired attributes including improved nutritional quality, organoleptic properties, safety, shelf-life and acceptability of locally fermented foods. Multifunctional scrutiny for the selection criteria should be constantly considered with ultimate health benefits (Akpi et al. 2020). The best starter culture reduced fermentation time and pH inhibited the pathogenic and toxigenic bacterial and fungal contamination hence improving the safety and shelf-life quality of the fermented condiments.

4 Conclusion

The fermentation process was best carried out in a moderately alkaline food environment which favors the growth and proliferation of the probiotic lactobacilli and better shelf-life quality. The probiotic lactobacilli served as potential starter cultures for possible fermentation, preservatives, and shelf-life quality of fermented condiments. This study affirms a positive contribution towards the enhancement of economic and sustainable development in Nigeria through ensuring proper nutrition, low-skilled women empowerment initiatives, and provision of employment opportunities in Nigerian food industries.

Acknowledgments The authors hereby appreciate all scientists in Biotechnology Research Cluster, Covenant University for their support through the period of this research work.

Funding The authors hereby appreciate financial supports from Nutricia Research Foundation, Netherland (Award 2020-T1) and publication support from Covenant University Centre for Research Innovation and Development (CUCRID) Nigeria.

Author's Contribution Background concept: SUO, YDO; Sample collection and experimental design: YDO, PAA, Manuscript preparation and editing: YDO, PAA; Manuscript reviewing: PAA, SUO, KOA; Supervision: SUO, KOA. All authors read and approved the final manuscript.

Competing Interests Authors have declared no competing interest.

References

Adebayo, F. O. (2018). Microbiological Profile of 'Ogiri' Condiment Made from Seeds of Watermelon (*Citrullus lanatus*). *Asian Journal of Agricultural and Food Sciences 1*: 1-9.

Adewumi, G. A., Grover, S., Isanbor, C., & Oguntoyinbo, F. A. (2019). Phylogenetics, Safety and In Vitro Functional Properties of *Bacillus* Species Isolated from Iru, a Nigerian Fermented Condiment. *Korean Journal of Microbiology and Biotechnology 47*(4): 498-508.

Afolabi, F. T., Abdulkadir, M., & Onilude, A. A. (2016). Isolation and Screening of Microorganisms Associated with Locust Bean (Iru) for the Ability to Ferment Soya Bean to Produce Soy Iru. *Frontiers in Microbiology 5*(2): 1-10.

Ahire, J. J., Jakkamsetty, C., Kashikar, M. S., Lakshmi, S. G., & Madempudi, R. S. (2021). In Vitro Evaluation of Probiotic Properties of *Lactobacillus plantarum* UBLP40 Isolated from

Traditional Indigenous Fermented Food. *Probiotics and Antimicrobial Proteins 7*(1): 1-12. https://doi.org/10.1007/s12602-021-09775-7.

Alan, Y. (2019). Culture fermentation of Lactobacillus in traditional pickled gherkins: Microbial development, chemical, biogenic amine and metabolite analysis. *Journal of Food Science and Technology 56*(8): 3930-3939.

Akpi, U. K., Nnamchi, C. I. and Ugwuanyi, J. O. (2020). Development of Starter Culture for the Production of African Condiments and Seasoning Agents. *Advances in Microbiology 10*(12): 599-622.

Atere, A. V., Adedeji, A., Akinmoladun, A. C., Oyetayo, V. O., & Akinyosoye, F. A. (2020). Local Condiment, Iru, obtained from the Fermentation of *Parkia biglobosa* Seed Substantially Reduced the Serum Cholesterol Level of Wister Rats. *Preventive Nutrition and Food Science 25*(2): 153. DOI: https://doi.org/10.3746/pnf.2020.25.2.153.

Balogun, O. D., & Omoloso, M. M. (2017). In Vitro Assessment of Microbial Post-Harvest Spoilage of Locust Beans. *International Journal of Innovative Research and Advanced Studies 4*(10): 66-68.

Braide, W., Azuwike, C. O., & Adeleye, S. A. (2018). The role of microorganisms in the production of some indigenous fermented foods in Nigeria. *International Journal of Advanced Research in Biological Sciences 5*(5): 86-94.

Chen, S., Cao, P., Lang, F., Wu, Z., Pan, D., Zeng, X., & Lian, L. (2019). Adhesion-related immunomodulatory activity of the screened *Lactobacillus plantarum* from Sichuan Pickle. *Current Microbiology 76*(1):29-36.

Chukwu, M., Ezeagwula, C. G., Nwakaudu, A. A., Oti, W., & Anyaogu, I. (2019). Microbial loads of ogiri-ahuekere condiment produced from groundnut seed (*Arachis hypogaea* Linn). *Agricultural and Food Science 6*(1): 114-119. https://doi.org/10.2139/ssrn.3517493.

Dosumu, O. O., Oluwaniyi, O. O., Awolola, G. V., & Oyedeji, O. O. (2012). Nutritional composition and antimicrobial properties of three Nigerian condiments. *Nigerian Food Journal 30*(1): 43-52.

Duru, I. A., & Maduka, T. D. (2021). Profiling and Comparison of Fatty Acids in the Oils from the seeds of egusi melon (*Cucumeropsis mannii* Naudin) and watermelon (*Citrullus lanatus* (Thunb.) Matsum. & Nakai). *World News of Natural Sciences 37*(1): 31-40.

Ezeokoli, O. T., Adeleke, R. A., & Bezuidenhout, C. C. (2018). Core bacterial community of soy-daddawa: Insights from high-throughput DNA metabarcoding. *LWT 97*: 61-66.

Fakoya, A., Adeleke, B. S., Adegbehingbe, K. T., Adejoro, D. O., & Jemilaiye, T. A. (2018). Effect of fermentation on the nutrient and antinutrient contents of lebbeck (*Albizia Lebbeck*) seeds. *Journal of Emerging Trends in Engineering and Applied Sciences 9*(5): 227-233.

Goel, A., Halami, P. M., & Tamang, J. P. (2020). Genome analysis of *Lactobacillus plantarum* isolated from some Indian fermented foods for bacteriocin production and probiotic marker genes. *Frontiers in Microbiology 11*(1): 40-52.

He, Q., Li, J., Ma, Y., Chen Q, & Chen, G. (2021). Probiotic Potential and Cholesterol-Lowering Capabilities of Bacterial Strains Isolated from *Pericarpium Citri* Reticulate 'Chachiensis'. *Microorganisms 9*(6):1224.

Ifesan, B. O. T., Akintade, A. O., & Gabriel-Ajobiew, R. A. (2017). Physicochemical and nutritional properties of *Mucuna pruriens* and *Parkia biglobosa* subjected to controlled fermentation. *International Food Research Journal 24*(5):1-24.

Khalil, E. S., Manap, A., Yazid, M., Mustafa, S., Alhelli, A. M., & Shokryazdan, P. (2018). Probiotic properties of exopolysaccharide-producing *Lactobacillus* strains isolated from tempoyak. *Molecules 23*(2): 398-418.

Mantzourani, I., Kazakos, S., Terpou, A., Alexopoulos, A., Bezirtzoglou, E., Bekatorou, A. and Plessas, S. (2019). Potential of the probiotic *Lactobacillus plantarum* ATCC 14917 strain to produce functional fermented pomegranate juice. *Foods 8*(1): 4-17.

Mulaw, G., Sisay, T., Muleta, D., & Tesfaye, S. (2019). *In vitro* evaluation of probiotic properties of lactic acid bacteria isolated from some traditionally fermented Ethiopian food products. *International Journal of Microbiology 1*: 7179514.

Musliu, A., Yusuf, M., & Adebisi, S. (2021). *In vivo* Evaluation of Soya Beans Flour Fermented with Lactic Acid Bacteria as a Potential Probiotic Food. *Frontiers in Microbiology 31*(4): 16-27.

Nwagu, T. N., Ugwuodo, C. J., Onwosi, C. O., Inyima, O., Uchendu, O. C., & Akpuru, C. (2020). Evaluation of the probiotic attributes of *Bacillus* strains isolated from traditional fermented African locust bean seeds (*Parkia biglobosa*), "daddawa". *Annals of Microbiology 70*(1): 1-15.

Nwanekwu, K. E. N., & Dike, K. S. (2018). Molecular Identification of *Bacillus* spp as Starter Culture for the Production of Ugba (*Pentraclethra macrophylla*), a Nigerian Fermented Condiment. *Nigerian Food Journal 36*(2): 83-87.

Ojewumi M. E. (2018). Effects of salting and drying on the deterioration rate of fermented *Parkia biglobosa* seed. *Journal of Nutritional Health & Food Engineering 8*(1):1-5.

Ojewumi, M. E., Omoleye, J. A., Ajayi, A. A., & Ekanem, G. P. (2021). Fermentation rate monitoring in the production of African protein-based condiments. In IOP Conference Series: *Environmental Earth Sciences 65*(1):012012. IOP Publishing.

Okechukwu, Q. N., Javed, F., & Ivantsova, M. N. (2019). Advancement of food processing biotechnology in developing countries. In AIP Conference Proceedings (Vol. 2174, No. 1, p. 020236). AIP Publishing LLC.

Okorie, P. C., Olasupo, N. A., Anike, F. N., Elemo, G. N., & Isikhuemhen, O. S. (2017). Incidence of enteric pathogens in ugba, a traditional fermented food from African oil bean seeds (*Pentaclethra macrophylla*). *International Journal of Food Contamination 4*(1): 1-8.

Olanbiwoninu, A. A., & Odunfa, S. A. (2018). Microbial interaction in selected fermented vegetable condiments in Nigeria. *International Food Research Journal 25*(1): 439-445.

Olasupo, N. A., & Okorie, P. C. (2019). African Fermented Food Condiments: Microbiology Impacts on Their Nutritional Values. In Frontiers and New Trends in the Science of Fermented Food and Beverages. *IntechOpen*. https://doi.org/10.5772/intechopen.73404.

Owusu-Kwarteng, J., Parkouda, C., Adewumi, G. A., Ouoba, L. I., & Jespersen, L. (2020). Technologically relevant *Bacillus* species and microbial safety of West African traditional alkaline fermented seed condiments. *Critical Reviews in Food Science and Nutrition 5*(2): 1-18.

Shukla, S., Park, J., Park, J. H., Lee, J. S., & Kim, M. (2018). Development of lotus root fermented sugar syrup as a functional food supplement/condiment and evaluation of its physicochemical, nutritional, and microbiological properties. *Journal of Food Science and Technology 55*(2): 619-629.

Ugwuanyi, J. O., & Okpara, A. N. (2019). Current Status of Alkaline Fermented Foods and Seasoning Agents of Africa. In New Advances on Fermentation Processes. IntechOpen. https://doi.org/10.5772/intechopen.87052.

Uzodinma, E. O., Mbaeyi-Nwaoha, I. E., & Onwurafor E. U. (2020). Suitability of bacterial fermentation and foil packaging of condiment from African mesquite (*Prosopis africana*) seeds for nutritional retention and commercialization. *African Journal of Microbiology Research 14*(7): 340-348.

Isolation and Molecular Characterization of Salmonella Serovars Distributed in Benue State, Nigeria

B. O. Okpa, G. M. Gberikon, S. Oranusi, and T. Ichor

1 Introduction

The genus *Salmonella belongs to the family Enterobacteriaceae,* they are facultative anaerobes, Gram-negative, flagellated rods that contain all the important species of *Salmonella* (Feasey et al. 2012). *S. enterica* is a vital food pathogen, it is presently divided into 2587 serovars (Grimont and Weill 2007), while the typhoidal *Salmonellae* are human pathogens and include *S. parayphi* and *S. typhi* (Uzzau et al. 2000). They are the leading causes of typhoid or enteric fever, which is prevalent in the sub-tropical and tropical countries, with 540 per 100,000 estimated annual incidences (Anyanwu et al. 2010; Dong et al. 2014). The worldwide incidence was plausible to reach up to 17 million cases (WHO 2018) accounting for about 600,000 deaths per annum (Udeze et al. 2010).

Salmonella species, subspecies, and serotypes causes salmonellosis, a leading cause of diarrhea globally, with high morbidity rate. It is often associated with contaminated foods and drinks, infected pets and reptiles (Lar et al. 2006). *S. enterica* serotypes Paratyphi and Typhi cause typhoidal salmonellosis, while Enteritidis and Typhimurium causes non-typhoidal salmonellosis (Ohad et al. 2014). The incidence of salmonellosis in several parts of Africa has been reported, Crump et al. (2004)

B. O. Okpa · G. M. Gberikon
Department of Microbiology, University of Agriculture, Makurdi, Benue State, Nigeria

S. Oranusi
Department of Biological Sciences, Covenant University, Ota, Ogun State, Nigeria

T. Ichor (✉)
Department of Microbiology, University of Agriculture, Makurdi, Benue State, Nigeria

Department of Biological Sciences, Covenant University, Ota, Ogun State, Nigeria
e-mail: tersagh.ichor@covenantuniversity.edu.ng

© The Author(s), under exclusive license to Springer Nature Switzerland AG 2022

A. O. Ayeni et al. (eds.), *Bioenergy and Biochemical Processing Technologies*, Green Energy and Technology, https://doi.org/10.1007/978-3-030-96721-5_27

reported 465 persons per 100,000 in East Africa, while 123 and 195 persons per 100,000 were estimated for West and South Africa respectively. Similarly, Ohad et al. (2014) observed the incidence of 725 persons per 100,000 in sub-Saharan Africa. Information on Salmonella isolation from Africa, is limited (Raufu et al. 2013), there is therefore, the need for more data on the Salmonella epidemiology to help estimate the incidence in Africa (WHO 2018). A limited number of Salmonella serovars which often vary from region to region, are known to cause human infections. In Africa, implicated serovars comprise S. typhimurium, S. enteritidis, S. Isangi, and S. concord (Feasey et al. 2012; Abdullahi et al. 2014 and Kingsley et al. 2009).

A report of Non-Typhoidal *Salmonella* (NTS) in chicken farms in the six geopolitical zones of Nigeria reported the presence of *S. kentucky, S. poona, S. elizabethville, S. iarochelle, S. agama* and *S. graz*, with *S. kentucky* as the most predominant (Raufu et al. 2013). Several other studies in chicken farms reported the presence of Non-Typhoidal *Salmonella* in various parts of Nigeria including *S. kentucky* and *S. hadarin* in Jos (Agada et al. 2014), *S. paratyphi* A, *S. gallinarum* and *S. typhinurium* in Awka (Anyanwu et al. 2010) and *S. hiduddify* in Maiduguri, Northeast Nigeria (Raufu et al. 2013).

In humans, Non-Typhoidal *Salmonella* including *S. apapa, S. dublin, S. enteritidis, S.* infantis, *S. jukestown, S. monaschaui, S. oritamerin* and *S. typhimurium* were reported in Ibadan with *S. typhimurium and enteritidis* as the most widespread (Fashae et al. 2010). In Lagos, Akinyemi et al. (2007) reported only *S. enteritidis* from stools of children under 5 years while in Abuja, Nigeria, three *Salmonella* serovars: *S. zanziba, S. brancaster* and *S. enteritidis* were recovered from children with acute gastroenteritis. In the northeastern sections of Nigeria, S. *eko*, S. hadar and *enteritidis* were reported as the predominant serovars from humans (Mamman et al. 2014). Similarly, some studies of Typhoidal *Salmonella* have reported various prevalence rates including 45.0% of *S. typhi* among other bacterial isolates in Zaria, Kaduna State (Adeshina et al. 2009), 27% for *S. paratyphi* A, 25% for *S. paratyphi* B, 13.7% for *S. paratyphi* C and 20% for *S. typhi* respectively in Biu, Borno State (Abdullahi et al. 2014), while in Benue State, Umeh and Agbulu, (2009) reported a prevalence rate of 57.6% for *S. typhi,* 26.3% for *S. Paratyphi* and 15% for the mixture of both serovars in Okpokwu Local Government Area.

Enteric fever is caused by the typhoidal Salmonella serotypes; it is an invasive, systemic disease that is often life threatening, with a very high mortality rate (Njunda and Oyerinde 1996; Crump et al. 2004). Pathogen spread is via the fecal-oral route; thus, it is prevalent in areas lacking clean water and with inadequate sanitation, which facilitates the pathogen spread (Njunda and Oyerinde 1996). Typhoidal Salmonella serotypes have average incubation period of 14 days with symptoms that persist for up to 3 weeks. Crump et al. (2004) highlighted the clinical indices of enteric fever to include related symptoms of salmonellosis, sustained fever, chills, headache, abdominal pain, diarrhea, constipation, nausea, anorexia, hepatosplenomegaly, rash, and bacteremia.

Dissemination Sites

The high morbidity rates of typhoidal Salmonella serotypes, compared to the non-typhoidal serotypes, has been associated with the high invasiveness of the deeper tissues and spread to the systemics. CDC (2008) observed a global death rate of 90,300 and 178,000 for typhoidal and non-typhoidal salmonellosis in 2015. Typhoid fever is widespread in Nigeria, affecting both the young and the old. Inadequate treatment/disposal of human wastes and indiscriminate antibiotics use are some factors responsible for its spread. Talabi (1994) and Akinyemi et al. (2005) reported increased morbidity and mortality due to Salmonella associated illnesses.

Non-typhoidal Salmonella infections have been reported in the immunocompromised, infants and young children, and the elderly. In Africa, disease burden is influenced by co-infection with human immunodeficiency virus (HIV) and malaria (Grimont and Weill 2007). Transmission sources, and modes for non-typhoidal Salmonella in Africa, is not clearly understood because of ineffective or nonexistent nationwide epidemiological surveillance systems (Barbiour et al. 2015). Salmonella exists for a long period as latent infection, and healthy carriage in food animals, with fecal shedding that contaminate foods during animal slaughter, and processing. It is also a major route of contamination for the environment, feed, and water (Mamman et al. 2014).

In Nigeria, and for most of Africa, the actual situation of salmonellosis is not very clear because of the non-availability of the necessary infrastructure for disease surveillance and diagnosis.

2 Materials and Methods

2.1 Study Area

Benue State is the study area, located in the North central, middle belt region of Nigeria, it lies in the lower river Benue trough at Longitude 7° 47″ and 10° 0″ East, Latitude 6° 25″ and 8° 8" North. Benue State has 23 Local Government Areas (LGAs), divided into three Senatorial zones, Benue North-East, Benue North-West, and Benue South. The landmass of Benue State is 34,059 square kilometers and had a population of about 4,253,641 in the 2006 census. The Tiv, Idoma, and Igede inhabit it predominantly. Benue State is bound to the North by Nasarawa State, Taraba to the East, to the south is Cross-River State, Ebonyi and Enugu states to the South-West, the republic of Cameroon on the Southeast and to the West is Kogi State.

The state has Guinea Savannah vegetation with flat to undulating topography and few hilly areas. It experiences two distinct seasons, the rainy (wet) season with annual 100-200 mm annual rainfall, from April to October. The dry season is from November to March. In the year, temperatures vacillate between 21–37 °C. River Benue is the main topographical feature in the state. It separates Makurdi, the capital city into two segments, the North and South Banks.

Table 1 Sample code and locations of *Salmonella* positive samples in cultural characterization

S/N	Sample code	Location	Result	S/N	Sample code	Location	Result
1	GBK 033	Gboko	Positive	10	KWD 026	Kwande	Positive
2	GBK 034	Gboko	Positive	11	KWD 023	Kwande	Positive
3	GBK 037	Gboko	Positive	12	KWD 038	Kwande	Positive
4	GBK 039	Gboko	Positive	13	KA 021	Katsina-Ala	Positive
5	MKD 014	Makurdi	Positive	14	KA 034	Katsina-Ala	Positive
6	MKD 015	Makurdi	Positive	15	KA 038	Katsina-Ala	Positive
7	OJU 002	Oju	Positive	16	KA 045	Katsina-Ala	Positive
8	OJU 015	Oju	Positive	17	OTK 005	Otukpo	Positive
9	OJU 017	Oju	Positive	18	OTK 014	Otukpo	Positive

2.2 *Sample Collection and Sampling Distribution*

Exactly 420 stool samples were collected aseptically as described by Cheesbrough (2002) from patients attending secondary health facilities in six Local Government Areas (LGAS), two from each of the three Senatorial districts. Table 1 shows the positive sample distribution, based on locations. Seventy samples of stool were collected from each of the six selected health care facilities. Two hundred and twenty-six samples were obtained from male patients while 194 samples were from female patients. Early morning stool samples was collected from each patient in a sterile wide-mouthed plastic bottle, properly labeled and placed in ice box and moved to the laboratory in the Department of Microbiology, Federal University of Agriculture, Makurdi, for examination as described by Cheesbrough (2002), Winn et al. (2005), WHO (2010).

2.3 *Isolation and Identification of* Salmonella *from Fecal Samples*

The isolation and identification of *Salmonella* was carried out following to the WHO global food borne infections network laboratory protocol, (2010). Exactly 5 g of each specimen was suspended in 45 ml of buffered peptone water, a pre-enrichment broth (Tm media, India) and 24 hours incubation at 37 °C. A 100 ul of the suspension was added to 10 ml of Rappaport-vassiliadis enrichment broth, (Tm media India) incubation was for 24 hours at 37 °C. A 1 ml of the suspension was then transferred to 10 ml of Selenite F broth (Tm media, India), and incubated at 42 °C for 24 hours. A loop full of each of the enrichment broth was streaked onto Brilliant Green Agar (BGA), and Xylose Lysine Deoxycholate (XLD) agar, incubation was for 24 hours at 37 °C. Cultural and morphological physiognomies was employed for preliminary Salmonella identification (Jinu et al. 2014).

Biochemical Characterization of Isolated *Salmonella* Species

Pure cultures of presumptive *Salmonella* isolates were further identified following standard biochemical tests, Gram staining, Triple sugar iron (TSI) agar, Citrate, Indole, Oxidase, Catalase, Urease, Methyl red and Voges-Proskauer tests, Motility test (Winn et al. 2005). API 20E biochemical test strip as a non-conventional biochemical test was used to further confirm *Salmonella* isolates.

2.4 *Molecular Characterization and Identification of* Salmonella *Serovars and Strains*

Extraction of Genomic DNA

The bacteria isolates confirmed by biochemical tests were sub-cultured on Luria-Bertani (LB) (Oxford, UK), incubated overnight at 37 °C and broth cultures were prepared. The boiling method was employed for the extraction of DNA from the bacterial cultures (Anejo-Okpobi et al. 2016). A 1.5 ml of the sample extracted was centrifuged at 10,000 rpm for 5 minutes. The supernatant was decanted, and the sediments washed in sterile deionized water twice. To the pellets, was added 200 μm of sterile deionized water, and vortexed to homogenize, boiled at 100 °C for 10 minutes in a water bath. This was vortexed and centrifuged at 12,000 rpm for 5 minutes. The DNA containing supernatant was stored at −20 °C in a test tube. The NanoDrop spectrophotometer was employed for the quantification and purity of the extracted DNA. The quantified DNA extract was prepared and used for Polymerase chain reaction (PCR) (Anejo-Okpobi et al. 2016).

Polymerase Chain Reaction (PCR)

PCR was employed for rapid and definite identification of *Salmonella serovars* using *Salmonella* specific primers as described by Anejo-Okpobi et al. (2016). Amplification was carried out in 50 μl reaction volumes containing; 25 μl of the DNA extract, 15 μl of nuclease free water, 25 μl Dream Taq Master Mix (Thermoscientific, USA) and 2.5 μl of each primer. The Amplification was done in 35 cycles at 95 °C for 5 minutes' initial denaturation followed by a 95 °C for 2 minutes' denaturation step, primer annealing, and extension was at 55 °C for 30 s and 72 °C for 1 min respectively, with the final extension at 72 °C for 10 min. The amplified DNA products from *Salmonella* specific-PCR were further analyzed. Aliquots 5 μl of reaction mixture were electrophoresed through 1.5% agarose gels stained with ethidium bromide; ultraviolet illumination was employed for visualization.

Sequencing

The purification of all the PCR products was with Exo sap and sent to Epoch Life science (USA) for Sanger sequencing (Hoffmann et al. 2013). The online blast search at http://blast.ncbi.nlm.nih.gov/Blast.cgi was used to identify the corresponding sequences.

Sequence Profiling and Identification of *Salmonella* Strains

The generated 16 s rRNA sequences and known 16 s rRNA sequences in the database of the NCBI (National Centre for Biotechnology Information) were compared, using the BLASTn (Basic Local Alignment Search Tool) algorithm. Similarity at between 95–100% of the sequences was used to define the identification at genus, species and serovar levels. The CLAUSTAL IV (Hoffmann et al. 2013) was used to align sequences together with the GeneBank derived reference sequences.

3 Results and Discussion

Cultural identification gave 18 positive presumptive *Salmonella* species out of the sample population of 420 patients. Table 1 presents the sample codes and locations of the positive samples. The organisms possessed characteristic pinkish colonies and black dots at the center on the XLD agar and whitish-green colonies on BGA.

Table 2 shows the results of biochemical tests using conventional methods, and API 20E strips respectively. *Citrobacter sp.* was used as the control, and the result confirmed the 18 presumptive positive isolates from cultural method as *Salmonella* species.

Conventionally, both cultural and biochemical methods of characterization yielded accurate preliminary identification at the generic level within the limit of their diagnostic resolutions. For instance, all the 18 *Salmonella* positive cases were congruent in both cultural and biochemical methods. They were also confirmed to be true in the molecular analysis. However, the molecular approach was more specific as sequencing data led to precise *Salmonella* identification at the sub species level including serovars and strains, consummating in the determination of phylogenetic relationship among them. Similar view was previously upheld among epidemiologists (Arlet et al. 2006).

Table 3 gives the prevalence of *Salmonella* infection in Benue State and across the six LGAs. Eighteen positive samples out of 420 population size were found with mean prevalence of 4.29%. Gboko and Katsina-Ala LGAs recorded the highest number of positive cases with 4 each, representing 22.2% of the positive cases and prevalence of 5.71%. Oju and Kwande LGAs had 3 (16.7%) cases each and prevalence of 4.29%. Makurdi and Otukpo recorded the least number of cases (2 cases each representing 11.1% of the positive cases) and prevalence of 2.89%. In the

Table 2 Biochemical tests and identification of *Salmonella* species using conventional and API test kit

Conventional tests	Gram staining	Triple sugar iron agar	Oxidase	Catalase	Citrate	Methyl red	Urease		Indole	Voges Proskauer											Identification
1–18	–	+	–	+	+	+	–		–	+											*Salmonella* spp
API biochemical tests	O-nitrophenyl-D-g alactopyranoside	Argine dihydrolase	Lysine decarboxylase	Orthine decarboxylase	Citrate	Hydrogen sulphide	Urease	Tryptophan deaminase	Indole	Voges Proskauer	Gelatinase	Glucose	Mannose	Inositol	Sorbitol	Rhamnose	Sucrose	Melibiose	Amygdalin	Arabinose	Identification
1–18	–	+	+	+	+	+	–	–	–	+	–	+	+	–	+	+	–	+	–	+	*Salmonella* spp
19	+	+	+	+	+	+	+	+	+	+	+	+	+	–	+	+	+	+	+	+	*Citrobacter* spp

1–18: Samples number 1–18 of *Salmonella* isolates; 19: Samples number 19 of *Citrobacter spp*

Table 3 Prevalence of *Salmonella* infections in different locations of Benue State

General Hospital	Sample size	No. and percentage of positive sample	Prevalence (%)
Gboko	70	4 (22.2%)	5.71
Makurdi	70	2 (11.1%)	2.86
Oju	70	3 (16.7%)	4.29
Adikpo	70	3 (16.7%)	4.29
Katsina-Ala	70	4 (22.2%)	5.71
Otukpo	70	2 (11.1%)	2.86
Total	**420**	**18**	**4.29 (mean)**

$\chi^2_{(5)}$ (location and prevalence) = 1.89, $P = 0.863 (P > 0.05)$

study area, prevalence of *Salmonella* infection did not depend on locations of sampling since no significant association was established between the two variables ($\chi^2 = 1.89$, $P > 0.05$).

Prevalence and serovars distribution in relation to demographic parameters have been reported, (Adeshina et al. 2009; Udeze et al. 2010; Jacob and Cohen 2016). Based on this report, the mean prevalence of *Salmonella* cases in Benue State was 4.29%. This is lower than the 5% permissible limits of the WHO (2017). However, Gboko and Katsina-Ala recorded higher prevalence of 5.71%, above the WHO permissible limits. All the six LGAs have equal chances of *Salmonella* infection, because the predisposing factors highlighted are equally present in the six locations of the study area. For instance, all locations are facing the challenges of unsafe drinkable water, poor food handling among farmers or vendors and poor sanitary conditions. Poultry and other poultry products thought to be the foremost means of transmission of *Salmonella* (EFSA 2013; Kaniz et al. 2014) are commonly seen. All the locations have similar market structures where fruits and vegetables are sold in the open.

Salmonella infection is highly prevalent in tropical regions and in populations that lack access to safe water and adequate sanitation (Abioye et al. 2017; CDC 2008). These are the prevailing factors in the six studied locations. Thus, the low mean prevalence rate recorded in this study, could be because the subjects were hospital patients attending the selected health care facilities, studies that focus on achieving total seroprevalence rate must therefore capture not only hospital patients but also populations of those not attending the hospitals.

This study reveals that *Salmonella* infection in the area of study was caused by two species: *S. enterica* and *S. bongori* Table 4. The latter had 5.6% occurrence with prevalence of 0.24% while the former was the dominant type with 94.4% occurrence and prevalence of 4.05%. Hence, occurrence of *Salmonella* infection was significantly tied to species type ($\chi^2 = 78.85$, $P < 0.05$). This result is consistent with previous findings Amber et al. (2006) and Hsiao et al. (2016). According to Brenner et al. (2000), 99% of Salmonella infections in man is associated to *S. enterica* subsp. *enterica*.

Raufu et al. (2013) opined that most enteric infections in humans are caused by more than one serovars of a given species, which may vary from time to time and from country to country. In the present work, seven distinct serovars of *S. enterica and* a serovar of *S, bongori are identified,* significant association was established between occurrence of *Salmonella* infection and causative serovars ($\chi^2 = 57.93$, $P < 0.05$). It suggests a huge diversity in the genomics and physiological adaptation of the microbe in the host. This has a huge implication in disease treatment and control because wider genetic make-up may widen the chances of multi drug

Table 4 Prevalence of *Salmonella* infections based on serovars

Salmonella serovars	Frequency and percentage of occurrence	Prevalence %
S. enterica Agona	1 (5.56%)	0.24
S. enterica Paratyphi B	1 (5.56%)	0.24
S. enterica Heidelberg	2 (11.1%)	0.48
S. enterica Typhi	2 (11.1%)	0.48
S. enterica Typhimurium	6 (33.33%)	1.43
S. enterica Enteritidis	4 (22.22%)	0.95
S. enterica Huaian	1 (5.56%)	0.24
S. bongori	1 (5.56%)	0.24
Total	**18 (100%)**	**4.29**

$\chi^2_{(7)}$ (serovars and occurrence of infection) = 57.93, $P = 0.000 (P < 0.05)$

resistance (MDR) and ability to develop different mechanisms for virulence. This view aligns with previous reports on salmonellosis as a bacterial infection caused by more than one *Salmonella* species, with many subspecies, and serotypes (Lar et al. 2006; Raufu et al. 2013).

Figure 1 shows the phylogenetic relationships among 18 *Salmonella* strains identified using plasmid gene sequencing data. Dendrogram formed 2 main clusters (numbered 36 and 81) and 2 sub clusters. The first sub cluster of main cluster (numbered 57) comprised four strains isolated from different locations: *S. enterica* Heidelberg-MG663473.1 sourced from Gboko, *S. enterica* Typhimurium-JQ22 8518.1 sourced from Katsina-Ala, *S. enterica* Typhimurium-CP014981.1 sourced from Makurdi and *S. enterica* Typhimurium-CP023166.1 sourced from Kwande. The *S. enterica* Typhimurium-CP023166.1 sourced from Kwande was a unique strain that showed wider genetic variability but closely related to the check strain (*S. bongori* strain KU060291.1). The second main cluster (numbered 81) comprised 8 strains belonging to 4 Enteritidis, 2 Typhi, 1 Huaian and 1 Typhimurium cutting across all locations except Gboko.

From the dendrogram, *S. enterica* Typhimurium-MH196335.1 was a divergent strain identified, it diverged from the source of the main clusters (88) bearing no close relationship with any existing check strain. It was isolated from a 20-year-old female patient in Adikpo General Hospital. The most unique of all the 18 strains identified was the *S. enterica* serovar Agona strain 392869-2 isolated from a

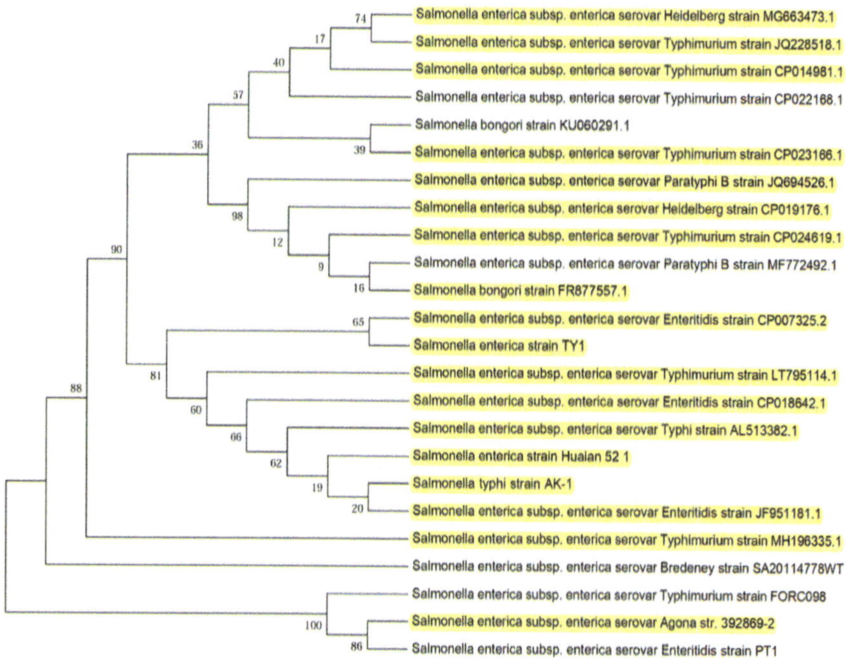

Fig. 1 Phylogenetic construction of *Salmonella* strains from Plasmid gene sequencing

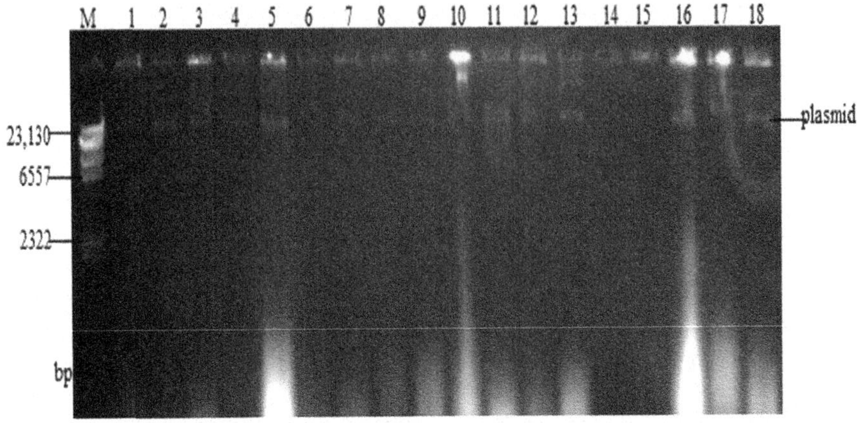

Fig. 2 Amplified gel image of Plasmids from *Salmonella* strains (23,130 bp). M = DNA Ladder, 1 = *S. enterica* Agona-392,869-2, 2 = *S. enterica* Paratyphi B-JQ694526.1, 3 = *S. enterica* Heidelberg-MG663473.1, 4 = *S. enterica* Heidelberg-CP019176.1, 5 = *S. enterica* Typhi-AK-1, 6 = *S. enterica* Typhimurium-CP014981.1, 7 = *S. enterica* Enteritidis-CP007325.2, 8 = *S. enterica* Typhi-AL513382.1, 9 = *S. enterica* Typhimurium-CP024619.1, 10 = *S. enterica* Typhimurium-MH196335.1, 11 = *S. enterica* Typhimurium-CP023166.1, 12 = *S. enterica* Enteritidis-JF951181.1, 13 = *S. enterica* Huaian-H52.1, 14 = *S. bongori*-FR877557.1, 15 = *S. enterica* Typhimurium-LT795114.1, 16 = *S. enterica* Typhimurium-JQ228518.1, 17 = *S. enterica* Enteritidis-TY1, 18 = *S. enterica* Enteritidis-CP018642.1

50-year-old male patient in Gboko General Hospital. It bore a close relationship with a known check *Enteritidis* strain.

Figure 2 presents the gel image of Plasmids from the *Salmonella* strains, the bands made up of 23,130 base pair.

It is also likely that every serovar is pathologically distinct, since there are evidence on the capability of every serovar to acclimate itself to the environment inside its host in a unique way (Thiennimitr et al. 2011). These adaptations are ascribed to copious virulence factors, and other microbial physiognomies of a precise serovar which makes it survive in the host. This may account for why *Salmonella* originates a wide array of human diseases which include enteric fever, gastroenteritis, bacteremia, and systemic infections (Monteville and Matthews 2008; Andino and Hanning 2015).

Apart from the epidemiological implication of diverse serotypes, the cost implications in the identity of the broad spectrum of all the serovars affecting a given population is huge. This can only be consummated at the molecular level as implicated in this work, thus changing the routine conventional method of diagnosis in public health and making treatment difficult.

In conclusion, this work reported that the serovars that infect the sample population are *Salmonella enterica.* This finding was corroborated Kramarenko et al. (2014), Yang et al. (2016), Ed-dra et al. (2017). The prevalence rate of each serovar in this work is less than 5%. This result deviates from previous findings in Nigeria reporting higher prevalence rates based on serovar types. Some workers reported

45.0% prevalence of *S. typhi* (Raufu et al. 2013). Anejo-Okpobi et al. (2016) published 27% for *S. paratyphi* A, 25% for *S. paratyphi* B, 13.7% for *S. paratyphi* C and 20% for *S. typhi* among human subjects in Jos Plateau State. Umeh and Agbulu (2009) reported 57.6% for *S. typhi,* 26.3% for *S. paratyphi,* and 15% for the mixture of both serovars in Okpokwu Local Government Area of Benue State.

Acknowledgments The authors acknowledge the financial support of Covenant University in the publication of this manuscript.

Conflict of Interest This article is our original research work and has not been submitted anywhere for this purpose and the result has not been published anywhere. The consent of the Authors has been sought and approval obtained before this submission, hence the Authors declare no conflict of interest.

Ethical Approval The protocol for this research was submitted and the ethical approval sought and obtained from the Benue State Hospital Management Board.

Authors' Contributions This work was carried out as collaboration with all authors. Authors Okpa, B.O. and Ichor, T. worked on the study design, laboratory, statistics, and draft manuscript. Author Gberikon, G.M. was the supervisor under whose guide the entire execution of the work and corrections were effected to this present standard. Author Oranusi, S. ensured that all the anomalies in the drafted manuscript were corrected, effected, and assisted in managing the literature searches. The authors declare no competing interest.

Ethical Consideration Approval was obtained of the management consent from the involved General Hospitals in the six Local Government Areas, through the office of the Executive Secretary to Benue State Hospitals Management Board Makurdi.

References

Abdullahi M, Olonitola S.O, Umoh V.J, et al. Antibacterial Resistance Profile, and PCR Detection of Antibiotics Resistance Genes in Salmonella Serovars, Isolated from Blood Samples of Hospitalized Subjects in Kano, Northwest Nigeria. *British Microbiology Research Journal.* 2014. 5(3): 245-256

Abioye JOK, Bulus S, Adogo LY. Prevalence of *Salmonella* Typhi infection in Karu LGA of Nasarawa State, Nigeria, *Journal of Advances in Microbiology*. 2017. 6(2):1-8

Adeshina G, Osuagwu N, Okeke C, et al. Prevalence and Susceptibility of *Salmonella* Typhi and *Salmonella* Paratyphi in Zaria, Nigeria. *International Journal of Health Research*. 2009. 2(4): 353-369

Agada GOA, Abdullahi IO, Aminu M, et al. Prevalence and Antibiotic Resistance Profile of Salmonella Isolates from Commercial Poultry and Poultry Farm-handlers in Jos, Plateau State, Nigeria. *British Microbiology Research Journal*. 2014. 4(4):462-479.

Ajiboye RM, Solberg OD, Lee BM et al. Global spread of mobile antimicrobial drug resistance determinants in human and animal *Escherichia coli* and *Salmonella* strains causing community-acquired infections. *Clinical Infectious Diseases*. 2009. 49, 365–371.

Akinyemi KO, Smith SI, Oyefolu AO, et al. Multidrug resistance in 2005.

Akinyemi KO, Phillip W, Beyer W, et al. In-vitro antimicrobial susceptibility Patterns of Salmonella enterica serovars and emergence of Salmonella phage type DT071 in a suspected community-associated outbreak in Lagos, Nigeria. *Journal of Infection in Developing Countries*. 2007. 1:48-54.

Amber H, Trevor T, Hye JS et al. Interaction between *Salmonella* and Schistosomiasis: A Review. *PLoS Pathogens*. 2006. 12(12):e1005928.

Amenu D. Antimicrobial Resistance for Enteric Pathogens isolated from acute gastroenteritis patients and Antibiotic Resistance. *Review of Science Technology*. 2014. 16: 709-715.

Andino A and Hanning I. *Salmonella enterica*: Survival, colonization, and virulence differences among serovars. *Science World Journal*. 2015. 52, (01):7-9.

Anejo-Okpobi JA, Isa SE Audu O, et al. Isolation and polymerase chain reaction detection of virulence invA gene in *Salmonella spp*. From poultry farms in Jos. *Nigeria Journal of Medical Microbiology in the Tropics*. 2016. 18:98-102

Anyanwu AL, Fasina FO, Ajayi OT, et al. Antibiotic Resistant *Salmonella* and *Escherichia coli* Isolated from Day-Old Chicks, Vom, Nigeria. *African Journal of Clinical and Experimental Microbiology*. 2010. 11(1): 129136

Ao TT, Feasey NA, Gordon MA, et al. Global burden of invasive nontyphoidal *Salmonella* disease. *Emerging Trend in Infectious Disease*, PMCID: PMC4451910. 2010.

Arlet G, Barrett TJ, Butaye P, et al. *Salmonella* resistant to extended-spectrum cephalosporins: prevalence and epidemiology. *Microbial infection*. 2006. 8(7):1945-54.

Asrat D. Shigella and Salmonella serogroups and antibiotic susceptibility patterns in Ethiopia. *East Mediterranean Health Journal*. 2008. 14:760-767.

Barbiour EK, Ayyash DB, Alturkistni W, et al. Impact of sporadic reporting of poultry *Salmonella* serovars from selected developing countries. *Journal of Infectious Diseases*. 2015. 9:1-7

Brenner FW, Villar RG, Angulo FJ, et al. *Salmonella* nomenclature. *Journal of Clinical Microbiolology*. 2000.38: 2465-2467

Centre for Disease Control (CDC). U.S. Department of Health & Human Services. Accessed at: www.hhs.gov. 2008.

Cheesebrough M. District Laboratory Practice in Tropical Countries. Tropical Health Technology Publishers, Great Britain. 2002. 40-56pp.

Crump JA, Luby SP, Mintz ED. The global burden of typhoid fever. *Bulletin of World Health Organisation*. 2004. 82: 346-353.

Dong P, Zhu L, Mao Y, et al. Liang R., Niu L., Zhang Y. Prevalence and profile of *Salmonella* from samples along the production line in Chinese beef processing plants. *Food Control*. 2014. 38, 54–60.

Ed-dra A, Filali FR, Karraouan B, et al. Prevalence, molecular and antimicrobial resistance of *Salmonella* isolated from sausages in Meknes, Morocco. *Microbial Pathogens*. 2017. 105: 340–345.

EFSA. Scientific Report of EFSA and ECDC: The European Union summary report on trends and sources of zoonoses, zoonotic agents and food-borne outbreaks in 2011. *EFSAJournal*. 2013 11:3129.

Fagbamila I, Kabir J, Abdu P, et al. Antimicrobial screening of commercial eggs and determination of Tetracycline residue using two microbiological methods. *International Journal of Poultry Science*. 2010. 9: 959-962.

Fashae K, Ogunsola F, Aarestrup FM, et al. Antimicrobial susceptibility and serovars of Salmonella from chickens and humans in Ibadan, Nigeria. *Journal of Infectious Disease in Developing Countries*. 2010. 4(8):484-494.

Feasey NA, Dougan G, Kingsley RA, et al. Salmonella disease: An emerging and neglected tropical disease in Africa. *Lancet*. 2012. 379:2489-2499.

Food and Agricultural Organization (FAO). FAOSTAT. Available: http://www.fao.org. 2015.

Institut Pasteur, Paris, France. 2007.

Hoffmann M, Muruvanda T, Allard M W, et al. Genome sequence of a multidrug-resistant *Salmonella enterica* Serovar Typhimurium var. 5- strain isolated from chicken breast. *Genome Announcement*. 2013. 1:e01068–13

Hsiao A, Toy T, Seo HJ, et al. Interaction between Salmonella and Schistosomiasis: A review. *PLos Pathology*. 2016. 12(2): e1005928.

Hyeon JY, Chon JW, Hwang IG, et al. Antibiotic resistance, and molecular characterization of *Salmonella* serovars in retail meat products. *Journal of Food Protocol*. 2011. 74:161–166.
Igwe NN and Agbo, E.A. Incidence of co-infection of enteric Salmonella and Schistosoma in Kachia LGA of Kaduna State, Nigeria. *International Journal of Tropical Medicine and Public Health*. 2014. 3(1): 12-17
Jacob NJ, and Cohen MB. Update on Diarrhoea. *Paediatrics in Review*. 2016. 37(8): 313-322
Jinu M, Agarwal RK, Sailo B, et al. Comparison of PCR and conventional cultural method for detection of Salmonella from poultry blood and faeces. *Asian Journal of Animal Veterinary Advances*. 2014. 9: 690-701
Jones FT. A review of practical Salmonella control measures in animal feed. *Journal of Applied Poultry Resources*. 2011. 20: 102-113.
Kaniz F, Mahfuzur R, Suvomoy D, et al. Comparative analysis of multi-drug resistance pattern of Salmonella spp. isolated from chicken faeces and poultry meat in Dhaka city of Banghladash. *Journal of Pharmacy and Biological Sciences*. 2014. 9: 147-149.
Kingsley RA, Msefula CL, Thomson, NR, et al. K.E. Epidemic multiple drug resistant *Salmonella* Typhimurium causing invasive disease in sub-Saharan Africa have a distinct genotype. *Genome Research*. 2009. 19: 2279–2287
Kosek M, Bern C, Guerrant RL. The global burden of diarrhoeal disease, as estimated from studies published between 1992 and 2000. *Bulletin of World Health*. 2003.
Kramarenko T, Nurmoja I, Karssin A, Meremae K, Horman A, Roasto M. The prevalence and serovar diversity of *Salmonella* in various food products in Estonia. *Food Control*. 2014. 42, 43–47.
Lar PM, Omojevwe ME, Onah JA. Mixed infections of *Schistosoma* and *Salmonella* in the Federal Capital Territory, Abuja. *Journal of Natural Sciences*. 2006. 2(10): 1119-1104
Mamman PH, Kazeem HM, Raji MA et al. Isolation and characterization of *Salmonella* Gallinarum from outbreaks of fowl typhoid in Kaduna State, Nigeria. *International Journal of Public Health and Epidemiology*. 2014. 3: 082-088.
Monteville T, Matthews K. Salmonella species In Food Microbiology. ASM Press. 2008. p. 97-112.
Muhammad M, Muhammad LU, Ambali AG, Mani, A.U., Azard, S. and Barco, L. Prevalence of *Salmonella* associated with chick mortality at hatching and their susceptibility to antimicrobial agents. *Veterinary Microbiology*. 2010. 140: 131-135.
Njunda AL, Oyerinde JP. *Salmonella* typhi infection in *Schistosoma* infected mice. *West Africa Journal of Medicine*. 1996. 15(1): 2430
Ohad G, Erin CB, Guntram AG. Same species, different diseases: how and why typhoidal and non typhoidal Salmonella enterica serovars differ. *PMC Frontier in Microbiology*. 2014. 5: 391
Orji MC, Onuigbo HC, Mbata TI. Serological survey of mycoplasmosis and pullorum disease in Plateau State of Nigeria. *International Journal of Infectious Diseases*. 2005. 9(2): 86-89
Raufu I, Bortolaia V, Svendsen CA et al. The first attempt of an active integrated laboratory-based *Salmonella* surveillance programme in the North-Eastern region of Nigeria. *Journal of Applied Microbiology*. 2013. 115: 1059-1067.
Scherer CA, Miller SI. Molecular pathogenesis of Salmonellae. In *Principles of Bacterial Pathogenesis Principles of Bacterial Pathogenesis*, Groisman, E. A. (Ed). Academic Press, United States of America. 2001. pp:265-316.
Talabi HA. Medical aspects of typhoid fever in Nigeria. *Nigerian Postgraduate Medical Journal*. 1994. 1:51-56.
Thiennimitr P, Winter SE, Winter MG, et al. Intestinal inflammation allows *Salmonella* to use ethanolamine to compete with the microbiota. *Proceeding of National Academy of Science*. 2011. 108:17480–17485
Udeze AO, Abdulrahman F, Okonkolo A. Seroprevalence of Salmonella typhi and Salmonella paratyphi among the first year students of University of Ilorin, Ilorin, Nigeria. *Middle-East Journal of Scientific Research*. 2010. 6(3):257-262
Umeh EU, Agbulu C. Distribution pattern of salmonella typhoidal serotypes in Benue State. *The International Journal of epidemiology*. 2009. 8(1): 1-7

Uzzau S, Brown DJ, Wallis T, et al. Host adapted serotypes of *Salmonella enterica*. *Epidemiology of Infectious Diseases*. 2000. 125(2):229-255.

Winn W, Allen S, Janda W et al. Koneman's Color Atlas and Textbook of Diagnostic Microbiology. 6th Edition. New York, USA: Lippincott Wilkins. S. 2005. 165- 171

World Health Organization (WHO). Typhoid Facts Sheet. www.who.int/mediacentre/factsheets/typhoid. 2017.

World Health Organization (WHO). Typhoid Facts Sheet. www. who.int/mediacentre/ factsheets/typhoid. 2018.

Yang X, Huang J, Wu Q et al. Prevalence, antimicrobial resistance and genetic diversity of *Salmonella* isolated from retail ready-to-eat foods in China. *Food Control*. 2016. 60. 50–56.

Phytochemical Screening of *Citrullus colocynthis* (L.) Schrad and its Cytotoxic Activity Against Cervical Cancer Cells

J. A. O. Olugbuyiro, J. O. Bamidele, and A. A. Fatokun

1 Introduction

For decades, cancer, a major killer disease has remained a notable leading cause of mortality across the globe. It has been a major health concern affecting mostly, but not limited to, the aged populations. According to recent reports from the International Agency for Research on Cancer (IARC), over 19 million cancer cases and a total of about ten million mortality cases resulting from cancer have been documented (IARC 2021) as of 2020. According to these new estimates, within 5 years, nearly 50 million people are alive after a previous cancer diagnosis. A further report projected that by 2040, the number of new cases will have increased to 28 million, most especially among low or medium income countries (GLOBOCAN 2020a). According to the GLOBOCAN fact sheet, three cancer types are currently most prevalent in Nigeria, one of which is cervical cancer (GLOBOCAN 2020b). Despite the fact that a number of treatment modalities exist and there have been recent advances, to date, there are no adequate therapeutic approaches that counter mortality from cancer while also significantly prolonging survival period for metastatic cancer. Chemotherapy, a conventional treatment modality, has been associated with certain side effects that in some cases lead to a deteriorated state of health in patients. Of further concern is the inability of novel synthesized chemotherapeutic drugs to yield expected results despite the exorbitant cost of their development

J. A. O. Olugbuyiro (✉) · J. O. Bamidele
Department of Chemistry, College of Science & Technology, Covenant University, Ota, Nigeria
e-mail: joseph.olugbuyiro@covenantuniversity.edu.ng

A. A. Fatokun
Centre for Natural Products Discovery (CNPD), School of Pharmacy and Biomolecular Sciences, Faculty of Science, Liverpool John Moores University, Liverpool, UK

© The Author(s), under exclusive license to Springer Nature Switzerland AG 2022

A. O. Ayeni et al. (eds.), *Bioenergy and Biochemical Processing Technologies*, Green Energy and Technology, https://doi.org/10.1007/978-3-030-96721-5_28

(Solowey et al. 2014). These challenges have given rise to the dire need to develop from plants, novel and affordable anticancer drugs with high efficacy and side effects.

Herbal medicines have been and still continue to be utilized as the primary source of medical care in developing countries. Multiple studies have identified plant species having anticancer characteristics, with a particular emphasis on those utilized in herbal medicine in developing nations (Ochwang'I et al. 2014; Greenwell and Rahman 2015). Over the years, cancer has been prevented and treated using medicinal plants, and according to Cragg (1998), much of the conventional anticancer drugs in clinical usage have their origin in plants and natural products. Report showed that over 80%, especially in developing countries, still rely on herbals and over 60% of cancer patients use traditional herbs in cancer treatment (Madhuri and Pandey 2008). This in part could be due to the unsatisfactory result from chemo and the inability to treat resistant and metastatic cancers. Furthermore, it has also been shown that the risk of developing cancer or other deadly diseases could be lowered by regularly taking plants and food sources rich in phytochemicals. One of such important plant species is *Citrullus colocynthis*, traditionally documented for its pharmacological activities.

Citrullus colocynthis (L.) Schrad is a Nigerian medicinal plant that grows in arid, sandy soils. It is native to tropical Africa and India. The potentials of phytochemicals to prevent and cure diseases necessitated its usage in ancient practices, as demonstrated by ancient medicinal practices (Fadeyi et al. 2013; Mann 2002). *C. colocynthis* is reported to be useful traditionally in treating cancer, cold, jaundice, leprosy, cough, diabetes, toothache, asthma, bronchitis, wound, gastrointestinal disorders, and bacterial infections (Ponsankar et al. 2018; Ríos et al. 2005). Its fruit has been of interest in studies because of its diverse pharmacological properties such as cytotoxic, antidiabetic, anti-inflammatory, antilipidemic, antimalaria, and antioxidant properties (Hussain et al. 2014).

Previous studies on *C. colocynthis* showed that it is rich in secondary metabolites (otherwise referred to as phytochemicals) such as flavonoids, phenols, alkaloids, glycones, glucosides, and steroids (Ramakrishna et al. 2019; Uma and Sekar 2014; Nora et al. 2015). Studies have also shown that extracts of *C. colocynthis* can inhibit malignant cell's development (Shaukat et al. 2017). The anticancer properties of this plant is said to be from its accumulation of cucurbitacin glycosides.

Cucurbitacins are one of the constituents of *C. colocynthis* identified to play a crucial part in the inhibition of the growth and replication of cancerous cells (Abdulridha et al. 2019). Li and colleagues observed that the growth, proliferation, metastatic movement, and development of cancer were inhibited by cucurbitacin E from *C. colocynthis,* through the induction of elongation translation factor phosphorylation and apoptotic induction. Similarly, cucurbitacin B from the same plant induced death in HepG-2 cancer cells through caspase induction, cell cycle inhibition, and Bax2/Bcl apoptotic pathway induction (Li et al. 2006). However, so far, no literature has reported the cytotoxic effect of *C. colocynthis* methanol and ethyl acetate extracts on cervical cancer cells. This research study therefore is intended to determine the anticervical cancer potential of *C. colocynthis*, a Nigerian medicinal plant which is diversely used as a condiment and for soup and stew.

2 Materials and Methods

2.1 Identification and Preparation of Plant Sample

Citrullus colocynthis was collected from a garden in Ilaro, Ogun State, a South-West zone of Nigeria. The plant material for this research was identified and authenticated. The sample with the voucher number Cc/Bio/H839 was submitted to an herbarium at the Federal Research Institute of Nigeria, Ibadan. The plant material was washed to ensure it was clean and clear of dirt and foreign matters. The whole fruit of the plant was cut into pieces to increase the surface area, air-dried for some days and soaked immediately in the appropriate solvent.

2.2 Extraction and Concentration of Plant Sample

Extraction of plant sample was carried out using successive extraction according to the procedure described by Olugbuyiro, Odugbesan, Rotimi, and Uchechukwu (Olugbuyiro et al. 2017). The defatted fruit was soaked in ethyl acetate and methanol successively. The extracts were then strained and filtered over a drying agent. This was followed by filtrate concentration with a rotary evaporator at 45 °C.

2.3 Phytochemical Analysis

This was conducted according to the standard procedure described by Olugbuyiro, Moody, and Hamann (Olugbuyiro et al. 2013) to qualitatively identify the secondary metabolites present in the plant samples.

2.4 Cell Culture

This was carried out using a human cervical cancer cell line (HeLa), as previously reported (Omondi et al. 2020). Dulbecco's Modified Eagle Medium (DMEM) was used to culture the cells along with the following supplements: Foetal calf serum (10%), glutamine (2 mM), and Antibiotic-antimycotic solution of 1%. Cells were seeded into 96-well plates at 5 x104 cells/ml (100 μl/well), and the cell culture plates were kept at 37 °C and 5% CO_2 in a humidified incubator.

2.5 *Cytotoxicity Assay*

Twenty-four hours after seeding cultures were treated with up to 500 μg/ml (each in triplicate) of each extract for 48 h (stock concentrations of extracts were prepared in DMSO, while working dilutions were prepared in growth medium). MTT [3-(4,5-dimethylthiazol-2-yl)-2,5-dipenyltetrazolium bromide] assay was then used in quantifying the cells' viability as previously described (Omondi et al. 2020). After incubating for 3 h with MTT, each well's contents were aspirated, and DMSO was used to solubilize the formazan that resulted. The plate was then gently shaken on a plate shaker for 5 min and absorption was read at 570 nm wavelength on a microplate reader (CLARIOstar, BMG Labtech, Germany). The negative control wells' average viability was set to 100% and the average viability of each treatment was normalized to it. The Inhibitory Concentration (IC_{50}), defined to be the concentration corresponding to 50% cell viability, was calculated from the MTT assay data.

Statistical Analysis of Data
Data were expressed as mean triplicate values± standard error of the mean. The data was analyzed using GraphPad Prism 9.2.0.

3 Results and Discussion

3.1 *Phytochemical Screening of* Citrullus colocynthis

The result of the phytochemical screening of the whole fruit of *Citrullus colocynthis* is presented in Table 1. The result shows phenols, tannins, flavonoids, cardiac glycosides, saponins, triterpenoids, steroids, and alkaloids. These phytochemicals were

Table 1 Summary of the phytochemical screening of *C. colocynthis*

Phytochemical	*C. colocynthis*
Tannins	+++
Phenols	+++
Flavonoids	+++
Saponins	+
Cardiac glycoside	+++
Steroid	+++
Alkaloids	+++
Triterpenoids	+++

Key: +++ = highly intense, + = less intense

found to be abundantly present in the plant extract, while anthraquinones were absent. The presence of these phytochemicals agreed with the report of Ramakrishna et al. (2019) who reportedly noted that flavonoids, glycosides, tannins, phenols, alkaloids, and steroids are present in the fruit pulp of *C. colocynthis* methanol extract. Phenolic compounds are one of the most abundant constituents of plant secondary metabolites. They are reported to exhibit numerous biological activities, one of which is against cancer as they possess anti proliferation, anti-angiogenic, antimetastatic, and antiapoptotic properties. They also offer protection against cardiovascular diseases and exhibit anti-aging as well as anti-inflammatory properties (Uma and Sekar 2014). Alkaloids have been reported to play important roles as anticancer agents, and have been in use as anticancer drugs over a long period of time, an example of which is the vinca alkaloid, vincristine. They have different anticancer mechanisms by which they operate. They prevent topoisomerase activity of enzymes that gets involved in mimicking of DNA; they can induce apoptosis and inhibit p53 gene expression (Meybodi et al. 2017). The presence of certain phytochemicals in plants confers particular medicinal activities on them. The anticancer properties of *C. colocynthis* are therefore attributed to the presence of one or more of these phytochemicals.

3.2 *Cytotoxic Effects of Ethyl Acetate and Methanol Extracts of* C. colocynthis *Against Cervical Cell Line (HeLa)*

The results of the MTT assay (Fig. 1) show that both *C. colocynthis* ethyl acetate (CCE) and methanol (CCM) extracts reduced HeLa cell viability as the concentration increased. However, CCE had nearly thrice more potency than CCM with IC_{50} (μg/ml) values of 33.95 ± 3.39 and 101.59 ± 24.37, respectively (Fig. 2). This finding indicates the higher efficacy of ethyl acetate extract of *C. colocynthis* on the

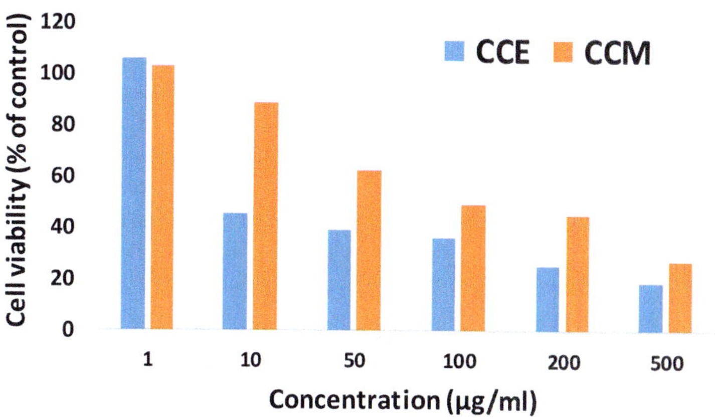

Fig. 1 Effects of CCE and CCM extracts of *C. colocynthis* on the viability of HeLa cells

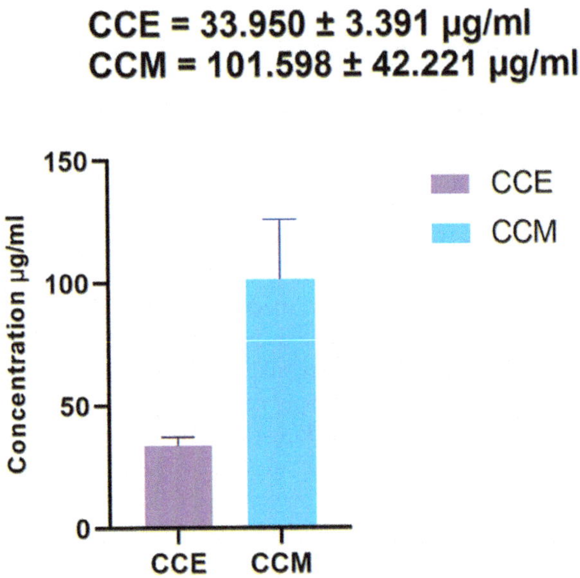

Fig. 2 IC_{50} values for the cytotoxic effects CCE and CCM extracts of *C. colocynthis* against HeLa cell. Results are expressed as mean ± SEM

HeLa cell line. Previous studies have also reported that *C. colocynthis* extracts had a cytotoxic effect on cancer cell lines. In a study carried out by Perveen et al. (2020), the cytotoxic effects of the methanol extract and ethyl acetate fraction of *C. colocynthis* was reported. These authors demonstrated the cytotoxic action of *C. colocynthis* methanol and ethyl acetate extract on the human breast cancer cell line (MCF-7). Similar to our result, the authors found out that both extracts reduced the viability of MCF-7 in a concentration-dependent manner but with the ethyl acetate fraction being more potent with IC_{50} values (µg/ml) of 30 ± 4.15 and 12 ± 0.46, respectively.

In another report, the anticancer potential of *C. colocynthis* ethanol extract was also documented. The result of the study showed that *C. colocynthis* ethanol extract induced cytotoxicity in a dose-dependent manner on the breast (MDA-MDB-21) and colon (HT-29) cancer cell lines used in the study (Bourhia et al. 2020). The authors reported the IC_{50} values (µg/ml) of the ethanol extract on the MDA-MDB-21 and HT-29 cancer cell lines as 170.34 and 132. 31. These studies and more (Muhkerjee and Patil 2012; Rezai et al. 2017) have shown that the various extracts of this medicinal plant exhibit cytotoxicity towards the cell lines tested. Our report showed that *C. colocynthis* is rich in phytochemicals as shown in Table 1. The extracts of this plant has also been reported to exhibit antioxidant activities (Benariba et al. 2013; Hussain et al. 2014). The cytotoxicity of this plant extracts could therefore be associated with its phytochemical constituents and antioxidant activities.

4 Conclusion

Cancer is a global burden that has posed a threat to the whole world owing to its increasingly alarming rate. In Nigeria, cervical cancer continues to rank as one of the most reported cancer cases. Traditionally, medicinal plants have been well and long used in treatment of cancerous diseases. The present research study investigated the phytochemical constituents and cytotoxic (potential anticancer) activity of *C. colocynthis* extracts. The two extracts studied were found to lower HeLa cell viability in a concentration-dependent manner with the ethyl acetate extract being more potent. The findings indicate that *C. colocynthis* could serve as a potential source of drugs for the treatment of cervical cancer. However, further studies are in progress for a detailed characterization of the cytotoxic effects and the associated cellular and molecular mechanism, and also for the isolation of the active principles from the plant.

Acknowledgement The authors thankfully acknowledge Covenant University, Nigeria for financial support towards the publication of this article.

References

Abdulridha MK, Al-Marzoqi AH, Ghasemian A (2019) The anticancer efficiency of *Citrullus colocynthis* toward the colorectal cancer therapy. J. Gastrointest Canc 51: 439– 444.

Benariba N, Djaziri R, Bellakdhar W, Belkacem N, Kadiata M, Malaisse WJ, Sener A (2013) Phytochemical screening and free radical scavenging activity of *Citrullus colocynthis* seeds extracts. Asian Pac J Trop Biomed 3(1):35-40.

Bourhia M, Messaoudi M, Bakrim H, Mothana RA, Sddiqui NA, Almarfadi OM, El Mzibri M, Gmouh S, Laglaoui A, Benbacer L (2020) *Citrullus colocynthis* (L.) Schrad: Chemical characterization, scavenging and cytotoxic activities. Open Chem 18(1): 986-994.

Cragg GM (1998) Paclitaxel (Taxol®): A success story with valuable lessons for natural product drug discovery and development. Med Res Rev 18(5):315-331.

Fadeyi SA, Fadeyi OO, Adejumo AA, Okoro C, Myles EL (2013) In vitro anticancer screening of 24 locally used Nigerian medicinal plants. BMC Complement Altern Med 13(79): 1-9

Global Cancer Observatory (GLOBOCAN) (2020a) New global cancer data. https://publications.iarc.fr/Non-Series-Publications/World-Cancer-Reports. Accessed 8 Aug 2021

Global Cancer Observatory (GLOBOCAN) (2020b) Cancer today. https://gco.iarc.fr/today/home. Accessed 8 Aug 2021

Greenwell M, Rahman, PKSM (2015) Medicinal Plants: Their use in anticancer treatment. International J Pharm Sci Res 6(10): 4103-4112

Hussain AI, Rathore HA, Sattar MZA, Chatha SAS, Sarker SD, Gillani AH (2014) *Citrullus colocynthis* (L.) Schrad (bitter apple fruit): a review of its phytochemistry, pharmacology, traditional uses and nutritional potential. J Ethnopharmacol 155(1): 54-66

International Agency for Research on Cancer (IARC) (2021) Latest global cancer data. https://www.iarc.who.int/faq/latest-global-cancer-data-2020-qa/. Accessed 8 Aug 2021

Li B, Gao MH, Zhang XC, Chu XM (2006) Molecular immune mechanism of C phycocyanin from *Spirulina platensis* induces apoptosis in HeLa cells in vitro. Biotechnol Appl Biochem 43(3):155–164.

Madhuri S, Pandey G (2008) Some dietary agricultural plants with anticancer properties. Plant Archives 8:13-16

Mann J (2002) Natural products in cancer chemotherapy: past, present and future. Nat Rev Cancer 2:143-148

Meybodi NM, Mortazavian AM, Monfared AB, Sohrabvandi S, Meybodi FA (2017) Phytochemicals in cancer prevention: A review of the evidence. Iran J Cancer Prev 10(1): 1-8

Muhkerjee A, Patil SD (2012) Effects of alkaloid rich extract of *Citrullus colocynthis* fruits on Artemia salina and human cancerous (MCF-7 AND HEPG-2) cells. J Pharm Sci Tech 1(2):15-19.

Nora NB, Hamid K, Snouci M, Boumediene M, Abdellah M (2015) Phytochemical and antibacterial screening of *Citrullus colocynthis* of South-west Algeria. J Chem Pharm Res 7(5):1344-1348.

Ochwang'I DO, Kimwele CN, Oduma JA, Gathumbi PK, Mbaria JM, Kiama SG (2014) Medicinal plants used in treatment and management of cancer in Kakamega County Kenya. J Ethnopharmacol 151: 1040-1055.

Olugbuyiro JAO, Odugbesan D, Rotimi SO, Uchechukwu SL (2017) Antioxidant activity, DNA damaged protection and apoptosis induction potential of *Vernonia amygdalina* leaf extracts. Int J Pharm Sci Res 8(7): 3122-3127.

Olugbuyiro JAO, Moody J, Hamann M T (2013) Phytosterold from *Spondias mombin* Linn with antimycobacterial activities. Afr J Biomed Res 16(1):19–24.

Omondi RO, Bellam R, Ojwach, SO, Jaganyi D, Fatokun AA (2020) Palladium(II) complexes of tridentate bis(benzazole) ligands: Structural, substitution kinetics, DNA interactions and cytotoxicity studies. J Inorg Biochem 210(2020):1-13.

Perveen S, Ashfaq H, Ambreen S, Ashfaq I, Kanwal Z, Tayyeb A (2020) Methanolic extract of *Citrullus colocynthis* suppresses growth and proliferation of breast cancer cells through regulation of cell cycle. Saudi J Biol Sci (Article in Press). https://doi.org/10.1016/j.sjbs.2020.11.029.

Ponsankar A, Vasantha-Srinivasan P, Thanigaivel A, Edwin E-S, Selin-Rani S, Chellappandian M, Senthil-Nathan S, Kalaivani K, Mahendiran A, Hunter WB (2018) Response of Spodoptera litura Fab. (Lepidoptera: Noctuidae) larvae to *Citrullus colocynthis* L. (Cucurbitales: Cucurbitaceae) chemical constituents: larval tolerance, food utilization and detoxifying enzyme activities. Physiol Mol Plant Pathol 101:16–28

Ramakrishna D, Suvarchala V, Chaitanya G, Shasthree T (2019) Preliminary phytochemical screening of a medicinally important cucurbit *Citrullus colocynthis* (L.) Schard. Res J Chem. Environ 23(11):41 – 55.

Rezai M, Davoodi A, Asori M, Azadbakhht M (2017) Cytotoxic activity of *Citrullus colocynthis* (L.) schrad fruit extract on gastric adenocarcinoma and breast cancer cell lines. Int J Pharm Sci Rev Res 45(1): 175-178

Ríos JL, Escandell JM, Recio MC (2005) New insights into the bioactivity of cucurbitacins. Stud Nat Prod Chem 32:429–469

Shaukat A, Dostal A, Menk J, Church TR (2017) BMI is a risk factor for colorectal cancer mortality. Digest Dis Sci 62(9):2511–2517.

Solowey E, Lichtenstein M, Sallon S, Paavilainen H, Solowey E, Lorberboum-Galski L (2014) Evaluating medicinal plants for anticancer activity. The Sci World J 2014:721402

Uma C, Sekar KG (2014) Phytochemical analysis of a folklore medicinal plant *Citrullus colocynthis* L (bitter apple) J Pharmacog Phytochem 2(6):195-202.

Susceptibility Pattern of *Vibrio cholerae* isolated from surface water sources in Makurdi Local Government to commonly used antibiotics

Tersagh Ichor, E. T. Azua, and Grace Oyenike Bolaji

1 Introduction

Water is significant in every living thing and is very essential to life and its processes. High-quality of water for drinking is crucial for health safety of all humans and as a means through which other essential elements required for the well-being of man are acquired through consumption (Abulude et al. 2007). Water covers about 75% of the earth surface whereas the human body is composed of approximately 70% of it (Pant 2004).

The task of water purification, safety, and security has received considerable attention with appreciable progress made in Europe and America as compared to developing nations which are still grappling with the daunting challenge of disease burden with high populace placed at risk of water-borne diseases. In developing nations, surface water bodies are vulnerable to pollutants from substances of varying origins and pathogenic agents due to lack of protective cover hence, using such water for irrigation and other purposes may institute severe unrestricted health threats (Shuval 1990; Obi et al. 2004; Okoh et al. 2007; Igbinosa and Okoh 2009; Chigor et al. 2010; Gemelle and Schmidt 2012). The presence of pathogenic organisms such as bacteria, fungi, helminths, viruses, and protozoa render a water source unsanitary, non-potable and may possibly lead to diffusion of water-borne diseases to agricultural employees, swimmers, and the users of crops dampened with such dirtied waters (Shuval 1990; Freeman et al. 2015).

The consequence of polluted water, according to the WHO report is that, today, an estimated population of one billion individuals lack access to innocuous water

T. Ichor (✉) · G. O. Bolaji
Department of Microbiology, Joseph Sarwuan Tarka University, Makurdi, Nigeria

E. T. Azua
Department of Zoology, Joseph Sarwuan Tarka University, Makurdi, Nigeria

© The Author(s), under exclusive license to Springer Nature Switzerland AG 2022

A. O. Ayeni et al. (eds.), *Bioenergy and Biochemical Processing Technologies*, Green Energy and Technology, https://doi.org/10.1007/978-3-030-96721-5_29

intake and 2.4 billion to cleanliness (WHO/UNICEF 2010). It further opined that not as much as two third of the world populace use enhanced sanitation facilities. Surface water pollution thus remains a global disquiet responsible for high sickness and death rates occasioned by food and water-borne ailments (Prüss et al. 2002; WHO 2006).

Surface water sources are important for both home and industrialized events and the accessibility of good quality of water is a requisite for prevention of water related diseases and improved standard of life (Anhwange et al. 2012). Water is a significant multi-usage component for drinking, energy production, and irrigation (Hacioglu and Basaran 2009). Humans, who utilize natural water sources, are disposed to a number of water-borne health hazards ranging from biological, chemical, and physical ones. Among biological hazards are bacteria such as *Vibrio* species which is one of the widespread pathogenic organisms in surface water (Adeleye et al. 2010). *Vibrio parahaemolyticus* and *Vibrio vulnificus* are harmful to humans but the most important ones are *Vibrio cholera* (Nhung et al. 2007).

Vibrio cholerae was originally described as the cause of cholera by Pacini in 1854. However, *Vibrio cholera* (O1 and O139), the causal agent of cholera is a known and dangerous source of high morbidity and mortality globally especially in Sub-Saharan Africa. *Vibrio cholerae* is about 1.4–2.6 mm long. The bacterium reacts positively to oxidase, it is implicated in reduction of nitrate and has a single, sheathed, flagellum that is polar for motility. It has a capacity for respiratory and fermentative metabolism. Its growth is enhanced by addition of 1% sodium chloride, though it can grow in nutrient broth without salt which distinguishes it from other *Vibrio* species (Barumann et al. 2001).

The bacterium is autochthonous to aquatic environments and are transmitted to human via fecal-oral route (Daniels and Shafaie 2000). Humans get tainted with it through ingestion of raw or undercooked contaminated food stuffs such as legumes, fish, or cross contamination of ready-to eat-food and also through poor personal hygiene and sanitation including unhygienic practices like none or improper hand washing or drinking of contaminated water. It causes acute diarrhea sickness and is reportedly liable for more deaths yearly.

Previous research reports from endemic areas have affirmed a persistent replication of frequency of *Vibrio cholerae* and cholera disease (Mishra et al. 2012; Fooladi et al. 2013). Nigeria has from the time when the earliest manifestation of widespread cholera, reported that sporadic occurrences have been taking place. In late 2010, the Northern region of the country reported roughly 3000 infections and 781 deaths due to cholera which showed how vulnerable some of underprivileged communities were to the infectious agent with children inclusive. The outburst of the disease was credited largely to rain which washed sewages into surface water bodies including ponds, streams, and rivers that people depend on for drinking and other domestic purposes. The region was reportedly ravaged by the blight mostly in states such as Adamawa, Bauchi, Borno, Cross River, Gombe, Jigawa, Yobe, Osun, Taraba, FCT, and Rivers States. Although the scourge occurred and were documented in the affected states mentioned, facts emanating from epidemiological studies showed that Nigeria is at a peril with the assumption that the occurrence remained as a result of hypervirulent strains of the organism (WHO 2011).

There is thus, a subsistence of abandoned reservoirs of the causative agent, *Vibrio cholera* which the disease crops up in environments wherever the conditions are recognized as predominantly unsanitary subjecting inhabitants to nonexistence of hygienic water. Supply of treated water to Makurdi residents is undependable such that pipes habitually go without water pumped through it for longer times compelling several inhabitants to obtain water from natural water source like the nearby River Benue for domestic purposes including drinking. Therefore, there is a need to assess the natural water sources for the presence of *Vibrio cholerae* at water collection points.

Furthermore, the immense usage of antibiotics for prophylaxis due to prior cholera epidemic in the world ensued the emergence of strain of *Vibrio cholera* that resist several drugs (Mandal et al. 2012). Contemporary examination of the organisms' sensitivity to antibiotics should not be limited to only clinical isolates but also to isolates from the environment in areas that are endemic to the disease since the disease causative agent that are resistant to drugs can be found in the environment as its reservoir. The occurrence of *Vibrio cholerae* has always been associated exclusively with an epidemic (Mishra et al. 2012; Fooladi et al. 2013).

The aim of the study was to isolate and identify *Vibrio cholerae* from surface water and to determine its antibiotic susceptibility pattern in metropolis. The relevance of the study cannot be underscored since *Vibrio cholerae* are usually free living in marine, freshwater, and estuarine environments. They are pathogenic and can cause a great deal of infections which are often severe and dangerous. Its presence in water could however pose a degree of threat and hazards to human health. Results from the study would create awareness to stakeholders in the water industry including users, government authorities, regulatory agencies, and agencies involved in management and treatment of water sources for effective prevention of *Vibrio cholera* infection.

2 Materials and Methods

2.1 Study Area

The area is Makurdi, the capital city of Benue State with temperature maximum of 32 °C and a mean minimum of 26 °C (Tyubee 2009).

2.2 Sterilization of Material Used

Various glass wares used in the course of this work were thoroughly washed, air dried and sterilized in a pressure pot at 121 °C for 15 minutes. The wire loop used was flame red hot and allowed to cool before use. Work bench surface were also cleaned before and after analysis. Aseptic method was employed in the course of this work as much as possible. Collection of samples and analysis were done aseptically.

2.3 Sample Collection

Surface water samples were randomly obtained from 16 points in 9 different locations of Makurdi Town using 100 ml sterile bottles and were taken to Biology Laboratory of Joseph Sarwuan Tarka University for analysis within 6 hours of collection. These sites include Wurukum area, Modern Market, River Benue, North Bank, Wadata area, Kastsina – ala street area, Kanshio area, Pilla Village, Mbaikyo Village.

2.4 Media Preparations

The media, thiosulphate citrate bile salt sucrose agar (TCBS) was prepared according to the manufacturer's instructions. It was then sterilized by heating gently over a Bunsen flame and kept until it cooled to 50 °C after which it was dispensed into sterilized Petri-dishes (Acharya 2015).

2.5 Sample Analysis

Serial dilution method was employed. A total fivefold dilution was carried out by taking 1 ml of the original samples and add 9 ml of peptone water. This was repeated for five times serially for each sample in the different tubes and then shaken before 1 ml was taken from the fifth dilution sample and pipette into a Petri-dish (Umeh et al. 2007).

2.6 *Enrichment of* Vibrio cholerae *from Surface Water in Peptone Water, Plating on TCBS Medium and Identification of Isolates*

The water samples were cultured in Alkaline peptone water with a pH of 8.6 ± 0.2 which was in a measurement of 20 ml each and kept in the incubator for 4–6 hours to allow for maximum growth of *Vibrio cholerae*.

After 6 hours, the alkaline peptone water containing the samples was collected and cultured on Thiosulphate citrate bile salt sucrose medium given the total of 16 samples, the TCBS Agar used was 44 g which was liquified in 500 ml of distilled water and boiled. It was then left to cool to 47 °C and transferred to the Petri dishes. After the TCBS solidified, the inoculation process followed which was done with the aid of a wire loop and a Bunsen burner. The Petri-dishes were put into the incubator for 24 hours. Yellow colonies which were button shaped measuring 1–2 mm diameter was observed (Acharya 2015).

Subculture was carried out using Nutrient Agar 7.8 g which was dissolved in 500 ml of water and poured into clean and contaminants free Petri dishes for further use for the Biochemical tests.

2.7 *Gram Stain Technique*

The technique was employed as described by (Fawole and Oso 2004). Gram-positive colonies were purple whereas Gram-negative were either pink or red. Samples that were Gram-negative and comma shaped were further subjected to biochemical tests which included: Oxidase reaction, Voges Proskauer reaction, Indole reaction, Citrate test reaction, Motility test reaction and Methyl red test reaction for genus confirmation as described by (Acharya 2015).

2.8 *Antibiotics Susceptibility Test*

Sensitivity test was carried out on all positive isolates using the Kirby-Bauer disc diffusion method, as described by (Nhung et al. 2007). The positive colonies were inoculated on dried surface of the nutrient agar plate by streaking round the plate using a sterilized wire loop. The incubated plate was left to dry for about 5 minutes and six (commonly used antibiotics in severe cholera treatment) antibiotics discs comprising of Tarivid (10 microgram), Reflacine (10 mcg), Ciproflox (10 mcg), Augmentin (30 mcg), Gentamycin (10 mcg), Streptomycin (30 mcg), Ceporex (10 mcg), Nalidixic acid (30 mcg), Septrin (30 mcg), Amplicin (30 mcg), were then applied at 37 °C overnight. After the incubation, the inhibition zone diameter was measured using a transparent plastic ruler and interpreted according to the zone diameter interpretive chart of CLSI (CLSI 2007).

3 Results and Discussion

Studies on the susceptibility of Vibrio cholerae isolated from surface water sources in Makurdi Local Government to commonly used antibiotics was carried out. *Vibrio cholerae* is normal resident bacteria of water environments, therefore water can serve as its channel of transmission. Water is one of the most important resources in life and it is used for various purposes. Diseases associated with water-borne infections are often capable of causing diarrhea, gastroenteritis vomiting and even fever (Uphadaya et al. 2013).

3.1 *Bacteriological Analysis of Surface Water Sources for* Vibrio cholera

Bacteriological analysis of different surface water sources was undertaken. Table 1 shows the total bacteria count of stream water samples. Total bacteria count ranged from 4.1×10^5 to 1.0×10^5 Cfu/ml. Modern Market Stream have the highest bacteria count (4.1×10^5 Cfu/ml), Old GRA Stream have the least bacteria count of 1.0×10^5 Cfu/ml. Samples from Wurukum Stream and Mbaikyo Village Stream have the bacteria count of 2.2×10^5 Cfu/ml. Wadata Stream have a bacteria count of 1.8×10^5 Cfu/ml, while Northbank Stream have a bacteria count of 2.4×10^5 Cfu/ml. Kanshio Stream have a bacteria count of 2.3×10^5 Cfu/ml, while Pilla Village Stream have a

Table 1 Bacteriological content of stream water within Makurdi metropolis

Stream code	Total bacteria count (Cfu/ml)
S1	2.2×10^5
S2	2.2×10^5
S3	1.8×10^5
S4	2.4×10^5
S5	2.3×10^5
S6	2.1×10^5
S7	4.1×10^5
S8	1.6×10^5
S9	1.0×10^5

Keys: *S1* Wurukum Stream, *S5* Kanshio Stream, *S2* Mbaikyo Stream, *S6* Pilla Village Stream, *S3* Wadata Stream, *S7* Modern Market Stream, *S4* North Bank Stream, *S8* Kastinala Area Stream, *S9* Old GRA Stream

Table 2 Bacteriological content of pond water within Makurdi metropolis

Pond code	Total bacteria count (Cfu/ml)
P1	1.8×10^5
P2	6.0×10^5
P3	1.2×10^5
P4	1.7×10^5
P5	1.4×10^5
P6	1.3×10^5

Keys: *P1* Wurukum Pond, *P4* Kanshio Pond, *P2* Wadata Pond, *P5* Modern Market Pond, *P3* North Bank Pond, *P6* Kastinala Pond

Table 3 Bacteriological content of River Benue

Sample	Total bacteriological count (Cfu/ml)
River Benue	4.2×10^5

bacteria count of 2.1×10^5 Cfu/ml. Kastinala Area Stream have a bacteria count of 1.6×10^5 Cfu/ml.

Table 2 shows the total bacteria count of pond water samples. Total bacteria count vacillated from 6.0×10^5 Cfu/ml to 1.2×10^5 Cfu/ml. The highest bacteria count was obtained from Wadata Pond which is 6.0×10^5 Cfu/ml, while the lowest count was gotten from North Bank Pond (1.2×10^5). Wurukum Pond have a bacteria count of 1.8×10^5 Cfu/ml, while Kanshio Pond have a bacteria count of 1.7×10^5 Cfu/ml. Modern Market Pond have bacteria count of 1.4×10^5 Cfu/ml, while Katsina ala Pond have a bacteria count of 1.3×10^5 Cfu/ml.

Table 3 shows the total bacteria count of River Benue water at one point of water collection. The colony count is 4.2×10^5 Cfu/ml.

From the findings of this study, there is high level of *Vibrio cholerae contaminants* in surface waters sources may be ecologically normal but remains threatening since people use such water for other human related activities. The samples isolated showed the presence of *Vibrio cholerae* at different area which include Wadata, Wurukum, North Bank, Kanshio, Kastinala, Modern Market, Mbaikyo village, Pilla village, and River Benue. The morphological characteristics, biochemical and Gram reactions of the colonies grown from 16 different surface water samples within Makurdi metropolis were observed. A total of 15 out of 16 surface water samples were positive for *Vibrio cholerae* when cultural, biochemical, and Gram stain techniques were used for identification. The Cultural characteristics of *Vibrio cholerae* from surface water were determined on a selective media: Thiosulphate Citrate Bile Salt Sucrose Agar (TCBS) which it grew as large yellow mucoid colonies. The presumptive identification of *Vibrio cholerae* was based on its morphology, Gram staining, and the result of biochemical tests. Gram negative rods observed were presumed to be *Vibrio cholera*. The organism was Voges Proskauer positive, Indole positive, MR positive, Oxidase test positive, Citrate positive and Motility positive.

Previous work carried out in Hati has shown that *Vibrio cholerae* were present in surface waters of Hati (Kahler et al. 2015). The routine usage of surface waters that are polluted with discharged wastewater run-offs for either home, recreation or irrigation purposes is likely a major predisposing factor to the risk of infection by water related infectious agents including *Vibrio cholerae*.

High occurrence of *Vibrio cholerae* in surface water was previously reported in Bangladesh (Faruque et al. 2004). The high frequency of *V. cholerae* was attributed to their ability to tolerate high salt concentration. Caldini et al. (1997) reported that out of 150 *Vibrio spp* isolated from the Aron River basin in Italy, *Vibrio cholerae* was the most prevalent species which was credited to the fact that the river basin is completely freshwater. Uphadaya et al. (2013) also isolated these bacteria from water bodies in India. Studies have also shown the capacity of *Vibrio cholerae* to associate freely with a diversity of zooplankton, phytoplankton, blue-green algae (cyanobacteria) which protract its subsistence (Faruque et al. 2002; Farque and Mekalanos 2012).

Akoachere et al. (Akoachere et al. 2013) stated that the presence of toxigenic *Vibrio cholerae* in water during the non-outbreak period in Cameroun is positively correlated with the physico-chemical characteristics such as temperature, pH (alkaline) of the water. In this finding, temperature value ranged from 22.40 °C to 29.90 °C but the optimal temperature for *Vibrio cholerae* growth is 37 °C with possible growth at 10–43 °C. pH value ranged from 6.00 to 8.40, but optimal pH for *Vibrio cholerae* growth is 7.6 with possible growth at 5.0–9.6. In this study, one could opine that the physico-chemical conditions mainly pH and temperature of surface waters could be the features sustaining the perseverance of toxigenic *Vibrio cholera* in surface waters.

3.2 Antibiotic Susceptibility of Identified Vibrio cholerae *with Antibiotic Disc*

There are two major roles of antibiotic susceptibility test has two major functions in both human and veterinary medicine. It can be used primarily to guide on which antimicrobial can be chosen for effective treatment of a patient. It may also be used at a tool to survey and monitor antimicrobial resistance. Fifteen selected isolates were carried out to antibiotic susceptibility test against preselected antibiotics. Antibiotics used in this experiment were: Nalidixic acid, Reflacine, Gentamycin, Augmentin, Ciproflox, Septrin, Streptomycin, Amplicin, Ceporex, and Tarivid. The zones of inhibition measured in mean diameter of zone of inhibition in mm is shown in Table 4.

Table 4 Diameter of zone of inhibition (in mm) of identified *Vibrio cholerae* with antibiotic discs

	Mean diameter of zone of inhibition in mm														
Antibiotic discs	S1	S2	S3	S4	S5	S6	S7	S8	P1	P2	P3	P4	P5	P6	RB
Blank discs	0	0	0	0	0	0	0	0	0	0	0	0	0	0	0
OFX	22	14	22	12	22	22	20	24	22	23	20	19	22	25	28
PEF	16	18	20	16	20	20	21	26	23	22	20	21	23	18	23
CPX	30	20	18	10	18	22	24	22	22	24	20	20	22	18	21
AU	0	14	14	0	14	0	8	10	16	0	12	18	0	20	16
CN	18	16	16	12	16	16	18	23	22	21	18	20	22	21	20
S	18	16	12	16	5	10	14	18	18	20	18	14	10	12	18
CEP	18	0	12	0	12	8	10	0	12	18	10	16	14	12	8
NA	0	0	0	0	0	0	0	0	0	0	0	0	0	0	0
SXT	20	12	10	8	10	14	13	8	20	14	16	18	10	14	8
PN	16	0	14	8	14	0	18	20	18	19	17	0	0	8	0

Table 5 Antibiotic susceptibility of identified *Vibrio cholerae* with antibiotics discs

Antibiotic Discs	S1	S2	S3	S4	S5	S6	S7	S8	P1	P2	P3	P4	P5	P6	RB
Blank discs	0	0	0	0	0	0	0	0	0	0	0	0	0	0	0
OFX	S	S	S	S	S	S	S	S	S	S	S	S	S	S	S
PEF	S	S	S	S	S	S	S	S	S	S	S	S	S	S	S
CPX	S	S	S	S	S	S	S	S	S	S	S	S	S	S	S
AU	R	S	S	R	S	R	S	S	S	R	S	S	R	S	S
CN	S	S	S	S	S	S	S	S	S	S	S	S	S	S	S
S	S	S	S	S	S	S	S	S	R	S	S	S	S	S	S
CEP	S	R	S	R	S	S	S	R	S	S	S	S	S	S	S
NA	R	R	R	R	R	R	R	R	R	R	R	R	R	R	R
SXT	S	S	S	S	S	S	S	S	S	S	S	S	S	S	S
PN	S	R	S	S	S	R	S	S	S	S	S	R	R	S	R

Table 5 shows the susceptibility of the isolated ***Vibrio cholerae*** to antibiotics used. The results showed that all the isolates from different water sources were susceptible all the antibiotics except to Nalidixic acid which the organism resisted.

The percentage of isolates susceptible to the various antibiotic drugs with Tarivid, Reflacine, Ciproflox, Gentamycin, and Septrin had the highest percentage (93.8%) of isolates susceptible to the drugs, followed by Streptomycin having (87.5%), Ceporex (75%), Augmentin (62.5%) and Amplicin (62.5%) (Table 6).

Table 7 shows the percentage number of resistant isolates to antibiotic drugs. Nalidixic acid had the highest resistance percentage of (93.8%), followed by Augmentin (31.2%), Amplicin (25%), Ceporex (18.8%), and Streptomycin (6.3%).

Rehydration is an effective strategy that has reduced rate of mortality during cholera epidemics. The use of antibiotics as a means of shedding of the organism, administering treatment for severe illnesses and to reduce duration of staying in the hospital and morbidity (Mandal et al. 2012). However, the resistance of microorganism including *Vibrio cholerae* to antibiotics has become a serious health challenge

Table 6 Percentage distribution of antibiotic susceptibility of the *Vibrio cholerae*

Antibiotics (ug)	Number (%) susceptible to drugs *Vibrio cholerae* ($n = 16$)
OFX	15(93.8)
PEF	15(93.8)
CPX	15(93.8)
AU	10(62.5)
CN	15(93.8)
S	14(87.5)
CEP	12(75)
NA	0(0)
SXT	15(93.8)
PN	10(62.5)

Table 7 Percentage distribution of antibiotic resistance of *Vibrio cholerae* isolates from surface water

Antibiotics (ug)	Number (%) resistance to drugs
OFX	0(0)
PEF	0(0)
CPX	0(0)
AU	5(31.2)
CN	0(0)
S	1(6.3)
CEP	3(18.8)
NA	15(93.8)
SXT	0(0)
PN	4(25)

Keys: *S1* Wurukum Stream, *S2* Mbaikyo Village Stream, *S3* Wadata Stream, *S4* North Bank Stream, *S5* Kanshio Stream, *S6* Pilla Village Stream, *S7* Modern Market Stream, *S8* Kastinala Area Stream, *S9* Old GRA Stream, *P1* Wurukum Pond, *P2* Wadata Pond, *P3* North Bank Pond, *P4* Kanshio Pond, *P5* Modern Market Pond, *P6* Kastinala Pond, *S* Sensitivity, *R* Resistance, *OFX* Tarivid, *PEF* Reflacine, *CPX* Ciproflox, *AU* Augmentin, *CN* Gentamycin, *S* Streptomycin, *CEP* Ceporex, NA Nalidixic acid, *SXT* Septrin, *PN* Amplicin, *OFX* Tarivid, *PEF* Reflacine, *CPX* Ciproflox, *AU* Augmentin, *CN* Gentamycin, *S* Streptomycin, *CEP* Ceporex, *NA* Nalidixic acid, *SXT* Septrin, *PN* Amplicin

worldwide. All isolate where strongly resistant to Nalidixic acid which agrees with the finding of Khan et al. (2005) and Shrestha et al. (2010). Where severe cholera cases are reported, health workers need to be aware of such relevant information to take a wise decision regarding the best drug to be prescribed to patients. Isolated *Vibrio cholerae* demonstrated high sensitivity to a number of antibiotics including Tarivid, Reflacine, Ciproflox, Gentamycin, and Septrin which corroborates the report of (Mustapha et al. 2016). Some *Vibrio* strains were previously reported to be resistant to antimicrobials like tetracycline, co-trimoxazole, trimethoprim, and sulfamethoxazole. The struggle to these antimicrobials alongside other factors like virulence and the capability to trigger epidemics, put up vibrios as pathogens of public health interest (Mohale 2003).

Some dissimilarities in the antibiotics resistant drug were reported in these findings with regard to the findings of Khan et al. (2015) and Shrestha et al. (2010). These resistant drugs include Augmentin, Streptomycin, Ceporex, and Amplicin. However, Faruque et al. (2004) reported that the resistance pattern of *Vibrio cholerae* depends on the geographical location and the time of isolation. Moreover, the cause of resistance to drugs by the isolates could be explained by various reasons including the presence of antibiotics released in the sewage system by the connected hospitals and humans by several pathways which get into surface waters (Chikwendu et al. 2014). Therefore, antibiotics residues that persist in surface waters due to their misuse and release in surface waters could be the basis of occurrence of antibiotics resistant organism, a serious public health.

4 Conclusion

Vibrio cholerae is an etiological agent of cholera, a deadly diarrhea disease was isolated from surface waters in Makurdi town, Benue State during non-outbreak period. These isolates were obtained from streams, ponds, and river. The isolates demonstrated a strong resistance to Nalidixic acid followed by to Augmentin, Streptomycin, Ceporex, and Amplicin.

Vibrio cholerae are present and can persist in surface waters and can be isolated even during non-outbreak periods. Surface waters could therefore serve as reservoir for *Vibrio cholerae*. Therefore, the people who usually use this water for drinking, cooking, washing, and swimming purposes are at high risk of infection with *Vibrio cholerae*. Hygienic practice which can prevent water from being contaminated should be observed example, boiling. Antibiotic susceptibility test for infected persons and the use of appropriate antibiotics for treatment to curtail the shedding of the organism which keeps the spread of the disease in check and to treat severe illness (by reducing volume of diarrhea) and shorten the extent of sickness and hospitalization is recommended.

References

Abulude FO, Obidiran GO and Orungbemi S. Determination of physicochemical parameter and trace metal conducts of drinking water samples in Akure Nigeria. Elect J. of Environ. Agric. and Food Chem. 2007. 6(8). 2297–2303.

Acharya T. Tests for Bacterial Motility: Procedure and Results. 2015.

Adeleye IA, Daniels FV and Enyinnia VA. Characterization and Pathogenicity of *Vibrio spp.* contaminating seafood in Lagos, Nigeria. Inter. J. of Food Safety. 2010. 12: 1-9.

Akoachere, JFTK, Thomas NM, and Henry AN. Multi-drug resistant toxigenic *Vibrio cholerae* O1 is persistent in water sources in New Bell-Douala, Cameroon. BMC Infect Dis.2013; 13: 366. doi: https://doi.org/10.1186/1471-2334-13-366.

Anhwange BA, Agbaji EB, and Gimba EC. Impact assessment of human activities and seasonal variation on river Benue, within Makurdi Metropolis. Inter. J. of Sci. and Tech. 2012. 2(5). 248–254.

Barumann P, Furniss AL and Lee JI. Genus I, *Vibrios*. 8th ed, 35. 2001. 185-282.

C.L.S.I. Performance standards for antimicrobial susceptibility testing. 2007.

Caldini G, Neri A, Cresti S et al. High prevalence of *Vibrio cholera* non – O1 carrying heat stable enterotoxin- encoding genes among Vibrio isolates from a temperate climate river basin of Central Italy. J. Appl. Environ. Microbiol. 1997. 63: 2934 – 2939

Chigor VN, Umoh, VJ, Smith SI et al. Multidrug resistance and plasmid patterns of *Escherichia coli* O157 and other *E. coli* isolated from diarrhoeal stools and surface waters from some Selected sources in Zaria, Nigeria. Inter. J. of Env. Res. and Pub. Health, 2010 7(10) 3831-3841.

Chikwendu CI, Ibe SN and Okpokwasili GC. Multiple Antimicrobial Resistance in *Vibrio* spp isolated from River and Aquaculture Water Sources in Imo State, Nigeria. Brit. Microbiol. Res. J.2014 4(5). 560-569.

Daniels NA and Shafaie A. A Review of pathogenic Vibrio infections for Clinicians. Infect. Med. 2000 17(10). 665-685.

Farque SM and Mekalanos JJ. Phage-bacterial infections in the evolution of toxigenic *Vibrio cholerae*. Virulence, 2012 3(7) 556–65.

Faruque SM, Asadulghani KMN, Ghosh AN et al. RSI Element of *Vibrio cholera* Can propagate horizontally as a filamentous phage exploiting the Morphogenesis genes of CTX phage. Infect and immun. 2002. 70(1). 163-170.

Faruque SM, Chowdhury N, Kamruzzaman M et al. Genetic diversity and virulence potential of environmental *Vibrio cholerae* population in a cholera-endemic area. Proc. Natl. Acad. Sci. 2004. 101. 2123–2128. doi: 10.1073/pnas.0308485100

Fawole MO and Oso BA. Characterization of Bacteria: Laboratory Manual of Microbiology. 4th Edn., Spectrum Book ltd., Ibadan, Nigeria. 2004. PP:24-33.

Fooladi IAA, Islamieh I.D, Doust HR et al. Design of a multiplex PCR method for detection of toxigenic pathogenic in *Vibrio cholerae*. Asian Pac. J. of Trop. Med. 2013. 115-118.

Freeman N, Boncy J, Colwell RR et al. Environmental surveillance for toxigenic *Vibrio cholerae* in surface waters of Haiti. Am J Trop Med Hyg. 2015. 92(1) 118-25. doi: https://doi.org/10.4269/ajtmh.13-0601.

Gemelle ME, and Schmidt S. Microbiological assessment of river water used for the irrigation of fresh produce in a sub-urban community in Sobantu, South Africa.Food Res. Inter.2012 47(2). 300-305.

Hacioglu N and Basaran D. Monthly variation of some Physiochemical and Microbiological Parameters in Biga, Canakkale, Turkey. Afr. J. of Biotech. 2009 8(9).1929-1937.

Igbinosa EO and Okoh AI. Impact of discharge wastewater effluents on the physicochemical qualities of a receiving watershed in a typical rural Community. Inter. J. of Environ. Sci. and Tech. 2009 6(12) 175-182.

Kahler AM, Haley BJ, Chen A, Mull BJ, Tarr CL, Turnsek M, Katz LS, Humphrys MS, Derado G, Freeman N, Boncy J, Colwell RR, Huq A, Hill VR. Environmental surveillance for toxigenic *Vibrio cholerae* in Surface waters of Hati. Am. J. of Trop. Med. and Hyg. 2015 92(1). 118-125,

Khan SA, Nawaz MS, Khan AA et al. Molecular characterization of multidrug-resistant *Enterococcus* spp. from poultry and dairy farms: Detection of virulence and vancomycin resistance gene markers by PCR. Mol. Cell Probes. 2005; 19:27–34. doi: https://doi.org/10.1016/j.mcp.2004.09.001.

Khan W, Saha D, Ahmed S et al. Efficacy of Ciprofloxacin for treatment of cholera associated with diminished susceptibility to ciprofloxacin to *Vibrio cholerae* O1. PLoS ONE. 2015. 10:e0134921. https://doi.org/10.1371/journal.pone.0134921

Mandal J, Dinoop KP and Parija SC. Increasing antimicrobial resistance of *Vibrio cholerae* O1 biotype El Tor strains isolated in a tertiary-care Centre Medical Association. J Health Pop. Nutr. ***2012*** 49 (179) 232-6.

Mishra A, Taneja N and Sharma M. Environmental and Epidemiological Surveillance of Vibrio cholerae in a cholera-endemic region in India with Freshwater environs. J. of Appl. Microbiol. 2012. 112(1). 225-237

Mohale NG. Evaluation of the adequacy and efficiency of sewage treatment works in Eastern Cape. MSc. Thesis: Rhodes University, South Africa; 2003.

Mustapha A, Isa T, Bello HS et al. Resistance Profiles of Bacteria Isolated from Wastewater in the University of Maiduguri Teaching Hospital. J. of Biotech. Res. 2016 .2(7): 49-54,

Nhung PH, Ohkusu K, Miyasaka J et al. Rapid and Specific identification of 5 human pathogenic *Vibrio* species by multiplex Polymerase chain reaction targeted to DNA gene. Diag. Microbiol. and Infect. Dis. 2007. 59. 271-275.

Obi CL, Green E, Bessong PO, Devilliers B, et al. Gene encoding virulence markers among *Escherichia* coli isolate from diarrheic stool samples and river sources in rural Venda Communities of South Africa. Water SA, 2004. AJOL. 30 (1) 36–41.

Okoh AI, Odjadjare EE, Igbinosa EO et al. Wastewater Treatment Plants as a source of microbial pathogens in receiving watersheds. 2007. AJOL. 2007 6(25) 2932–2944.

Pant PR. Tailored media for the detection of E. coli and coliforms in the water samples. Journal of Tribhuvan University, 2004 24(1), 49–54.

Prüss A, Kay D, Fewtrell L et al. Estimating the burden of disease from water, sanitation, and hygiene at a global level. Environ. Health Persp. 2002 110 (5). 537-542.

Shrestha SD, Mellas ABR, Shakya G et al. Antibiotic Susceptibility patterns of *Vibrio cholerae* isolates. J. of the Nepal Med. Assoc. 2010. 49 (179): 232-6.

Shuval, HI. Wastewater irrigation in developing countries: Health Effects and Technical Solutions: a summary of World Bank Technical Paper Nost. Washington DC: USA, 1990, pp. 1-19.

Tyubee BT. The influence of ENSO and North Atlantic Sea surface Anomaly (SSTA) on extreme rainfall events in Makurdi, Nig. J. of Meteorol. Clim. Sci. 2009. 7. 28-33.

Umeh EU, Juluku JU and Ichor, T. Microbial Contamination of 'Naira' (Nigerian currency) Notes in circulation. Res. J. of Environ. Sci. 2007 1(6) 336-339.

Uphadaya A, Srivastava R and Skukla S. Cholera causing microorganism in different dynamic water bodies (a survey Study), J Adv Res Biosci, 2013 1(3).75-78.

World Health Organization (WHO). *Vibrio cholerae* in: WHO Guidelines for Drinking Water Quality, 2006.

World Health Organization and UNICEF (WHO/UNICEF). Progress on Sanitation and Drinking-Water; 2010 updated; WHO Press: Geneva, Switzerland. 2010. Pp. 1-60.

World Health Organization. EHEC outbreak: increased cases in Germany. Geneva, Switzerland. 2011. Available at http://www.euro.who.int/en/what-we-do/health-topics/emergencies/international-health-regulations/news/news/2011/06/ehec-outbreak-increasein-cases-in-germany.

Part III
Fuel Technology and Drilling Operations

Necessity for Nanofluids in Refrigeration Systems: An Overview

Mfon Udo, Sarah O. Akinkunmi, Sunday A. Afolalu, Ayodeji A. Noiki, Olabisi O. Yusuf, Moses E. Emetere, and Olusegun D. Samuel

1 Introduction

In recent times, dispersion of solid materials nanoparticles has become common phenomena to enhance thermal characteristics of working fluid in thermodynamics systems. Metal and metal oxides are renowned examples of nanoparticles. Thermal conductivity of nanofluid (a mixture of nanoparticles and the working fluid) is higher than the working fluid (Dincer and Kanoglu 2010). The first history of refrigeration can be traced back to the 1800s when ice from the polar region were cut into chunks and transported manually by a team over a long distance to be sold out to others who would keep the ice in their food 'storage rooms' with heat-insulated walls using sawdust. Scientists later detected the addition of certain salts to lower the temperature of the water. It would happen later on in 1755 when William Cullen attempted to produce ice mechanically by means of an air pump to lower the pressure of the liquid (water) in a closed apparatus. Many other investigations and experiments were carried out until Jacob Perkins in 1834 crafted the first vapour compression system using natural rubber as 'refrigerant' (Sharma 2018). Refrigeration can be described as a means of transferring heat from a

M. Udo (✉) · S. O. Akinkunmi · S. A. Afolalu · A. A. Noiki
Department of Mechanical Engineering, Covenant University, Ota, Nigeria
e-mail: mfon.udo@covenantuniversity.edu.ng

O. O. Yusuf
Department of Microbiology, Obafemi Awolowo University, Ile-Ife, Nigeria

M. E. Emetere
Department of Physics, Covenant University, Ota, Nigeria

O. D. Samuel
Department of Mechanical Engineering, Federal University of Petroleum Resources Effurun, Effurun, Delta –State, Nigeria

© The Author(s), under exclusive license to Springer Nature Switzerland AG 2022

A. O. Ayeni et al. (eds.), *Bioenergy and Biochemical Processing Technologies*, Green Energy and Technology, https://doi.org/10.1007/978-3-030-96721-5_30

low-temperature reservoir to a high-temperature reservoir. Sharma et al. (2018) identifies refrigeration as the process of cooling products and substances from a higher temperature within an enclosed space to a low temperature. Refrigeration is a high-demand process at both the industry level and the domestic level. The most common areas of application of refrigeration include cooling of food, medicine, artificial ice rinks, and air conditioning (Afolalu et al. 2018a). Most refrigeration systems are classified as vapour absorption systems. These systems include a condenser, evaporator, expansion or capillary tube and a compressor. Each is very vital to the running of the system. There are measures by which we ascertain the best performance of a refrigeration system and they include the Coefficient of Performance (COP), Energy consumption and Refrigeration capacity. The Coefficient of Performance is the measure of the effectiveness of the system and it depends on the other two mentioned above. It is mathematically expressed as the energy consumed divided by the refrigeration capacity (Taib et al. 2009).

Energy consumption has been found to be related to the economic growth of a country (Eggoh et al. 2011). Reducing the rate of energy consumption in a country starts from the smallest unit – the home (Fontana et al. 2013). Researchers have dedicated time to improve the efficiency of the home appliances while reducing energy consumption from the refrigerators, television, home stereos to home automation (as in the Amazon Echo Dot). The refrigerator is one of the highest consumers of electricity as it is constantly on and only goes off with a power cut. The various ways of reducing electricity consumption in refrigerators can be by ameliorating the material for the door and body using polyfoam or fiberglass, enhancing the heat transfer of condensers and evaporators by introducing ventilators or increasing the surface area and increasing the efficiency of the compressor which is the aim of this study. Improving the compressor's efficiency starts from friction reduction, which can be alleviated using a suitable lubricant (Ajayi et al. 2019a).

2 Types of Refrigeration System

2.1 Vapour Refrigeration System

Vapour compression refrigeration is classified under refrigeration using mechanical energy. The system is rated among the most important cycles because it has a wide range of heating and cooling temperatures as well as pressure (Sonnenrein et al. 2014). The temperature and pressure of the refrigerant vapour is amplified using a compressor. The refrigerant fluid is the working fluid for expansion and compression. A vapour compression refrigerator typically consists of an evaporator, a compressor, a condenser and a capillary or expansion valve. It sometimes contains auxiliary components (Prasad et al. 2009). The cycle works by taking in low atmospheric pressure vapour from the surroundings or from the closed chamber during the stroke known as suction stroke. The compressor then compresses the refrigerant to a high temperature, high pressure working fluid during the compression stroke, and then it is passed to the condenser which reduces the pressure and temperature and passes the heat to the coolant through the expansion valve. The expansion valve passes on the low temperature vapour into the evaporator at the storage chamber which absorbs the dry vapour and uses it to cool the objects around. The dry vapour is then recycled. Figure 1 shows a simple vapour compression cycle (Udayakumar et al. 2012).

2.2 *Vapour Absorption Refrigeration*

The vapour absorption system uses heat energy to run the system. Instead of the compressor, a pump is utilised here. The system consists of a generator, a condenser, an evaporator, a pump and an absorber (Maina and Huan 2015). The cycle works by passing ammonia solution or lithium bromide from the evaporator to the absorber which contains water. The mixture of the fluids results in a strong solution at high temperature. The heat is then reduced by passing cooled water through the pipe. The pump then pumps the solution to the generator where it is reheated. The reheating results in the evaporation of the ammonia from the solution; the ammonia later condenses back to liquid ammonia which is passed to the receiver (Secretariat 2000). Then, it is passed on to the expansion valve which expands the liquid at low pressure and temperature and it re-enters the evaporator in the closed chambers used to cool the surrounding object shown in Fig. 2 below (Lorentzen 1993).

Other types of refrigeration include evaporative cooling, thermoelectric refrigeration, cold air refrigeration; the first type of refrigeration known to mankind, and based on the usage; primary and secondary types of refrigeration.

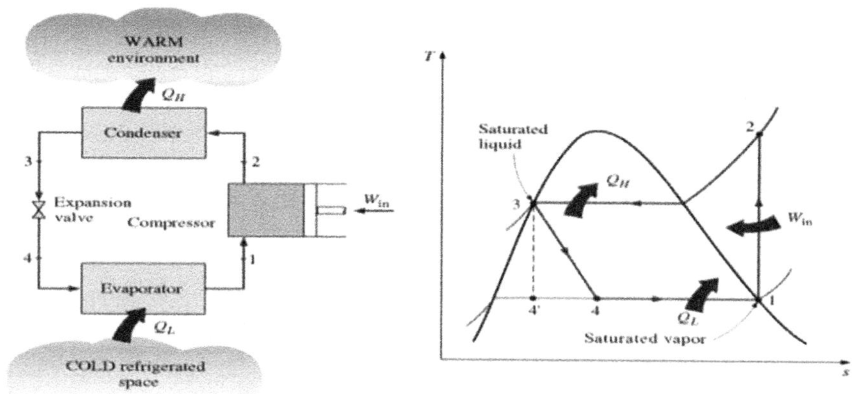

Fig. 1 Vapour compression cycle (Sonnenrein et al. 2014)

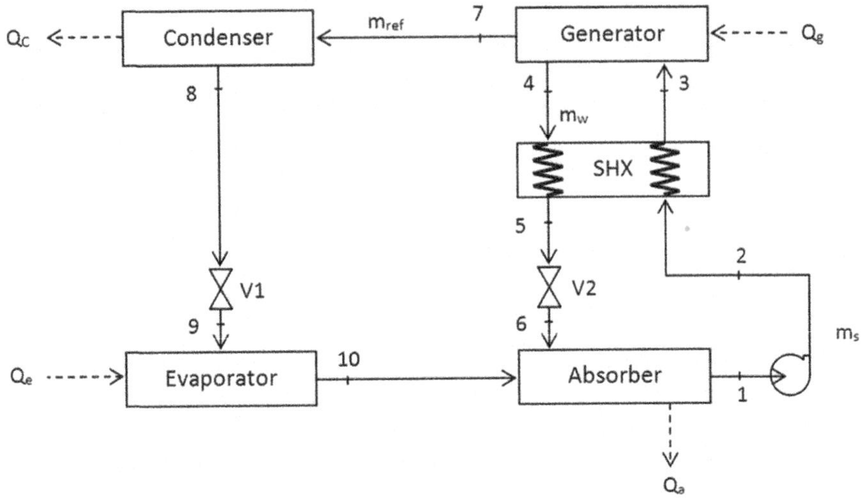

Fig. 2 Vapour absorption cycle (Lorentzen 1993)

3 Refrigerants and Its Roles

A refrigerant is a fluid substance basically in the liquid or gaseous state which brings out the refrigeration effect. These substances absorb heat at one place at a low temperature level and reject the same at some other place having higher temperatures and pressure. The common refrigerants are hydrocarbons and hydrofluorocarbons. The improvements made on refrigerants and the compressor likewise over the years have contributed to the wider use of refrigeration systems along with the development of improved systems and components of Refrigeration and Air conditioning systems. A refrigerant is a fluid that helps keep the refrigeration system in working condition. Using this, the first refrigerant would have been water as in the natural refrigeration system (Ajayi et al. 2019b). The first refrigerant used in a closed system was the product of the destructive distillation of rubber, ethyl ether which is highly dangerous and toxic, but this was the refrigerant used for more than a decade before other natural substances such as carbon dioxide, ammonia, and air later came into use. Out of these three natural substances, carbon dioxide (R744) was propounded to be a good refrigerant by Thaddeus S. Lowe in the late 1860s dwelling on toxicity and flammability factors, a while after he used it in balloons for the military. Different scientists also tested out carbon dioxide until Windhausen made a compressor just for the gas. Using his designs, other scientists built marine refrigerators and in the 1900s, the R744 was being used for air conditioning. The problems associated with it such as losses due to efficiency, sealing at high pressures and the ridiculous high components cost led to a search for better refrigerants that brought it our biggest problem today (Abioye et al. 2019).

Chlorofluorocarbons were introduced at the beginning of the 1930s. These substances had a vast application; from refrigerants in R and A systems to cleaning solvents and spray propellants (Navarro-Esbrí et al. 2013). The hoax that favoured the use of the substances was that they were eco-friendly. However, scientists later uncovered that the claim was false and that chlorofluorocarbons have caused suffocations as well as being a large contributor to destroying the ozone layer and causing the greenhouse effect. Out of the green gases, Chlorofluorocarbons (CFCs) causes about 20% of the climate change (Dalkilic and Wongwises 2010; (Padilla et al. 2010; Duan et al. 2015). In 1987, the Montreal Protocol was established banning any use of CFCs. They finally started phasing out. Unfortunately, some old systems using these substances as refrigerants have not been replaced. The conditions laid out for the new refrigerants are that they are eco-friendly, not reactive with the other materials in the system, do not cause corrosion, are not toxic and non-flammable and so on (Sethi et al. 2016).

The search for alternative refrigerants, therefore, was needed. Refrigerants with low Global Warming Potential (GWP) are being researched for. The change also requires a redesign of the domestic refrigeration systems to improve the efficiencies, for materials that would not easily corrode and leak the refrigerant gases and for components that can handle the new refrigerants being developed. Lorentzen et al. (1993) suggested the natural substances used before such as ammonia and carbon dioxide to replace CFCs. One of the alternative substances, R-134a or tetrafluoroethane which is non-toxic and non-flammable is projected to completely replace R-22 (chlorodifluoromethane) as a refrigerant in packaged air-conditioning units. However, the new system utilising R-134a will be expensive and larger to accommodate the larger tubing needed for the refrigerant to be effective. It also replaces R-12 (dichlorodifluoromethane) and R-400 in some system (Boyde et al. 2002). Later on, it was discovered to possess higher GWP than the other alternative refrigerants (GWP- 1430) and so other alternatives were sought to replace it. One of them is the R513A with a GWP of less than 600, and Navarro-Esbrí et al. (2013) investigated the refrigerant in a test bench and concluded that the replacement has a greater cooling capacity and mass flow rate. Although it consumed more in terms of power, the refrigerant has a higher coefficient of performance. The refrigeration system, however must be adjusted to accommodate the new refrigerant (Rizvi 2009). Another alternative refrigerant is R-123 in R and A systems, heat pump systems and centrifugal compressors to replace R-11. However, it has been reported to be toxic but not as much as R-11. It is required that the R and A workers have an exposure limit. Paraffin oils have been found to be compatible with the refrigerant. Dalkilic and Wongwises (2010) analysed mixtures of R290 and R600a with a proportion of 40 and 60 by percentage weight and found it suitable to replace R12. They also analysed R290 and R1270 as a blend and found it very suitable to replace R22. Padilla et al. (2010) proposed the use of R413a to replace R12 in a refrigeration system designed for the latter without any need to adjust or change the components. Zeotropic substances such as R-401A and R-401B have been accepted as alternatives in some applications. The main advantage of these substances is that they do not contain chlorine which is a major element in greenhouse gases that contribute to the depletion of the ozone layer (Karaosmanoğlu et al. 2013). The best alternative refrigerants proposed have been the natural substances (especially

Carbon dioxide) and the new synthetic substances; hydrofluorooleifins. Hydrofluorooleifins contain carbon, hydrogen and fluorine (Prasad et al. 2009). The two most commonly used; R1234yf and R1234ze (E) have GWPs close to the new refrigerants are reported to have low critical temperatures and pressures and very low flammability (Padilla et al. 2010). Lubricating oils are utilised in vapour compression refrigeration systems to serve as lubricants for the metal components, as sealants, as coolants, to dampen the vibrations of the compressors, reduce noise from the compressors and finally to also flush out unwanted particles (Adelekan et al. 2017; Bhattad et al. 2017). After the injection of unused lubricating oils into the refrigerating system, chemical reactions such as corrosion and the mixture with dirt and unwanted particles cause the oils to degrade which result in the reduction of the oil properties; the physical index such as viscosity and the chemical index. Thermal reactions also contribute in the degradation (Veera and Govindha 2018).

3.1 Nanofluid in Refrigeration

Bhattad et al. (2017) reviewed the importance of nanofluid in improving the coefficient of performance of refrigerators. It was stated that to improve the performance of refrigerators, two factors that need to change are system modification and redesign and/ or improving the heat transfer properties of the working fluids. After a lot of research, nanofluids were discovered to have the capacity. These nanofluids based on their area of application in refrigeration were classified as; nanorefrigerant, that is, nanoparticles immersed in refrigerant, nanolubricants, which are nanoparticles immersed in compressor oil as the base fluid. Scientists have investigated and concluded that the introduction of nanoparticles into the lubricating oil will significantly improve the refrigeration cooling capacity, coefficient of performance, energy consumption, second-law efficiency and the compressor's stability (Majgaonkar 2016; Ajayi et al. 2018; Fernando et al. 2018; Ajayi et al. 2019b). The other classifications are secondary fluids, coolants and secondary fluids with nanoparticles. The authors noted that while these fluids exit, literature has only been published on the first two. Veera and Govindha (2018) carried out investigation on improving the coefficient of performance with nanofluid but restricted their study to examine the better base lubricant for nanolubricants to mix in. They concluded that mineral oil was better than polyolester oil because of its compatibility with CFC refrigerants. It was noted that more basic research and applied research have to be explored to gain more insight in the field. Ajayi et al. (2018) experimented using titanium oxide nanorefrigerant resulting in lower temperature of the evaporator. Ajayi et al. (2019c) used diamond nanolubricant for the compressor and it resulted in higher COP and cooling capacity. There are many more literature that show nanofluid improving the performance of refrigerators.

4 Conclusion

Nanoparticles present significant properties from which numerous applications can be developed. The volume and high surface ratio make them suitable for different purposes in various fields, including promising qualities to overcome emerging challenges in refrigeration systems. In this study, we have examined the comparative advantage of nano based refrigerants over the conventional refrigerants, also a comprehensive assessment of the substantial impact of nano based refrigerants in refrigeration system particularly in the area of system performance enhancement. However, there are several ongoing researches in the area of nano based refrigerants geared to towards exploring the vast opportunities presented by nanomaterials and nanotechnology.

Acknowledgments We acknowledge the financial support offered by Covenant University in the actualisation of this research work for publication.

References

A. A. Abioye, D. O. Rotimi, O. O. Fasanmi, O. P. Abioye, S. A. Afolalu, T. F. Owoeye & C. C. Obuekwe (2019, December). Synthesis and Characterization of Selected Starch Nanoparticles as Matrix Reinforcements for Low Density Polyethylene. In Journal of Physics: Conference Series (Vol. 1378, No. 4, p. 042070). IOP Publishing

D. S. Adelekan, O. S. Ohunakin, T. O., Odunfa, M. K., R. O. Leramo, S. O Oyedepo & D. C. Badejo (2017). Case Studies in Thermal Engineering Experimental performance of LPG refrigerant charges with varied concentration of TiO_2 nano-lubricants in a domestic refrigerator. Case Studies in Thermal Engineering, 55–61.

S. A. Afolalu, A. A. Abioye, M. O. Udo, O. R. Adetunji, O. M. Ikumapayi, & S. B. Adejuyigbe (2018a). Data showing the effects of temperature and time variances on Nano-additives treatment of mild steel during machining. Data in Brief. 19(2018) 456–461

O. O. Ajayi, T. I. Okolo, E. Y. Salawu, F. T. Owoeye, D. K. Akinlabu, E. T. Akinlabi & S. A. Afolalu (2019a). Performance and Energy Consumption Analyses of R290/Bio-Based Nano lubricant as a Replacement for R22 Refrigerant in Air-Conditioning System. In Energy Technology, pp. 103–112. Springer, Cham.

O. O. Ajayi, C. C. Aba-Onukaogu, E. Y. Salawu, F. T. Owoeye, D. K. Akinlabu, S. A. Afolalu, & A. A. Abioye (2019b). Effect of Biomaterial (Citrullus Lanatus Peels) Nanolubricant on the Thermal Performance and Energy Consumption of R600a in Refrigeration System. In Energy Technology 2019 (pp. 91–102). Springer, Cham

O. O. Ajayi, O. F. Omowa, O. A. Omotosho, O. P. Abioye, E. T. Akinlabi & S. A. Afolalu (2018). Experimental Investigation of the Effect of ZnO-Citrus sinensis Nano-additive on the Electrokinetic Deposition of Zinc on Mild Steel in Acid Chloride. In TMS Annual Meeting & Exhibition (pp. 35–40). Springer, Cham

O. O. Ajayi, D. E. Ukasoanya, M. Ogbonnaya, E. Y. Salawu, I. P. Okokpujie, S. A. Akinlabi & F. T. Owoeye (2019c). Investigation of the Effect of R134a/Al2O3 –Nanofluid on the Performance of a Domestic Vapour Compression Refrigeration System. Procedia Manufacturing, 35, 112–117.

O. A., Akinlabi, E. T., Akinlabi, S. A., & Afolalu, S. A. (2018, March). Finite Element Modelling of Electrokinetic Deposition of Zinc on Mild Steel with ZnO-Citrus sinensis as Nano-Additive. In TMS Annual Meeting & Exhibition (pp. 199–211). Springer

A. Bhattad, J. Sarkar, & P. Ghosh (2017). Improving the performance of refrigeration systems by using nano fluids : A comprehensive review. Renewable and Sustainable Energy Reviews, 82,3656–3669

S. Boyde,, Randles, S. J., & Thompson, R. I. G. (2002). Reducing vibration with advanced lubricants. Industrial Lubrication and Tribology, 54(5), 209–214.

A.S. Dalkilic, S. Wongwises (2010). A performance comparison of vapour-compres- sion refrigeration system using various alternative refrigerants. Int. Commun. Heat Mass Transf. 37, 1340–1349

I. Dincer, & M. Kanoglu (2010). Refrigeration Systems and Applications (Second). Wiley and Sons Ltd.

H. Duan, D. Wang, & Y. Li (2015). Green Chemistry for Nanoparticle Synthesis. Chemical Society Reviews (February).

J. C. Eggoh, C. Bangake,, & C. Rault (2011). Energy consumption and economic growth revisited in African countries. Energy Policy, 39(11), 7408–7421.

D. Fernando, M. Pico, L. Ribeiro,, P. Smith, E. Pedone & B. Filho (2018). Performance evaluation of diamond nanolubricants applied to a refrigeration system Évaluation de la performance de nanolubrifiants à base de diamants appliqués à un système frigorifique. International Journal of Refrigeration, 100, 104–112.

L Fontana, V. Atella and D. M. Kammen (2013). Energy efficiency as a unifying principle for human, environmental, and global health. F1000Research 2013, 2(101).

F. Karaosmanoğlu,, S. Yüzer, H. Kerpiççi & E. Durak (2013), "The investigation of relationship between lubricating oil and refrigerant in refrigerator compressors", Industrial Lubrication and Tribology, Vol. 65 No. 6, pp. 456–465. https://doi.org/10.1108/ILT-02-2011-0017

G. Lorentzen (1993). Application of Natural Refrigerants, Proceedings of the Meeting of I.I.R. Commission B1/2, May 12–14, Ghent, Belgium. 55–64

P. Maina & Z. Huan (2015). A review of carbon dioxide as a refrigerant in refrigeration technology. South African Journal of Science, 111(9–10), 1–10.

A. S. Majgaonkar (2016). Use of Nanoparticles In Refrigeration Systems : A Literature Review Paper * Corresponding Author. International Compressor Engineering, Refrigeration and Air Conditioning, and High Performance Buildings Conferences, 1–10.

J. Navarro-Esbrí, J.M. Mendoza-Miranda, A. Mota-Babiloni, A. Barragan-Cervera, J.M. Belman-Flores (2013), Experimental analysis of R1234yf as a drop-in replacement for R134a in a vapor compression system, IJR, 36, 870–880.

M. Padilla, R. Revellin & J. Bonjour (2010). Exergy analysis of R413A as replacement of R12 in a domestic refrigeration system. Energy Conversion and Management. 51. 2195–2201Ajayi, O. O., Omowa, O. F., Abioye, O. P., Omotosho.

H. T. Prasad, P. K. Reddy & R. R. D. Reddy (2009). Exergy Analysis of Vapour Compression Refrigeration. International Journal of Applied Engineering Research, 4(12), 2505–2526.

S.Q.A. Rizvi (2009). A Comprehensive Review of Lubricant Chemistry, Technology, Selection, and Design; ASTM Stock Number: MNL59; ASTM International: West Conshohocken, MD, USA, 2009.

O. Secretariat (2000). The Montreal protocol on substances that deplete the ozone layer. United Nations Environment Programme, Nairobi, Kenya.

A. Sethi, E.V. Bercerra, S.F.Y. Motta, GWP Low (2016). R134a replacements for small refrigeration (plug-in) applications, IJR., 66, 64–72.

A. Sharma (2018). A Review on Various Methods of Transformer Protection (IJTSRD) 1108–1112.

G. Sonnenrein, A. Elsner, E. Baumhögger, A. Morbach, K. Fieback, J. Vrabec (2014). Reducing the power consumption of household refrigerators through the integration of latent heat storage elements in wire-and-tube condensers. International Journal of Refrigeration.

M.Y. Taib, M.S.M Sani, M. M. Noor & M Kadirgama (2009). Performance Analysis of a Domestic Refrigerator in Malaysia using Experimental Method.

V. Udayakumar, M. Shrestha & S. Suresh (2012). Thermodynamic Analysis of R134a-Dmac Vapor Absorption Refrigeration (Var) THERMODYNAMIC ANALYSIS OF R134A – DMAC VAPOR ABSORPTION REFRIGERATION (VAR) SYSTEM

R. K. Veera and R. N. Govindha (2018). Review on Applications of NanoFluids used in Vapour Compression Refrigeration System for Cop Enhancement IOP Conf. Ser.: Mater. Sci. Eng. 330 012112

Synthetic Heat Transfer Fluids: Alternative to Steam in Chemical Industries – A Review

A. Ayoola, S. Ogunlade, D. Vershima, O. Olomukoro, and N. Sonia

1 Introduction

A liquid or gaseous material required to transfer heat from one portion to another is a heat transfer fluid (Conrad and Gamble 2017). Heat transfer fluids transmit heat to a process stream. Heat transfer fluids transfer heat from a heat source to other heat demands (or cold streams) by circulation within a closed-loop system (Devi et al. 2020). The medium for transferring heat is circulated within a closed-loop system to supply heat to the end-user (Mukund and Santanu 2012). These fluids are applied in heating or cooling processes to obtain or sustain a specific temperature. They possess vital thermophysical properties such as high thermal conductivity, low viscosity, high heat capacity, low corrosivity, low cost, low flammability, low toxicity, and environment friendly (Hoffschmidt et al. 2012).

Heat transfer fluids are widely used in many conventional chemical and biochemical processes. In the brewing process, heat transfer fluids act as refrigerants for cooling during fermentation (Mehos et al. 2020). In the pharmaceutical manufacturing process, heat transfer fluids provide a high-temperature range for necessary chemical reactions and a low-temperature requirement for unit operations such as crystallization of final products (Mehos et al. 2020). The removal of impurities such as hydrogen sulphide (H_2S), water (H_2O), volatile organic compounds (VOCs) and carbon dioxide (CO_2) from gas in gas processing operations require heat transfer fluids to maintain temperatures that are suitable for the removal of these impurities. In the petrochemical and packaging industries, polymerization of polyesters, PET moulding and nylon processing require heat transfer fluids (Hoffschmidt et al. 2012).

A. Ayoola (✉) · S. Ogunlade · D. Vershima · O. Olomukoro · N. Sonia
Department of Chemical Engineering, Covenant University, Ota, Ogun State, Nigeria
e-mail: ayodeji.ayoola@covenantuniversity.edu.ng

© The Author(s), under exclusive license to Springer Nature Switzerland AG 2022

A. O. Ayeni et al. (eds.), *Bioenergy and Biochemical Processing Technologies*, Green Energy and Technology, https://doi.org/10.1007/978-3-030-96721-5_31

In chemical processing plants, steam, usually in the saturated form or of high quality, has been used for years as the primary heat source for process heating in heat transfer equipment such as heat exchangers, re-boilers and reactors. It is also used to clean equipment and generate electricity (Sinaga 2020a). Steam is used as a heat source because it comes from raw water, which is cheaper, less hazardous, and safe in its gaseous form to store and transfer energy (Sinaga 2020a).

In addition, the steam itself is beneficial as a process heating agent in chemical plants because the desired temperature can be obtained by pressure control. It is an efficient heat source with high throughput per unit mass of utility at a constant temperature because of its high heat of condensation, high heat transfer coefficient, non-flammability and non-toxicity. It is also an inert heating agent to many process fluids (Sinaga 2020b).

The main objective of this review is to identify alternative heat transfer fluids with more emphasis on synthetic heat transfer fluids, their properties, benefits, applications and limitations in chemical industries.

2 Literatures Gathering Approach

For this review, the information gathered was from databases that contributed to the topic. The given set of guidelines, procedures, inclusion and exclusion criteria, the literature of the studies and the chemistries of the substances to be studied were taken into full consideration. Articles were retrieved from databases such as *Scopus*, *Google Scholar*, *ResearchGate* and *ScienceDirect*. The keywords for the search include 'steam', 'heat transfer fluids', 'advantages and disadvantages of using steam as heat transfer fluid', 'alternative materials to use of steam as heat transfer fluids', 'chemistries of heat transfer fluids', 'mineral oils as heat transfer fluids', 'synthetic heat transfer fluids' and 'synthetic aromatic heat transfer fluids'. In addition, articles relevant to the study were obtained from reliable websites and resources that improved the overall structure of the review. Furthermore, the articles, journals and sample review works were obtained from 2011 to 2021, according to the ten-year limit stated in the inclusion and exclusion criteria.

Furthermore, the most recent updates on the chosen articles were utilised. The companies that work with these heat transfer fluids were consulted, and the chemistries of the various synthetic aromatic substances were taken into account and documented. Articles that discussed the effects of oxidation and thermal degradation of substances were gathered via the use of google search engine, thus enabling a more comprehensive outlook on the overall study.

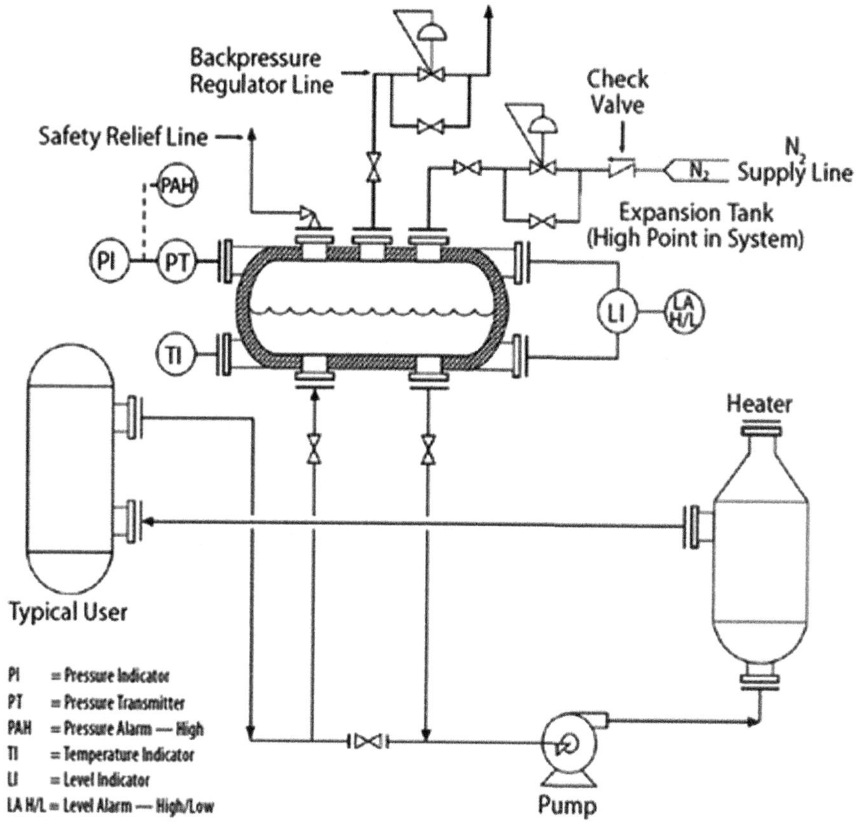

Fig. 1 Heat transfer fluid system and its main components (Lee 2016)

2.1 *Heat Transfer Fluid System*

A heat transfer fluid system is typically made up of the following components (Fig. 1):

- Heat transfer fluid
- Storage tank, which are used to store the fluid and fill the circuit
- Heater, which is the heat source such as furnace for heating and chiller for cooling
- The piping system, which include valves and pumps that transport fluid across various components such as user equipment
- User equipment, which are the units that require heat
- Filters, which help to remove particulates from the fluid
- Sampling line, which allows safe sampling of fluid for analysis and inspection
- Expansion tank**,** which is primarily added to enable expansion of fluid during heating and contraction when during cooling. It is also used to pressurise the system and vent off unwanted volatile organic compounds, gases and water

Heat transfer fluids act as supercritical fluids and experience phase change. Examples of heat transfer fluids are water, hydrocarbon oils, glycols, silicone oils, air and refrigerants. Heat transfer fluids according to their operating temperatures are classified as follows:

Low Temperature Fluids

They include silicone oils and hydrocarbons with a temperature range from below 0 °C to 100 °C.

Medium Temperature Fluids

They include glycols, such as ethylene and propylene glycol, can be used at temperatures below 0 °C (freeze protection) and 177 °C (with the presence of an additive).

High Temperature Fluids

These are heat transfer fluids been used at a temperature range of between 177 °C to 400 °C.

2.2 *Steam As a Heat Transfer Fluid: Its Operation and Limitations*

Steam is a widely used medium for the transfer of heat in many industrial systems. Steam is sent to the heat exchanger in the form of gas. The saturated steam transfers heat using latent heat, liberates much energy and then condenses to liquid. The condensed fluid exits the heat exchanger at a lower temperature. Figure 2 shows a

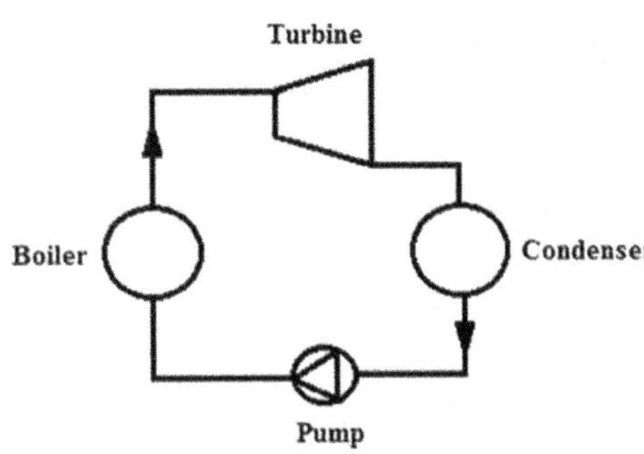

Fig. 2 Steam generation process (Ebrahimi and Torshizi 2012)

simple typical steam generation process which comprises of pumping unit, boiler, turbine and condenser.

In Fig. 2, the feed water at elevated pressure is first pre-heated before entering a boiler or series of boilers to superheat the steam to pass its dew point to produce high-pressure steam at correspondingly high temperatures. High-pressure steam can be used in plants for various operations. Part of it can also be converted using a series of valves and steam turbines to heat medium temperature steam, which can be used as a heat source for some process units and produce low-pressure steam to heat low-pressure processes. The low-pressure steam can be expanded inside the condensing turbines to produce shaft work and energy (Balamurugan and Samsoloman 2014). Despite its many advantages, steam also has its various disadvantages when operated within the processing plants.

Steam systems suffer from metal corrosion issues arising from the combined effects of corrosive agents such as air, hot water, salts and other reactive contaminants. The steam system promotes scale and deposits formation due to the minerals present in raw water supplies. The abrasion of steam also causes corrosion due to erosion. The steam system produces steam at a high-pressure level, which increases exponentially as the steam temperature increases. The high pressure and temperature pose safety hazards and can cause a severe burn, injury, hearing damage (due to prolonged exposure). High-pressure steam can lead to fatalities to plant and maintenance personnel due to invisibility to eyes and as a result of a pinhole leak in distribution piping system and catastrophic steam pipe ruptures due to erosion of pipe walls by steam condensate, corroded pipe walls due to acidic contaminants, overpressure effect of water hammer on of pipe. The steam system needs continuous maintenance of equipment such as steam traps, valves, condensate return pumps, expansion joints and water analysis and treatment, which are prone to failure and peculiar to the steam system only. The system also suffers from freezing, pipe bursting, and components damage in a cold climate. Steam systems require significant human effort for operation, supervision and maintenance. In some geographies, the local law mandated full-time licensed personnel to man the operation of high-pressure steam systems (Tanuma 2017).

Another challenge of steam is the environmental safety concern. Steam is produced from water that must be chemically treated to remove unwanted cations and anions that cause scale and corrode steam. The chemical treatment wastes present considerable environmental hazards since they cannot be discharged into sewers. Also, there is a need to discharge spent water into the environment, which is often regulated by law, and this means a provision is required for cooling if water is to be drained into sewers. Another setback for the steam system is the dependence on pressure regulation to control its temperature, which is not accurate and typically limited to ±10 °F error margin. As the system ages and corrosion takes its toll, temperature control practically becomes a severe issue (Tanuma 2017).

2.3 *Alternative Heat Transfer Fluids*

Heat transfer fluids are materials with specific heating, cooling and transport properties that are chemically stable at higher operating temperatures. They have their applications in various industrial processes such as oil and gas, pharmaceuticals, chemical manufacturing, biodiesel production, food and brewing processes. They are also used in concentrated solar power.

Unlike steam systems, alternative heat transfer fluid systems operate at low pressure. Pressure in some of the systems is limited to the pump discharge required for turbulent flow. Typical pump discharge pressures range from 35–65 psi and are a little higher for large systems. This low-pressure requirement makes the heat transfer system safer to handle even without a licensed supervisor that may be required in steam systems. Also, the low-pressure ratings positively impact system material selection and capital cost. The pressure rating favours wide industrial applications.

Unlike steam, alternative heat transfer fluids such as synthetic organic fluids and hot oils are easy to maintain. They do not have condensate return, blowdown, or water additives that make the steam system complex. Hot oil and synthetic fluid systems have been operated safely and efficiently for years with a minimal maintenance routine. Some of the alternative heat transfer fluids are enumerated below (Mehos et al. 2020).

Aliphatic Based Fluids

These are paraffinic or aliphatic based (hydrocarbon) fluids with an application temperature range between −10 to 300 °C. They are commonly called hot oils, and they belong to a group of chemically similar products such as mineral, white mineral, hydro-treated, paraffinic or petroleum oil. They have a high-temperature boiling point, which is a feature that permits them to be used in low pressured or non-pressured systems at a lower temperature range. At higher temperatures, they suffer a high rate of thermal degradation that aggravates operational issues and eventually costly fluid replacement (Balamurugan and Samsoloman 2014).

Hydro-Processed Organic Heat Transfer Fluids

These organic heat transfer fluids consist of the following types:

Hydro-Processed White Mineral Oil

It is food-grade mineral oil with a temperature range of − 30 to about 320 °C. It is thermally stable at a high temperature of 320 °C for long periods and possesses excellent heat transfer properties even at low temperatures (say around −45 °C). It is a colourless, odourless and non-toxic transparent liquid (Czaplicka et al. 2021).

Hydro-Processed Naphthenic Fluid

Hydro-processed Naphthenic heat transfer fluid is used in the liquid phase for cooling or heating at a temperature range of −5 to 320 °C. It is a light-yellow transparent liquid with good anti-coking characteristics (Tanuma 2017).

Inhibited Glycol-Based Fluids

The inhibited glycol-based is a mixture of water and glycols with operating temperatures below 175 °C (350 °F) whose properties and heat transfer performance are a function of the specific concentration of glycol in the fluid. They are usually in the form of either ethylene glycol or propylene glycol (Tanuma 2017). Ethylene glycol-based fluids have lower viscosity and better heat transfer properties, while propylene glycol-based products are less toxic. Propylene glycol-based fluids find applications in food processing industries. Glycols are used as heat transfer fluids because of their low corrosiveness, flammability and high pumping requirements. Inhibited glycol-based heat transfer fluid consists of water which provides heat transfer capacity. Ethylene or propylene glycol, in addition to heat capacity, helps to depress the freezing point of the fluid mixture. A corrosion inhibitor is added to the glycol-water mixture to protect the common metals of the construction of the system. The most important physical properties of ethylene and propylene glycol desired for their applications as heat transfer fluid are their relatively low vapour pressure, relative high boiling points and ability to lower the freezing point of water. Pure ethylene glycol has a freezing point of about 13 °C (9 °F), while pure propylene glycol freezes at about 20 °C (−4 °F) (Czaplicka et al. 2021).

Inhibited glycol-based heat transfer fluids are considered medium temperature thermal fluids. They are commonly used in the food industry for chilling and freezing in specific applications such as immersion freezing, cooling of liquid foods and fermentation cooling. They are also used in single fluid process heating and cooling and closed-loop, water-based heating, ventilation and air condition applications (Tanuma 2017).

Silicon-Based Heat Transfer Fluids

Silicon-based heat transfer fluids have good thermal stability. They are preferred for applications such as pharmaceutical processes that require heat transfer either at a low temperature of less than 29 °C or a high temperature of greater than 343 °C (Balamurugan and Samsoloman 2014).

2.4 *Synthetic Heat Transfer Fluids*

Synthetic heat transfer fluids are used to transfer heat at temperatures between −57 °C (70 °F) and 400 °C (750 °F). They have been established to have superior thermal stability over conventional hot oils, which means they will stay long in the system; less make-up volume is required, and operating cost is minimal. They are used in simple operating systems, unlike steam system, which typically comprises storage tank, piping, pump, expansion tank, filters, sampling line, user and heater. They operate at low pressure, making usage a safer option for indirect heating in chemical industries. Some synthetic heat transfer fluids are discussed below (Czaplicka et al. 2021).

Synthetic Heat Transfer Diaryl Alkyl

It is a synthetic heat transfer fluid with high-temperature stability, operating at a maximum temperature of 350 °C. It has a broad operating range of temperature of between −20 to 350 °C. Diary Alkyl has properties that are resistant to coking. During the continuous heating and cooling process, their properties remain intact. Due to the absence of coke formation, they produce a clean heating system, thereby sustaining the performance of the equipment (Czaplicka et al. 2021).

Synthetic Heat Transfer Biphenyl/Diphenyl Oxide Mixture

It is a transparent synthetic heat transfer fluid with excellent heat stability. It is applied in concentrated solar power (CSP) generation. It can operate within a temperature range of between 12 to 400 °C for liquid-phase heating systems and between 257 to 400 °C for vapour phase liquid systems at lower vapour pressures. The saturated vapour pressure at 400 °C is about 1.06 MPa. At low temperatures, it possesses lower viscosity. The heat transfer fluid's corrosivity rate is relatively low because of the low chlorine and sulphur content (Tanuma 2017).

Synthetic Heat Transfer Isopropyl Biphenyl Mixture

It is an aromatic synthetic heat transfer fluid suitable for low-pressure applications. It is a colourless transparent liquid with good heat stability that can operate at high-temperature heating systems for long periods. They operate at a temperature range of – 20 to 330 °C. They can resist any form of chemical degradation in a material through the air (oxidation) at high temperatures and will not cause corrosion in the heating system (Tanuma 2017).

Synthetic Heat Transfer Alkyl Substituted Aromatic Fluid

It is an aromatic synthetic heat transfer fluid that operates within a temperature range of −75 to 315 °C and is applied in both liquid and gaseous phases. It is a light yellowish transparent liquid. At low temperatures, it has excellent heat transfer and flow properties. It has the best heat transfer coefficient at low temperatures relative to other refrigerating fluids (Balamurugan and Samsoloman 2014).

Synthetic Heat Transfer Alkylbenzene

It belongs to the alkyl-substituted aromatic fluid category. It is a light yellowish transparent liquid. This heat transfer fluid has high thermal stability and has a low tendency to undergo oxidation. It works at the operating temperature range of −25 to 300 °C without any carbon deposits. It is an economical synthetic heat transfer fluid. It can easily flow (low viscosity) at low temperatures and is more suitable in cold regions (Czaplicka et al. 2021).

The Synthetic Aromatic Hydrocarbon Mixture

It is an organic heat transfer fluid with excellent thermal stability at a temperature range between −10 to about 380 °C. It has good anti-coking properties with a low saturated vapour pressure of 0.62 MPa at its maximum temperature of 380 °C. It has a low tendency to resist flow at lower temperatures (Balamurugan and Samsoloman 2014).

2.5 *Benefits of Synthetic Heat Transfer Fluids*

Synthetic heat transfer fluids have some advantages over steam, mineral oils and other natural thermal fluids. Synthetic heat transfer fluids have been established as viable alternatives to water or steam — and the most preferred option at temperatures above 200 °C.

Synthetic fluids, such as Dowtherm, offer long-term benefits even as an economical replacement for mineral oils in applications with moderate temperature requirements above 260 °C (500 °F).

They provide a safer operation with less monitoring because they operate at lower pressures than steam as heat transfer mean. For instance, for an application temperature of 343 °C (650 °F), steam has a vapour pressure of 13,790 kPa (2000 psi) which is 200 times higher than that of a synthetic thermal fluid that is approximately 71–86 kPa (Conrad and Gamble 2017) (Table 1).

Table 1 The synthetic heat transfer fluid system versus steam system (Pirobloc 2019)

	Steam	Alternative heat transfer fluid
Pressure at 350 C (Psi)	2610	150
Thermal heater efficiency	Less efficient	30% more efficient
Flash/Blowdown/De-aerator losses (%)	6–14/3/2	Not applicable
Service life	Shorter	Longer
Capital cost	High	Less
Flexibility	Less	Flexible
Simplicity	Less	Simple
Safety concern (Explosion risk due to pressurisation)	High	No
System corrosion issues	High	No
Cost of operation	Higher	Lower
Fuel consumption	Higher	Lower
Dedicated workforce needed	Yes	Not required
Maintenance	Frequent	Less
Chemical regulation (for fluid)	Required	Not required

2.6 *Limitation of Synthetic Heat Transfer Fluids*

The major limitation of synthetic heat transfer fluids is their inability to operate beyond a particular temperature of below 400 °C. Most synthetic heat transfer fluids function optimally at temperatures below 400 °C (Balamurugan and Samsoloman 2014). Due to aromatic-based chemistries of most synthetic heat transfer fluids, there is a high tendency for some fluids to have hazardous degradation by-products, which will require special regulatory permits for handling, shipping and transportation (Conrad and Gamble 2017).

3 Conclusion

Synthetic organic fluids transfer heat at temperatures between −57 °C (70 °F), and 400 °C (750 °F) have been established to have superior thermal stability over conventional hot oils and steam. They are the most preferred option at temperatures above 200 °C (392 °F). They are aromatic derived chemistries. They have good chemical stability at the operating temperature range. They also have low-temperature pumping characteristics and improved heat transfer properties over conventional hot oil. Further, they are non-corrosive, the desired property contributing to system efficiency, extending equipment life, and improving operating economics.

Information provided by thermal fluid heater manufacturers has shown that efficiencies of heat transfer fluid systems, when adequately operated, can be as much as

5–8% higher than steam systems. Further, synthetic organic heat transfer fluids do not require water treatment and are fouled less due to the lower heat flux. Furthermore, unlike steam systems, synthetic heat transfer fluid systems operate at low pressure, with some at atmospheric pressure, which makes it a safer option than steam.

References

Balamurugan S, Samsoloman DP (2014). Investigation of shell and tube heat exchangers by using a design of experiment. Journal of Heat and Mass Transfer Research 1(2):59-66 https://doi.org/10.22075/JHMTR.2014.180

Conrad E, Gamble PE (2017). Process Heating. In: Organic heat transfer fluid oxidation and its prevention. https://www.process-heating.com/articles/92366-organic-heat-transfer-fluid-oxidation-and-its-prevention. Accessed 26 Aug 2021

Czaplicka, N, Grzegorska A, Wajs J, Sobczak J, Rogala A (2021). Promising nanoparticle-based heat transfer fluids- Environmental and techno-economic analysis compared to conventional fluids. International Journal of Molecular Sciences 22(17):1-37. https://doi.org/10.3390/ijms22179201

Devi NP, Rao CS, Kumar KK (2020). Thermal performance of nanofluids in heat transfer loops. IOP Conference Series: Materials Science and Engineering 981:1-8. https://doi.org/10.1088/1757-899X/981/4/042029

Ebrahimi, M, Torshizi, SEM (2012). Optimization of power generation from a set of low-temperature abandoned gas wells, using organic Rankine cycle. Journal of Renewable and Sustainable Energy 4(6):1-13. https://doi.org/10.1063/1.4768812

Hoffschmidt B, Alexopoulos S, Göttsche J, Sauerborn M, Kaufhold O (2012). High concentration solar collectors. Comprehensive Renewable Energy 3: 165–209. https://doi.org/10.1016/b978-0-08-087872-0.00306-1

Lee LW (2016). Process Heating. In: Expansion tank design considerations for synthetic organic heat transfer fluid systems. https://www.process-heating.com/articles/91582-expansion-tank-design-considerations-for-synthetic-organic-heat-transfer-fluid-systems. Accessed 28 Aug 2021

Mehos M, Price H, Cable R, Kearney D, Kelly B, Kolb G, Morse F (2020). Concentrating solar power best practices study. National Renewable Energy Lab, Colorado. https://doi.org/10.2172/1665767

Mukund B, Santanu B (2012). Targeting minimum heat transfer fluid flow for multiple heat demands. Computer-Aided Chemical Engineering 31:675–679. Paper presented at the 11th international symposium on process systems, Singapore, 12 July 2012. https://doi.org/10.1016/b978-0-444-59507-2.50127-X

Pirobloc (2019). Thermal fluid versus steam: A capex-opex comparison. Thermal engineering blog. https://www.pirobloc.com/en/blog-en/thermal-fluid-versus-steam-capex-opex-comparison/#eficiencia. Accessed 28 Aug 2021

Sinaga, N. (2020a). Heat transfer enhancement. Project presented at conference on heat transfer, Diponegoro University, Semarang, Indonesia, June 2020 https://doi.org/10.13140/RG.2.2.20471.04008

Sinaga N (2020b). Heat transfer enhancement of heat exchangers. Project presented at conference on heat transfer, Diponegoro University, Semarang, Indonesia, June 2020. https://doi.org/10.13140/RG.2.2.31336.90885

Tanuma T (2017). Advances in steam turbines for modern power plants. In: Zheng X (ed) Introduction of new sealing technologies for steam turbines, Woodhead Publishing, Sawston, p. 307–320. https://doi.org/10.1016/b978-0-08-100314-5.00014-2

Gas Condensate Reservoir Developmental Techniques

Yuven Thelma Nchila, Fred T. Ogunkunle, Josephs E. Rachael, Oluwasanmi A. Olabode, and Christian N. Dinga

1 Introduction

During production from a gas-condensate reservoir, condensates dropout of the gas and build up around the wellbore when the pressure of the reservoir drops below the dew point line. A phenomenon sometimes referred to as condensate banking. This often results in productivity loss of both the gas and condensates. The condensate banking effect become more severe especially in gas condensate reservoirs with very low permeability. Also, the production of associate formation water and possibility of entrapment occurring around the region of the wellbore can worsen the blockage phenomenon (Ayub & Ramadan, 2019).

Depending on the pressure behavior of a gas reservoirs, most of the rich gas reservoir performance can mainly be divided into three time periods: (i) when the overall reservoir pressure is higher than dew point; (ii) when just the pressure of the near well drops below the dew point pressure; and (iii) when overall reservoir pressure is lower than dew point (Ayub & Ramadan, 2019). No immediate treatment is needed for the first category because the gas can be produced as a single phase. Nevertheless, to maximize recovery, adopting some sort of pressure maintenance is preferable to be able to extend the length of time for the natural depletion process of the reservoir. This is because when condensate gas reservoirs are depleted only by natural depletion all or some of the following disadvantages could be apparent:

(i) Valuable loss of condensates into the reservoir formation
(ii) Well productivity decline which must be offset by installing compressors and/or drilling more wells

N. Y. Thelma (✉) · F. T. Ogunkunle · J. E. Rachael · O. A. Olabode · D. N. Christian
Department of Petroleum Engineering, Covenant University, Ota, Nigeria
e-mail: yuven.nchilapgs@stu.cu.edu.ng

© The Author(s), under exclusive license to Springer Nature Switzerland AG 2022

A. O. Ayeni et al. (eds.), *Bioenergy and Biochemical Processing Technologies*, Green Energy and Technology, https://doi.org/10.1007/978-3-030-96721-5_32

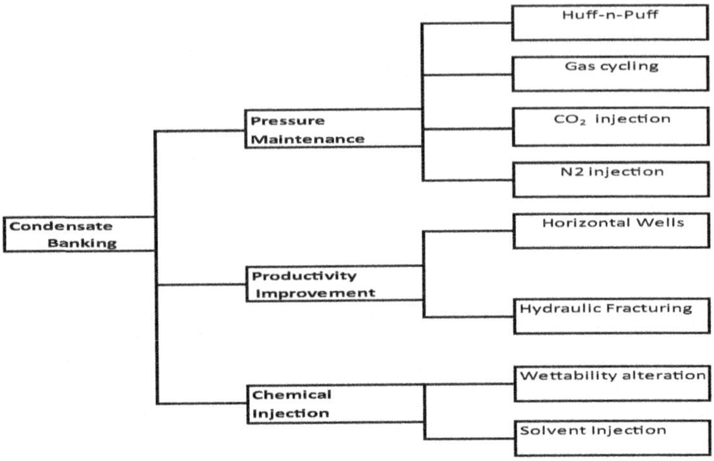

Fig. 1 Condensate reservoir development techniques

(iii) Decline in the loading efficiencies of the plant during the project life

Gas condensate bearing reservoirs are becoming more due to increase in production depth, pressure, and temperature (Orodu et al., 2012). There are several development techniques that can be applied to develop gas condensate reservoirs to improve condensate recovery or prevent condensate formation. Common practices to prevent the damage caused by condensate banking are to reduce draw down pressure and maintaining the bottomhole flowing pressure higher or to change the gas condensate phase behavior. This study describes the various development techniques that have been applied in the developing gas condensate reservoirs as summarized in Fig. 1. The advantages and disadvantages each development option will be highlighted.

2 Gas Condensate Reservoir Development Techniques

2.1 Pressure Maintenance

In the pressure maintenance approach, the goal is maintaining the reservoir pressure higher than dew point line to improve the recovery of condensates. Gases such as Liquefied Natural Gas (LNGs), CO_2, and N_2 are the common gases used for this approach. There are two pressure maintenance schemes: Full pressure maintenance has to do with continuous injection of gas and at the same time producing condensates from the reservoir with the purpose of preventing the pressure of the reservoir from dropping below the dew point pressure. With the partial pressure maintenance method, the same well is used as the injector well and the producer well. The injection of gas into the reservoir via the well occurs during the huff period and the production fluid production of fluid occurs during the puff period. This aims at revaporization of the condensates that have drop out of the gas. When injected gas comes in contact with condensates, processes that may occur include reservoir fluid displacement by the injected gas, revaporization of liquid dropouts, and PVT behavior change of the reservoir fluid.

Gas Cycling

Gas cycling is the re-injection of gas produced back into the reservoir formation. The produced gas is stripped of liquids and then the dry gas is reinjected back into the reservoir. With gas cycling techniques, the goal is to maintain the pressure of the reservoir at a value higher above the dew point line thereby slowing down pressure decline and inhibiting condensate formation (Adel et al. 2006; Nasiri et al. 2015). The gas cycling process also helps in the vaporization of condensates (at least intermediates) that might have been form back into the gas phase (Sayed and Al-Muntasheri 2016). It is estimated that around 60–80% of condensate liquids cannot be produced by natural energy of the reservoir. This can significantly be reduced to about 30–40% through gas recycling (Mohamadi et al. 2020). If considering condensate recovery by cycling gas process, it is important to consider factors such as vertical efficiency, sweep area, and condensate vaporization (Sayed and Al-Muntasheri 2016). The most significant factor that affects the sweep efficiency is the location of the injection and producing wells in the reservoirs. This determines the distribution of the injected gas and the production of wet gas among the producing wells.

Several reports have been reported in literature on the numerical and field applicability of gas cycling technique some of which are discussed below. The use of simulation studies can permit the prediction of optimum scenarios to develop condensate reservoirs and to maximize the condensates liquid recovery using the gas cycling approach. This is because making use of simulation considers many parameters such the petrophysical and geological data.

Siddiqui et al. (2014) used a numerical computational simulator in optimizing the parameters governing the gas cycling processes using stochastic optimization algorithm. They used Differential Evolution and Covariance Matrix Adaptation Evolution techniques to optimize the production/injection rates, well locations, and the reinjected gas fraction. The field average oil saturation and Net Present Value of the gas cycling project were used separately as the objective functions. Their results indicated that if parameters are optimized simultaneously rather than sequentially, the Net Present Value can be increased significantly. This suggest that well placement optimization alongside operational parameters is very important. The results also showed that optimization of production rates, sales volumes, and well placements significantly reduce the reservoir condensate saturation rate.

Higher injection of gas has showed higher condensate recovery factor than the normal depletion (El-Banbi et al. 2000).

The Arun field in North Sumatra in Indonesia experience significant decline in well production after 10 years since production began. Well transient pressure tests performed indicated that the condensate saturation around the wellbore was responsible for the decrease in well delivery (Rahimzadeh et al. 2016). PVT reports of the formation fluid show that the fluid in the formation was lean gas reservoir system with optimum liquid dropout of 1.1%. A massive lean gas cycling project was then conducted to remove the condensate bank near the wellbore and to maintain reservoir pressure higher than the dew point value. Lean gas was injected at the reservoir periphery to displace(sweep) the liquid condensates towards the producing well.

Although gas cycling may seem to be an ideal solution for retrograde formation, there are several concerns which could affect this technique of operation. Increase gas prices makes this technique unfavorable (Sayed and Al-Muntasheri 2016). Also, high initial investment is required for the gas compression and injection. The cycling gas process couple with subsequent periods of blow-down may result in prolong project life with underusage gas plant at the final stage and high cumulative operation cost. However, cycling of condensate reservoirs at declining reservoir pressure conditions is advantages from an economic and recovery point of view. This is because gas cycling at declining pressure, both the wet gas displaced from the swept area and the unswept portions by gas expansion are recovered simultaneously at the same time. The presence of any liquid condensates in the swept area is revaporized by the injected dry gas. Hence, it is important to evaluate the economics of adopting the gas cycling technique considering both the advantages and disadvantages as mentioned in this paragraph.

N_2 Injection

Due to economic drawbacks of reinjecting produced dry gas into the reservoir for condensate recovery, nitrogen has been adopted as an alternative to dry gas. Nitrogen gas of minimum miscible pressure is injected into a reservoir to free the hydrocarbons trapped in the reservoir formation. Nitrogen forms miscible slugs at very high pressures that helps to sweep oil and gas from areas that are difficult to reach. This method is one of the preferred because it is cost-effective and sustainable. Secondly, its inert chemical properties prevent combustion downhole and no corrosion effect on pipelines. During nitrogen injection, the nitrogen overrides the other reservoir fluids because of the difference in densities between the displaced fluid and displaying fluid. Oil displacement is considerably improved if CO_2 is added to the injection of nitrogen and natural gas (Fig. 2).

Mogensen and Xu (2020) modelled and investigated miscible nitrogen injection potential carbonate reservoir containing volatile oil. The simulation studies involved performing several sensitivity studies on the modelled. In their study miscible nitrogen injection was reported to be the viable option with potential of increasing the condensate liquid recovery by about 20%. Also, significant improvement in the sweep efficiency was observed because of N_2 miscibility with the oil under the reservoir conditions.

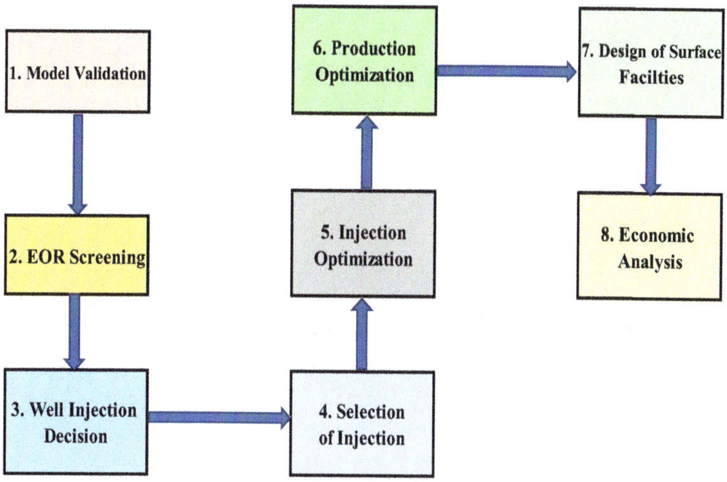

Fig. 2 Nitrogen injection workflow for EOR (Canchucaja and Sueiro 2018)

Canchucaja and Sueiro (2018) carried out a feasibility study of injecting nitrogen in a multilayered lean gas condensate reservoir located in the Peruvian Jungle. A key factor in the injection of nitrogen in this field was the difference in the rock properties of the producing formations between the multilayered reservoirs. It was noted that the minimum miscibility pressure (MMP) which is the pressure at which the injected N_2 undergoes a first contact miscibility process irrespective of its concentration was slightly lower than the fracture pressure. Nitrogen was then injected above the MMP by avoiding the multiple contact miscibility process in order not to alter the rock properties of the reservoir when surpassing the fracture pressure. A novel conceptual engineering approach was proposed for enhance oil recovery as shown in Fig. 3. Several observations were made in this study:

(i) Field recovery did not respond linearly with an increase nitrogen injection and hence the decision to inject N_2 will have to be greatly dependent on surface facility constraints and economic estimations.
(ii) Nitrogen injection provoked a series of effects that varies at different reservoir formation zones: Increased liquid dropout occurred in the nitrogen mixing zone and pressure support in another zone much farther away from the producing well because of the partial pressure voidage replacement being pervasive causing revaporization to occur.
(iii) The construction of an air separation and N_2 injection units are a major challenge for N_2 project because of the high cost, high energy demand, and considerable increase in flexibility of operation.

Davarpanah et al. (2019) simulated the injection of nitrogen to enhance productivity from a gas condensate reservoir in Iran and compared different injectivity scenarios. The objective of the study was to substitute natural gas with nitrogen for elevating the original pressure of the reservoir there by reducing the amount of

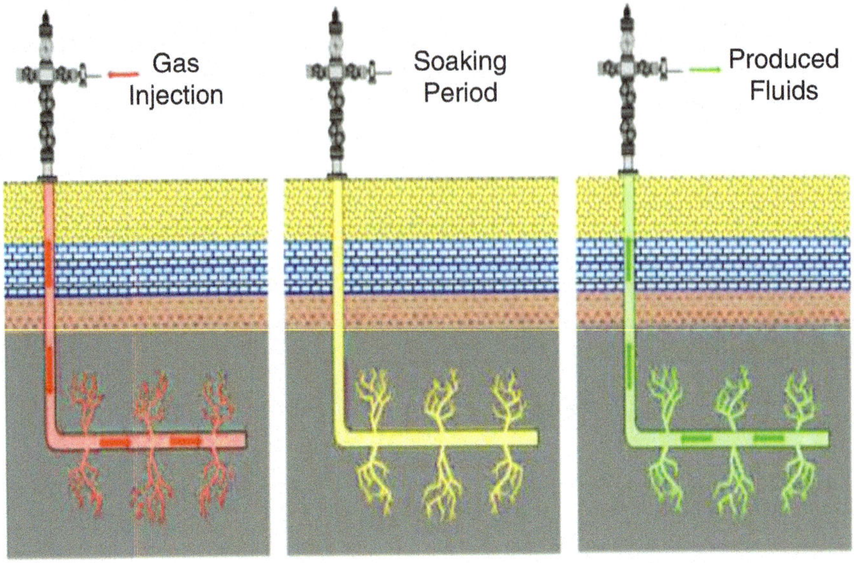

Fig. 3 Huff and puff technique for enhance oil recovery (Pankaj et al. 2018)

valuable gas use for injection processes. The effect of the injected nitrogen on the reservoir pressure, amount of gas contaminated by the injected gas mixture and the volume of condensates extracted were studied. Five million cubic meters of nitrogen was injected for a period of 16 months to attain the original reservoir pressure which was about one 8.3% of the volume of natural gas injected. This represents a potential breakthrough for the oil and gas industry. Small amounts of heavy components present in the gas produced with nitrogen injection was observed and it is completely dependent to the nitrogen dew point pressure which increases the injection and production time and decrease the concentration of nitrogen in the reservoir fluid mixture.

Several factors play a major role when considering nitrogen for gas condensate recovery on an economic standpoint. One very important factor has to do with the cost of energy needed to produce and compress nitrogen. Typically, large nitrogen generators are used for injection applications which require approximately 15–15.5 brake-horsepower-per-hour per 1000 cubic feet of nitrogen at pressures of about 5000 psig. Also, the location of these facilities has a pronounced effect on the accessibility of plant erection, maintenance, and operation personnel as well as the protection of the equipment against extreme weather conditions. The effect of these items differs, and each location must be evaluated on a case-by-case basis.

Nitrogen has a low density making it favorable for gas assisted gravity drainage, proving sufficient vertical relief to the reservoir and help in maintaining a gravity stable displacement. Due to the inert chemical nature of nitrogen gas, pipe corrosion is not an issue contrary to other gas such as CO_2 and H_2S. Nitrogen non-flammable posing no health risk.

However, nitrogen injection in a gas reservoir can sometimes raise the dew point pressure leading to early condensate dropout and loss of valuable liquids in the formation. Another concern is the early gas breakthrough which can occur when nitrogen is injected into the formation because of its high mobility ratio.

Carbon Dioxide Injection

Carbon dioxide injection is amongst the most widely used gas injection for EOR. The method is dependent on the carbon dioxide's ability to dissolve the oil in the reservoir. The oil displacement mechanism by CO_2 depends on whether, the displacement process is miscible or immiscible (Alagorni et al. 2015). For miscible displacement, CO_2 at high reservoir pressures and temperatures mixes with the condensate causing condensate swelling, reduction in the condensate viscosity and promoting the vaporization of the condensates thereby maintaining a single gas phase (Narinesingh and Alexander 2014). The displacement of oil by CO_2 injection occurs because of an increase in the oil volumes because of the CO_2 enrichment. This provides an effective oil displacement and additional washing out of residual oil. In miscible oil displacement process, CO_2 dissolves in the Oil. During the miscible displacement, the residual oil saturation is reduced to nearly zero, the viscosity increases, its mobility decreases, and the surface tension on the oil and water interface reduce. The oil displacement factor is lower in immiscible than in miscible displacement resulting in the occurrence of three-phase filtration in the formation which is characterized by an increase in the filtration resistance. That is, the residual oil saturation can remain in the reservoir after mixing. The use of immiscible oil displacement by CO_2 is associated with the lower cost of the process and the low injection pressure required. The oil swells, the viscosity reduces, and the mobility and capillary water penetration by porous media improve.

Su et al. (2017), using a simulation model, performed experimental investigations of CO_2 injection and water flooding for a gas condensate reservoir located in East region of China. They reported that CO2 injection proved to be more effective than the water flooding increasing the productivity by a factor of 1.39. Fath and Dashtaki (2016) in a case study, evaluated parameters on carbon dioxide injection process to determine the optimum condition for CO_2 injection in a gas condensate reservoir. Their results showed that pressure and rate of injection play a major role in the determination of the best condensate recovery. The application of CO_2 injection is preferable for use in fields with an occurrence depth of about 7000 m and high formation depths and low oil viscosity. Current technologies include but are not limited to Cyclic injection of CO_2, joint injection of carbon dioxide and surfactants: Continues CO_2 injection; injection of water saturated with CO_2: alternating plugs of CO_2 and water to reduce fingering.

Factors that influence the application of CO_2 include the lower formation sweep as compared to other conventional methods, well corrosion, CO_2 utilization, and incomplete mixability with oil and light hydrocarbons. However, the main problems in using CO_2 method are the availability of the sources close to the field, CO_2

separation from oil, and regeneration for future injection and its transportation. CO_2 injection provide a temporal solution and must be frequently repeated in the field. CO_2 injection in sandstones and carbonate formations changes the clay content, porosity as well as permeability of the formation.

Huff and Puff

The huff and puff process involves the injection of gas under high pressure into the partially depleted area of the reservoir with fracture networks having a low pressure. Only one well is used both for injection and production. This process usually starts with the injection of a recovery enhancement fluid such as CO_2 in to the well and the fracture networks. The solvent moves into fracture/matrix interface during the injection period by a pressure gradient mechanism. This is then followed by a shut-in period (soaking time) which allows for the gas injected to dissolve in the oil (condensate), causing its volume to swell, reducing its viscosity. The well is then put back into production after the injection and soaking period and the well response is monitored. There are several mechanisms for huff and puff processes reported in literature: oil swelling, repressurization, diffusion, oil viscosity reduction, relative permeability hysteresis, and interfacial tension reduction (Sanaei et al. 2018). The dominant mechanism in this method is the re-vaporization of the retrograde condensate into the flowing gas phase.

The injected gas migrates into the rock matrix during the early soaking period and help re-pressurize the depleted area and the limited area around the fracture network. More gas is transported into the matrix as the soaking period continues causing the swelling of oil, reduction in viscosity and the interfacial tension. The condensate liquid swelling leads to a slight increase in the matrix pressure creating a local pressure gradient and hence helps the oil extraction via microfractures. During the puff period (production period), the oil extracted further mixes with the gas injected and migrates towards the producing well more easily.

Wan and Mu (2018) investigated CO_2 huff and puff effect on remedying the condensate banking in Eagle Ford shale gas condensate reservoir. They considered the effect of Nano-pore confinement on the gas and oil productivity performance. Rich gas and lean condensates scenarios were compared. Their results showed that the two-phase regions shrink with increasing total content of carbon dioxide in the system. The condensate falls out of the gas decreased with increase addition of CO_2 into the reservoir fluid. They also reported that CO_2 showed to be more effective in improving the recovery of rich gas condensates than in the recovery of lean gas condensates.

Cyclic CO_2 injection was applied by Yang et al. (2019) in other to mitigate the condensate banking affect. They went further to investigate the role of multicomponent adsorption and the role of geomechanics in the development of the Eagle Ford shale condensate reservoir during primary and CO_2 enhance recovery. They reported that the impact of multicomponent adsorption on the productivities of the well are limited for both processes. This is because, the reservoir pressure is still higher even after a production period of 10 years whereas the small variation in the geomechanical model leads to almost double initial gas rate.

Sheng (2015) extended the investigation of the effect of huff and puff to evaluating condensate liquid recovery potential in shale gas condensate reservoirs. Their studies suggested that the huff and puff injection of gas can improve liquid condensate recovery more than primary depletion and gas flooding. They also suggested that huff and puff technique become more appreciable when the bottom hole flowing pressure is low or when the initial reservoir pressure is closer to the dew point.

Ayub et al. (2019) used a computational simulator for CO_2 huff and puff injection technique and its potential advantage in gas condensate removal. They finalized that CO_2 is comparatively more efficient in mitigating condensate banking around the wellbore as compared to regions farther away from the wellbore.

Sanaei et al. (2018) carried out a comprehensive study of gas cycling process and developed a numerical compositional model of the Middle Bakken tight oil reservoir to simulate CO_2 gas cycling considering various operational parameters. Their results obtain from the sensitivity study of the operational parameters indicated as follows:

I. There is an injection rate that is optimum for gas cycling operation. And that lower gas injection rate does not increase sufficiently the pressure and hence recovery of oil unlike at higher injection rates.
II. The number of cycles has a significant effect on the gas cycling effectivity. The volume of the injected gas increases with increased number of cycles and consequently increasing the volume of oil recovered by the injected gas volume.
III. There is an optimum duration of the injection period depending on the reservoir pressure around the wellbore region and the surface gas injection rate.
IV. Commencing the gas cycling process too soon in the life of the well negatively affects the effectiveness of the process.

CO_2 is the most used fluid for miscible displacement because it is less expensive than liquefied petroleum and it reduces the viscosity of the oil.

Nevertheless, the most significant limitation of this method has to do with the fact that although the injection of gas reduces or removes the condensate around the wellbore, condensates can potentially accumulate again when production resumes (Hwang, 2011). Another setback for gas injection techniques in shale gas oil reservoirs is the early gas break through (Wan and Mu 2018). A possible season for this is that the geomechanics process during injection helps to reopen the pre-existing natural fractures and recover the impaired conductivity of hydraulic fractures (Yang et al. 2019).

2.2 *Production Improvement of Condensate Reservoirs*

Productivity improvement method focuses mainly on the creation of higher permeability flow paths or in increasing the contact area between the wellbore and the reservoir thereby reducing the pressure drop and enhancing the productivity of the well. This technique potentially alters the flow of gas from radial to lineal flow and as a result leading to a delay in the condensate banking. Drilling of horizontal wells and hydraulic fracturing are the commonly deployed approach to this method.

Use of Horizontal Wells

Horizontal wells are high-angle wells often drilled to enhance the reservoir performance. There were relatively limited horizontal well drilling activities until 1980 when the first major trust of horizontal well drilling started (Joshi 2003). Today these wells have been widely adopted worldwide presenting an improved and even a more efficient method of condensate banking mitigation. Horizontal wells increase reservoir and well contact area there by improving the well productivity by spreading the pressure drop over that area. Hence, horizontal wells reduce the pressure drop and hence delaying the condensate banking near the wellbore in gas condensate reservoirs. A lesser drawdown pressure is achieved if the length of the horizontal well is longer (Sayed and Al-Muntasheri 2016).

Miller (2010) used numerical model to investigate the mitigation of gas condensate banking phenomenon in a giant gas condensate reservoir in Northeast China by studying the application of horizontal wells. They used two well models; a horizontal well model and with cartesian coordinates and a vertical well model with radial coordinates. Their objective was to determine the fractions of gas produced in a horizontal well caused by reduction in condensate banking and increased formation contact. Their results showed a smaller drawdown pressure for horizontal wells, and the magnitude of oil saturation build up around the wellbore in horizontal wells is smaller than that for vertical wells. The increased in productivity for horizontal wells even after the dew point shows that the PI increase for horizontal wells is proportional to the condensate reducing ability of the horizontal wells around the wellbore.

Olabode and Egeonu (2017) used a 3D reservoir simulation model to optimized production from a condensate reservoir system in the Niger Delta. They demonstrated the performance of vertical and horizontal wells and horizontal well length effect on condensate productivity. Results indicated that horizontal wells showed better performance in productivity as compared to vertical wells and that horizontal wells showed a slower decline in the pressure drawdown.

Horizontal wells help reduce the water coning in oil and gas reservoirs because of reduce in the draw down pressure; provide larger well and reservoir contact area there by improving the well productivity and potential for accessing multiple potentially isolated fault compartments.

There are several setbacks in the use of horizontal wells. Liquid loading can potentially be problematic if liquid accumulates in the fractures or wormholes (oil and water) as this may affect the productivity of gas. Also, horizontal wells can delay condensate formation around wellbore but eventually the condensates are expected to form again when the pressure declines again. Even though the condensates will still be form, it will take a longer time and the well productivity will remain higher for a longer period. Therefore, the benefits of considering horizontal wells must be weight against its cost longer (Sayed and Al-Muntasheri 2016).

Hydraulic Fracturing

The application of fractured horizontal wells is another method of reducing the impact of condensate blocking. Hydraulic fracturing causes improvement of condensate flow by creating fractures in the formation connecting the reservoir and the wellbore. It involves the inducement of fractures by pumping fluid under high pressure, higher than the fracture pressure into the target reservoir rock formation. The wells in hydraulically fractured reservoirs are operated at rate higher than those of vertical wells and so the higher-pressure drop will then be distributed on a larger area thereby delaying dew point attainment in gas condensate reservoirs. Several factors are responsible for the performance of hydraulic fracturing treatment such as type of injection fluid, formation properties, composition of the hydrocarbon fluids.

Hydraulic fracturing has its limitations as this technique does completely condensates accumulated in areas with lower formation pressures than the dew point. It helps delay the time when the pressure reaches the dew point but will not totally prevent condensate banking (Kerunwa et al. 2020). Hydraulic fracturing does not however generate conduits past the condensate saturation region or at least not for a longer period. The saturation around the wellbore will increase just like it did before if the surface pressure falls below the dew point pressure.

The optimization of the hydraulic fracturing fluid will help to minimize the fracture damage and enhance the performance after treatment (Rahim et al. 2012). It should be noted that this method is not a permanent technique as increase production with time occurs, drawdown increases, and the chances of condensate formation again is high.

2.3 Chemical Method of Condensate Reservoir Development

Chemical method of condensate reservoir development involves the injection of chemicals and solvents to the formation to alter the formation wettability thereby minimizing the rate of condensate blocking. This technique has been applied in well simulation for enhancement of productivity allowing for deeper penetration and hence higher treatment durability (Lopez et al. 2021).

Wettability Alteration Method

Wettability alteration is a novel approach gaining grounds recently and is becoming more attractive to researcher in the industry. The gas condensate reservoir rocks are mostly naturally liquid wetting. This technique changes the wetting ability of the treated region, decreasing the liquid saturation and increasing the preferential relative permeability of the gas. Altering the wettability of the reservoir rock from a liquid wetting to preferential gas wetting can help increase the condensate mobility and the gas relative permeability as confirmed by Sheydaeemehr et al. (2014). In an experiment carried out by Kewen and Firoozabadi (2000) the rock wetting ability was changed by chemical treatment with solutions FC722 and FC759 under laboratory conditions.

Franco-Aguirre et al. (2018) synthesized and evaluated the effectiveness of nanofluid to change the sandstone wetting state in a tight gas reservoir. They synthesized functional particles using nanoparticles of silicon dioxide and the commercial surfactant Silnyl-FSJ. The changes in wettability in sandstone with oil-wet and water-wet states were evaluated by spontaneous imbibition tests and contact angle measurements for different nanofluid formulation and nanoparticle dosage. It was observed that the nanofluid containing 500 mg per liter of SSY5 nanoparticles summered in 0.46% solution of SY achieved better performance in the static evaluation of wettability change. Also, the contact angle changes from 123° to 115° and from 0° to 93° respectively for the oil-wet samples.

Franco-Aguirre et al. (2018) conducted a dynamic assessment of the selected nanofluids at reservoir conditions of 212°°F with a pore pressure of 4500 pi and confining pressure of 6500 psia. It is noted that after the injection of the nanofluid, the mobility of oil and water increases, which indicates the transition of the system from a liquid wetting to a gas wetting state. This is confirmed by a sharp decrease in the saturation of oil and water. In addition, they observed an increase in condensate liquid recovery by 68%, which indicates that the use of nanoparticles can significantly reduce the energy consumption required the production of gas and oil because of changes in the wettability of the formation (Fig. 4).

Ali et al. (2019) evaluated the effect of wettability on gas and condensate production by investigating the optimizing wettability conditions near the wellbore. They used numerical simulation technique to examine a set of wetting conditions. Their

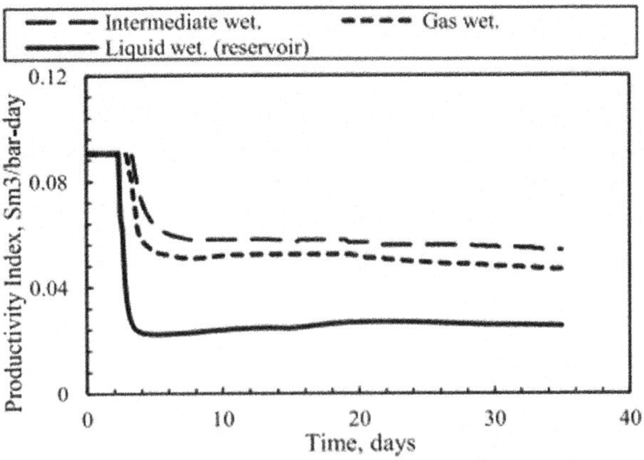

Fig. 4 Productivity improvement for different wetting states (Sheydaeemehr et al. 2014)

results indicate the possibility to achieve an optimal wetting state by maximizing delivery from a gas well. They also noted that completely reversing the conditions of wettability in the reservoir can improve productivity compared to the initial wetting state, however, can lead to a decrease in the volume of condensate recovered in some cases.

Wettability alteration techniques is a novel approach not widely applied in oil fields. Even though the theory is promising due to its effectives, flexibility of design, and durability, there are several concerns that are associated with this technique. That is, risk of formation damage and permeability reduction.

Application of Solvents

The purpose of solvent application in gas condensate reservoir systems is to dissolve the liquid condensate thereby reducing the interfacial tension between the gas condensate interfaces. This ultimately enhances the effective permeability of gas thereby improving the gas mobility. Generally, alcohols such as methanol are used as solvents to improves condensate recovery due to their multi-contact-miscibility abilities (AL-Anazi et al. 2005).

Du et al. (2000) carried out an experiment to evaluate the applicability of methanol in restoring the relative permeability of gas in retrograde gas reservoir. Their results showed that methanol increases the relative permeability of the gas by a factor of 2.5 depending on the saturation of the water in the reservoir. The methanol miscible displacement of the condensate and water phases was reported to be responsible for the increase in the gas relative permeability. This investigation presents an approach in the use of cheap alcohols in improving production in gas wells that have been hindered by condensate liquid build.

Correa et al. (2009) conducted an experiment in which methanol, methylene, and propanol were used as solvents in simulating a gas condensate reservoir at pressures below the dew point in removing condensate liquids and increase the effective permeability of gases near wellbore. They noted that all solvents increase the effective permeability of the gas after the removal of condensates. Methanol was reported to be more effective than propanol and methylene in removing condensate and increasing gas permeability.

They observed that all the solvents improved the gas effective permeability after removing the condensate banking. Methanol was reported to be more effective in condensate removal and improves gas permeability than propanol and methylene chloride.

Kumar et al. (2006) carried out a numerical simulation study of chemical treatment of a single well for a gas condensate reservoir. The results of the simulation showed that chemical treatment has the potential to greatly improve productivity from a gas condensate reservoir at a low-cost relative to increased revenue when considering condensate treatment only around the wellbore. They also reported that considering condensate bank treatment for a radius of about 20 ft can increase the fluid productivity by a factor of up to 1.8 if the treated zone relative permeability increases by a factor of 2.

Solvents such as alcohols (methanol, ethanol, and isopropyl alcohol) are effective in the remediation of liquid build up around the wellbore in carbonate and sandstone reservoirs. However, the use of solvents is a temporary treatment.

Solvents such as methanol can be mixed with other solvents to improve its miscibility in hydrocarbons and water formations, as well as reducing the cost of purchasing pure solvent. In addition, knowledge of phase behavior of the methanol-water mixtures phase behavior, which can be applied under varying reservoir conditions, especially at high temperatures and different compositions, is crucial for this approach.

3 Summary

The design of a pressure maintenance project involves the determination of the type and quantity of the fluid to be used of the injection. In a reservoir where the dew point pressure is higher, some sort of full pressure maintenance is required to prolong the life span of the reservoir in the gas phase. When early liquid condensate drop out is expected in the reservoir a partial pressure maintenance approach is preferable to help in the vaporization of the liquid condensates so that the reservoir can be produced as a single gas phase.

The selection of a gas for the cycling process should consider greatly the economics of the project. The increasing price and high demand for methane can make methane gas cycling impractical economically especially during the early and midlife of the production field. Application of dry gas cycling at the late stage of field production should then be considered. The gas cycling project design for condensate recovery desires correct and reliable information of the phase behavior and thermodynamics characteristics of the reservoir fluid and the mixtures generated during the cycling process.

The use of nonhydrocarbon gases such as CO_2 and N_2 have been suggested and widely reported in literature. It is important to evaluate the mode of application of the selected injection technique. Nitrogen gas being an inert gas and is just slightly miscible with the reservoir fluid at very high pressures and hence is preferable for pressure maintenance purposes. The injection of N_2 into a gas reservoir raises the dew point pressure of the reservoir fluid/injected gas mixture (Canchucaja and Sueiro 2018). However, the composition of the injected nitrogen gas should be taken into consideration. In some cases, the nitrogen might not be a good idea as the nitrogen may contaminate the reservoir fluid there by increasing the purification cost. For reservoirs with low heterogeneity N_2 injection can be a viable option if operational parameters such as production and injection rates are maintained constant (Sayed and Al-Muntasheri 2016).

The use of horizontal wells and hydraulic fracturing aims at increasing the contact surface area between the well and the reservoir. This will delay the accumulation of condensates thereby improving the productivity of the reservoir. This however is temporary solution to the condensate banking problem. Drilling horizontal wells might be economical and hence cost should be an important factor when considering its applicability.

The use of chemicals in gas condensate reservoir aims at altering the wettability from a liquid wetting to a preferential gas wetting and to reduce the interfacial tension between the gas and the condensate interface. Chemical's solvents such as alcohols have proven to be effective in the condensate banking removal. However only laboratory application is reported in literature. Alcohol solvents will provide a major economic breakthrough for improve condensate recovery and hence further research is required for its field scale applicability especially for high temperature reservoirs.

4 Conclusion

Gas condensate fields are very resourceful when natural gas prices are high as the natural gas market is constantly expanding. Gas condensate fields are an invaluable resource when natural gas prices are high as the natural gas market is constantly expanding. Similarly, they are valuable at low natural gas prices due to the production of liquids. The development of gas condensate reservoirs is like the development of dry gas reservoirs. However, two significant differences: the flow of condensate liquids near the wellbore and the significant production of liquid during the life of the reservoir greatly reduces the productivity of the gas. This review describes the various options for developing gas condensate reservoirs and highlights the setbacks of each technique.

Acknowledgement The authors would like to appreciate Covenant University management for providing an enabling environment to carry out this research, and assistance in publication.

References

Adel, H., Tiab, D., and Zhu, T. (2006). *Effect of gas recycling on the enhancement of condensate recovery, Case study*: Hassi R'Mel South Field, Algeria. SPE Paper 104040.

Al-Anazi, H. A., Solares, J. R., & Al-Faifi, M. (2005, April). *The impact of condensate blockage and completion fluids on gas productivity in gas-condensate reservoirs*. In *SPE Asia Pacific Oil and Gas Conference and Exhibition*. OnePetro.

Alagorni, A. H., Yaacob, Z. Bin, & Nour, A. H. (2015). *An Overview of Oil Production Stages Enhanced Oil Recovery Techniques and Nitrogen Injection*. *6*(9). https://doi.org/10.7763/IJESD.2015.V6.682

Ali, N. E. C., Zoghbi, B., Fahes, M., Nasrabadi, H., & Retnanto, A. (2019). The impact of near-wellbore wettability on the production of gas and condensate: Insights from experiments and simulations. *Journal of Petroleum Science and Engineering*, *175*, 215-223.

Ayub, M., & Ramadan, M. (2019). Mitigation of near wellbore gas-condensate by CO2 huff-n-puff injection: A simulation study. *Journal of Petroleum Science and Engineering, 175*, 998-1027

Canchucaja, R., & Sueiro, M. (2018). *Feasibility of Nitrogen Injection in a Multi-layered Lean Gas Condensate Reservoir. SPE Russian Petroleum Technology Conference.* https://doi.org/10.2118/191652-18rptc-ms

Correa, T., Tiab, D., & RESTREPO, D. P. (2009). Evaluation of solvents efficiency in condensate banking removal. *Dyna, 76*(157), 163-171.

Davarpanah, A., Mazarei, M., & Mirshekari, B. (2019). A simulation study to enhance the gas production rate by nitrogen replacement in the underground gas storage performance. *Energy Reports, 5*, 431-435.

Du, L., Walker, J. G., Pope, G. A., Sharma, M. M., & Wang, P. (2000, October). Use of solvents to improve the productivity of gas condensate wells. In *SPE annual technical conference and exhibition*. OnePetro.

El-Banbi, A. H., Aly, A. M., Lee, W. J., & McCain, W. D. (2000, April). Investigation of waterflooding and gas cycling for developing a gas-condensate reservoir. In *SPE/CERI Gas Technology Symposium*. OnePetro.

Fath, A. H., & Dashtaki, N. B. (2016). Evaluation of effective parameters on CO2 injection process in a gas condensate reservoir: A case study. *Energy Sources, Part A: Recovery, Utilization, and Environmental Effects, 38*(24), 3680-3686.

Franco-Aguirre, M., Zabala, R. D., Lopera, S. H., Franco, C. A., & Cortés, F. B. (2018). Interaction of anionic surfactant-nanoparticles for gas-Wettability alteration of sandstone in tight gas-condensate reservoirs. *Journal of Natural Gas Science and Engineering, 51*, 53-64.

Hwang, J. (2011). Gas injection techniques for condensate recovery and remediation of liquid banking in gas-condensate reservoirs (Doctoral dissertation).

Kerunwa, A., Princewill, O. A., Nkemakolam, C. I. (2020). Economic Evaluation of Hydraulic Fracturing in a Gas Condensate Reservoir Operating below Dewpoint. Journal of Yangtze Gas & Oil (14)5, 73-86. https://doi.org/10.4236/ojogas.2020.53007

Kewen, L., & Firoozabadi, A. (2000). Modeling gas-condensate relative per-meabilities and the effect of wettability change to gas wetness. *Society of Petroleum Engiceers Jouroal, 3*(2), 139-149.

Kumar, V., Bang, V. S. S., Pope, G. A., Sharma, M. M., Ayyalasomayajula, P. S., & Kamath, J. (2006, January). Chemical stimulation of gas/condensate reservoirs. In *SPE Annual Technical Conference and Exhibition*. Society of Petroleum Engineers.

Joshi, S. D. (2003, May). Cost/benefits of horizontal wells. In *SPE western regional/AAPG Pacific section joint meeting*. OnePetro.

López, D., Zabala, R. D., Matute, C., Lopera, S. H., Cortés, F. B., & Franco, C. A. (2021). Well injectivity loss during chemical gas stimulation process in gas-condensate tight reservoirs. *Fuel, 283*, 118931.

Miller, N. (2010). Increasing well productivity in gas condensate wells in Qatar's North Field (Doctoral dissertation, Texas A & M University).

Mogensen, K., & Xu, S. (2020). Comparison of three miscible injectants for a high-temperature, volatile oil reservoir-With particular emphasis on nitrogen injection. *Journal of Petroleum Science and Engineering, 195*, 107616.

Mohamadi, M., Azin, R., Osfouri, S., & Zendehboudi, S. (2020). Experimental and modeling investigation of non-equilibrium condensate vaporization in porous systems: Effective determination of mass transfer coefficient. Fuel, 262, 116011.

Narinesingh, J., & Alexander, D. (2014). CO2 enhanced gas recovery and geologic sequestration in condensate reservoir: a simulation study of the effects of injection pressure on condensate recovery from reservoir and CO2 storage efficiency. *Energy Procedia, 63*, 3107-3115.

Nasiri Ghiri, M., Nasriani, H. R., Sinaei, M., Najibi, S. H., Nasriani, E., & Parchami, H. (2015). Gas injection for enhancement of condensate recovery in a gas condensate reservoir. *Energy Sources, Part A: Recovery, Utilization, and Environmental Effects, 37*(8), 799-806

Olabode, O. A., & Egeonu, G. I. (2017). Effect of horizontal well length variation on productivity of gas condensate well. *Int J Appl Eng Res, 12*(20), 9271-9284.

Orodu, O. D., Ako, C. T., Makinde, F. A., & Owarume, M. O. (2012). Well deliverability predictions of gas flow in gas-condensate reservoirs, modelling near-critical wellbore problem of two phase flow in 1-dimension. *Brazilian Journal of Petroleum and Gas, 6*(4).

Pankaj, P., Mukisa, H., Solovyeva, I., & Xue, H. (2018). *Enhanced Oil Recovery in Eagle Ford: Opportunities Using Huff-n-Puff Technique in Unconventional Reservoirs. SPE Liquids-Rich Basins Conference - North America.* https://doi.org/10.2118/191780-ms

Sanaei, A., Abouie, A., Tagavifar, M., & Sepehrnoori, K. (2018, September). Comprehensive study of gas cycling in the Bakken shale. In *Unconventional Resources Technology Conference, Houston, Texas, 23-25 July 2018* (pp. 2057-2071). Society of Exploration Geophysicists, American Association of Petroleum Geologists, Society of Petroleum Engineers.

Rahim, Z. Al-Anazi, H. Kanaan, A. (2021). Productivity increase through Hydraulic Fracturing in Coventional and Tight Gas Reservoirs-Expectation Versus Reality. Presented at the SPE Middle East Uncoventional Gas Conference and Exhibition, Abu Dhabi, 23-25 January. SPE-153221-MS.

Rahimzadeh, A., Bazargan, M., Darvishi, R., & Mohammadi, A. H. (2016). Condensate blockage study in gas condensate reservoir. *Journal of Natural Gas Science and Engineering, 33*, 634-643.

Sanaei, A., Abouie, A., Tagavifar, M., & Sepehrnoori, K. (2018, September). Comprehensive study of gas cycling in the Bakken shale. In Unconventional Resources Technology Conference, Houston, Texas, 23-25 July 2018 (pp. 2057-2071). Society of Exploration Geophysicists, American Association of Petroleum Geologists, Society of Petroleum Engineers.

Sayed, M. A., & Al-Muntasheri, G. A. (2016). Mitigation of the Effects of Condensate Banking: A Critical Review. SPE Production & Operations, 31(02), 085–102. https://doi.org/10.2118/168153-pa.

Sheng, J. J. (2015). Increase liquid oil production by huff-n-puff of produced gas in shale gas condensate reservoirs. *Journal of Unconventional Oil and Gas Resources, 11*, 19-2

Sheydaeemehr, M., Sedaeesola, B., & Vatani, A. (2014). Gas-condensate production improvement using wettability alteration: a giant gas condensate field case study. *Journal of Natural Gas Science and Engineering, 21*, 201-208.

Siddiqui, M. A. Q., Alnuaim, S., & Khan, R. A. (2014). *Stochastic Optimization of Gas Cycling in Gas Condensate Reservoirs. Abu Dhabi International Petroleum Exhibition and Conference.* https://doi.org/10.2118/172107-ms

Su, Z., Tang, Y., Ruan, H., Wang, Y., & Wei, X. (2017). Experimental and modeling study of CO2-Improved gas recovery in gas condensate reservoir. *Petroleum, 3*(1), 87-95.

Yang, S., Wu, K., Xu, J., Li, J., & Chen, Z. (2019). Roles of multicomponent adsorption and geomechanics in the development of an Eagle Ford shale condensate reservoir. *Fuel, 242*, 710-718.

Wan, T., & Mu, Z. (2018). The use of numerical simulation to investigate the enhanced Eagle Ford shale gas condensate well recovery using cyclic CO2 injection method with nano-pore effect. *Fuel, 233*, 123-132.

Index

© The Editor(s) (if applicable) and The Author(s), under exclusive license to Springer Nature Switzerland AG 2022
A. O. Ayeni et al. (eds.), *Bioenergy and Biochemical Processing Technologies*, Green Energy and Technology, https://doi.org/10.1007/978-3-030-96721-5

D

Q

R